Nanotechnology Applications in Soil Science:

From Remediation to Precision Agriculture

Abhishek Singh
https://orcid.org/0000-0002-0049-6910
Yerevan State University, Armenia

Vishnu D. Rajput
https://orcid.org/0000-0002-6802-4805
Southern Federal University, Russia

Rupesh Kumar Singh
https://orcid.org/0000-0002-2536-1967
University of Minho, Portugal

Karen Ghazaryan
https://orcid.org/0000-0002-8973-2595
Yerevan State University, Armenia

Tatiana Minkina
Southern Federal University, Russia

IGI Global
Scientific Publishing
Publishing Tomorrow's Research Today

Vice President of Editorial	Melissa Wagner
Director of Acquisitions	Mikaela Felty
Director of Book Development	Jocelynn Hessler
Production Manager	Mike Brehm
Cover Design	Jose Rosado

Published in the United States of America by
IGI Global Scientific Publishing
701 East Chocolate Avenue
Hershey, PA, 17033, USA
Tel: 717-533-8845 | Fax: 717-533-7115
Website: https://www.igi-global.com E-mail: cust@igi-global.com

Library of Congress Cataloging-in-Publication Data

No LCCN (CIP Data Pending)
ISBN13: 9798369370063
Isbn13Softcover: 9798369370070
EISBN13: 9798369370087

British Cataloguing in Publication Data
A Cataloguing in Publication record for this book is available from the British Library.

Table of Contents

Detailed Table of Contents

Abhishek Singh, Yerevan State University, Armenia
Karen Ghazaryan, Faculty of Biology, Yerevan State University, Armenia
Rupesh Kumar Singh, Universidade de Trás-os-Montes e Alto Douro, Vila Real, Portugal
João Ricardo Sousa, Universidade de Trás-os-Montes e Alto Douro, Vila Real, Portugal
Tatiana Minkina, Academy of Biology and Biotechnology, Southern Federal University, Rostov-on-Don, Russia
Vishnu D. Rajput, Academy of Biology and Biotechnology, Southern Federal University, Rostov-on-Don, Russia

Soil is an important part of the Earth's ecosystem because it supports farming, nutrient cycling, and the health of the environment. But practices of modern farming and industry have damaged soil to such an extent that it is no longer as fertile or as stable, imperiling both food security and the stability of ecosystems. Nanotechnology have gained importance in the context of soil science as a novel approach for new avenues to enhance soil, remediate pollution, and manage land in an environment-friendly manner. This study presents an in-depth meta-analysis of 252 publications (2010–2025) extracted from Scopus to evaluate the current state, research trends and future prospects for nanotechnology applications in soil science.

Nanotechnology has become an important technological advancement in increasing soil functionality, crop productivity, and the long-term sustainability of agricultural production systems. In this chapter, we highlight how soil physicochemical properties (pH, texture, organic matter and salinity) determine the fate, transport and availability of nanomaterials for uptake by plants, translocated through the plant and influence physiological responses to nanomaterials. A mechanistic insight of how nanomaterials modulate use efficiency of nutrients, redox balance, and plant growth will also be provided. Overall, our chapter highlights the importance of developing a mechanistic, soil-mediated understanding of the nanomaterials to maximize agronomic benefits while minimizing risks to the environment and human health so that nanotechnology may be incorporated into soil-based agricultural systems in a sustainable manner.

Biochar is a substance high in carbon that is produced by thermochemically converting biomass in areas with limited oxygen. It has garnered significant attention because it can address climate change, soil degradation, and waste management simultaneously. The chapter begins with the history of biochar in Amazonian Dark

Earths (Terra Preta) and then proceeds to discuss its complex physicochemical properties, including surface area, porosity, elemental composition, and functional groups. We examine various methods for producing biochar, including gasification, hydrothermal carbonization, slow pyrolysis, and fast pyrolysis. Different methods produce biochar with varying qualities that are best suited for different purposes. It also shows how the choice of feedstock changes the properties of biochar. This technology can be applied in various ways, including cleaning up polluted soils, storing carbon, treating and purifying water, enhancing agricultural productivity, and even storing energy.

Biochar is a carbon-rich substance made by pyrolyzing organic biomass in an environment with no oxygen. It has become a viable soil amendment that might greatly improve the chemical characteristics and fertility of soil. This chapter provides a comprehensive analysis of biochar's diverse effects on soil chemistry, encompassing changes to pH, cation exchange capacity, nutrient retention, and overall soil fertility measurements. Through examination of current research findings, we analysed the mechanisms by which biochar alters soil chemical dynamics, its interaction with native soil components, and the ensuing consequences for agricultural production and environmental sustainability. The conversation explores the role of biochar in carbon sequestration, nutrient cycling, and improving soil conditions. It also discusses how the impacts of biochar can vary depending on the feedstock sources, production settings, and soil types.

As nanotechnology advances, the deployment of nanomaterials in various applications increases, inevitably leading to their release into the environment, particularly soil. By elucidating the mechanisms underlying microbial responses to nanomaterials, researchers can develop strategies to mitigate potential adverse effects and harness the beneficial roles of nanomaterials in soil environments. Moreover, the chapter explores the concept of a "balancing act" in microbial-nanomaterial interactions, highlighting the delicate equilibrium between nanomaterial-induced stress and microbial adaptation. It discusses how microbial communities may undergo shifts in composition and function in response to nanomaterial exposure, potentially affecting nutrient cycling, soil fertility, and ecosystem resilience. By advancing our understanding of these interactions, we can develop sustainable practices for the safe deployment of nanomaterials in agriculture, remediation, and other soil-related applications, ensuring the preservation of soil ecosystem health and function.

This chapter focuses on nano-enabled nutrient delivery systems as emerging opportunities in agricultural technology that can address some of the current issues of nutrient use in agriculture. Different forms of nanomaterials such as nanofertilizers, nanosensors, nanopesticides, nanoemulsions, and nanoclays are described about their uses and how they work. The chapter focuses on the mode of utilization of nano-nutrient by the plant through root absorption, foliar absorption, and cellular uptake pathway of nano-particles. These systems show better uptake of the nutrients compared with the basic fertilizers hence reducing pollution while increasing the yields of crops. Case studies shows how nano-enabled systems are a better solution than traditional ones. This chapter lays out the opportunities of nano-enabled nutrient delivery systems in agriculture sector, but also the need to further investigate the valorization of these systems.

Omkar Singh, Department of Agriculture, KVCOS, Swami Vivekanand Subharti University, India
Shivangi Singh, KVK Bijnor, Sardar Vallabhbhai Patel University of Agriculture and Technology, Meerut, India
Uday Pratap Shahi, Sardar Vallabhbhai Patel University of Agriculture and Technology, Meerut, India
Vaishali Singh, Sardar Vallabhbhai Patel University of Agriculture and Technology, Meerut, India
Krishna Kumar Singh, KVK Bijnor, Sardar Vallabhbhai Patel University of Agriculture and Technology, Meerut, India
Sakshi Singh, Faculty of Biology, Yerevan State University, Armenia
Tatevik Derdzyan, Faculty of Biology, Yerevan State University, Armenia
João Ricardo Sousa, Center for the Research and Technology of Agroenvironmental and Biological Sciences, Inov4Agro, Universidadede Trás-os-Montes e Alto Douro, Quinta de Prados, Vila Real, Portugal

Innovative soil amendment research is driven by the need for sustainable agriculture to provide global food security and reduce environmental damage. Biochar, made from organic feedstocks that promote soil health and sequester carbon. Bulk characteristics may limit its efficacy. Nanotechnology can modify biochar's functioning. Submicrometer particle size, large surface area, and unique physicochemical properties make nano-biochar (NBC) a significant breakthrough. Nano-biochar in sustainable agriculture is covered in this chapter. NBC production and characterization are explained, along with its properties compared to regular biochar. In this chapter, we discuss NBC's many uses as a soil conditioner, remediator, agrochemical transporter, and plant stress reducer. A heavy metal contaminated soil remediation case study shows NBC's practicality and efficacy. Finally, the chapter discusses nano-biochar's role in next-generation sustainable agriculture systems.

and highly productive contributor with most of its publication trends leaning toward contamination sources, ecological risk, and human health impacts.

Foreword

It is with great pleasure that I extend my appreciation for the outstanding work presented in the edited volume "Nanotechnology Applications in Soil Science: From Remediation to Precision Agriculture" published by IGI Global Scientific Publishing. This book offers a comprehensive and timely exploration of the transformative role nanotechnology plays in advancing soil science and modern agriculture.

As global challenges such as soil degradation, contamination, and the increasing demand for sustainable food production intensify, nontechnology presents immense opportunities for soil remediation, nutrient management, and precision farming techniques. However, realizing this potential requires and in-depth understanding of the science behind nanomaterials and their practical applications in diverse soil environments.

This book provides thoughtful analyses and evidence-based insights into the latest advancements, techniques, and applications of nanotechnology in soil health restoration and crop productivity enhancement. The chapters are contributed by internationally recognized experts from leading research institutions, combining multidisciplinary knowledge to address the scientific and practical aspects of nanotechnology in agriculture and soil management.

I sincerely appreciate the dedication and scholarly commitment of the authors in assembling this important work. Their collaborative efforts have produced a resource that is not only information but also instrumental in guiding future research and sustainable practice sin soil science and in precision agriculture. Dr. Abhishek Singh (Yerevan State University, Armenia), Dr. Vishnu D. Rajput (Southern Federal University, Russia), Dr. Rupesh Kumar Singh (Universidade de Tras-os-Montes e Alto Douro, Portugal), Dr. Karen Ghazaryan (Yerevan State Universiy, Armenia), and Prof. Tatiana Minkina (Southern Federal University, Russia) have demonstrated remarkable leadership and cooperation in bringing together diverse expertise to create this valuable volume.

This edited book will serve as a vital reference for researchers, policymakers, and industry professionals striving to harness nanotechnology for improve soil management and sustainable agricultural and environment development in the future. I am confident that this volume will inspire innovation, informed decision-making, and collaborative efforts to address soil and crop challenges for the benefit of global food security and environmental health in the future.

Rafael Barkhudaryan
Yerevan State University, Armenia

Preface

INTRODUCTION

Environmental deterioration driven by climate change, land degradation, pollution, and declining soil fertility has become one of the most pressing challenges of the 21st century. These interconnected threats endanger ecosystems, agricultural productivity, and human well-being, urging scientists and policymakers to develop innovative and sustainable solutions. In recent years, nanotechnology has emerged as a transformative field capable of addressing these global concerns with remarkable precision and efficiency. Among the various nano-enabled approaches, biogenic nanoparticles, engineered nanomaterials, nano-enabled soil amendments, and nano-biochar have shown immense potential for soil remediation, fertility enhancement, and sustainable agricultural development.

Nanotechnology Applications in Soil Science: From Remediation to Precision Agriculture provides a comprehensive exploration of how nanoscale innovations are reshaping soil science and environmental management. This book brings together interdisciplinary insights from soil scientists, agronomists, environmental researchers, microbiologists, and nanotechnologists. It highlights both the opportunities and risks associated with nanotechnology in soil ecosystems, emphasizing its role in nutrient delivery, soil quality enhancement, pollutant removal, microbial interactions, and real-time precision monitoring. By examining these interactions, the book underscores the importance of integrating nanotechnology with sustainable practices to build climate-resilient and productive agricultural systems.

The chapters compiled in this volume follow a structured progression from foundational knowledge to advanced applications. Part I introduces the rise of nanotechnology in soil science, presenting global research trends, fundamental concepts, and key developments in biochar and nanomaterial production. Part II examines how nanomaterials and soil amendments; such as biochar, nano-enabled fertilizers, algal nanomaterials, and microbial systems; contribute to soil health improvement, nutrient cycling, and ecosystem stability. Part III shifts the focus to nano-biochar and its wide-ranging applications in sustainable agriculture and soil remediation, presenting innovative strategies for improving plant growth and restoring degraded land. Part IV show with recent advancements in nanosensors and diagnostic tools that enable precision soil monitoring and data-driven agricultural decision-making and Part V concludes recent meta-analysis-based monitoring for soil and health monitoring. This book serves as a valuable resource for researchers, agronomists, environmental engineers, policymakers, and students seeking to understand the evolving role of nanotechnology in soil science. By bridging scientific evidence, practical applications, and policy perspectives, it promotes responsible innovation and informed decision-making. We extend our sincere appreciation to all contributing authors whose expertise and dedication have shaped this volume. Their collective efforts make this book a meaningful contribution to the field. It is our hope that this work will inspire further research, support sustainable solutions, and contribute to a healthier environment and a more resilient agricultural future.

ORGANIZATION OF THE BOOK

Chapter 1: The Rise of Nanotechnology in Soil Science: A Bibliometric Meta-Analysis of Global Research Trends (2010–2025)

Soil is an important part of the Earth's ecosystem because it supports farming, nutrient cycling, and the health of the environment. But practices of modern farming and industry have damaged soil to such an extent that it is no longer as fertile or as stable, imperiling both food security and the stability of ecosystems. Nanotechnology have gained importance in the context of soil science as a novel approach for new avenues to enhance soil, remediate pollution, and manage land in an environment-friendly manner. This study presents an in-depth meta-analysis of 252 publications (2010–2025) extracted from Scopus to evaluate the current state, research trends and future prospects for nanotechnology applications in soil science.

Chapter 2: Nanotechnology from Soil to Shoot: Nanomaterial Dynamics for Soil Health and Sustainability

Nanotechnology has become an important technological advancement in increasing soil functionality, crop productivity, and the long-term sustainability of agricultural production systems. In this chapter, we highlight how soil physicochemical properties (pH, texture, organic matter and salinity) determine the fate, transport and availability of nanomaterials for uptake by plants, translocated through the plant and influence physiological responses to nanomaterials. A mechanistic insight of how nanomaterials modulate use efficiency of nutrients, redox balance, and plant growth will also be provided. Overall, our chapter highlights the importance of developing a mechanistic, soil-mediated understanding of the nanomaterials to maximize agronomic benefits while minimizing risks to the environment and human health so that nanotechnology may be incorporated into soil-based agricultural systems in a sustainable manner.

Chapter 3: Introduction to Biochar: Composition, Production, and Applications

Biochar is a substance high in carbon that is produced by thermochemically converting biomass in areas with limited oxygen. It has garnered significant attention because it can address climate change, soil degradation, and waste management simultaneously. The chapter begins with the history of biochar in Amazonian Dark Earths (Terra Preta) and then proceeds to discuss its complex physicochemical properties, including surface area, porosity, elemental composition, and functional groups. We examine various methods for producing biochar, including gasification, hydrothermal carbonization, slow pyrolysis, and fast pyrolysis. Different methods produce biochar with varying qualities that are best suited for different purposes. It also shows how the choice of feedstock changes the properties of biochar. This technology can be applied in various ways, including cleaning up polluted soils, storing carbon, treating and purifying water, enhancing agricultural productivity, and even storing energy.

Chapter 4: Biochar as a Soil Amendment: Chemical Properties and Fertility Enhancement

Biochar is a carbon-rich substance made by pyrolyzing organic biomass in an environment with no oxygen. It has become a viable soil amendment that might greatly improve the chemical characteristics and fertility of soil. This chapter provides a comprehensive analysis of biochar's diverse effects on soil chemistry, encompassing changes to pH, cation exchange capacity, nutrient retention, and overall soil fertility measurements. Through examination of current research findings, we analysed the mechanisms by which biochar alters soil chemical dynamics, its interaction with native soil components, and the ensuing consequences for agricultural production and environmental sustainability. The conversation explores the role of biochar in carbon sequestration, nutrient cycling, and improving soil conditions. It also discusses how the impacts of biochar can vary depending on the feedstock sources, production settings, and soil types.

Chapter 5: Exploring Microbial Responses to Nanomaterials in Soil Environments: Balancing Act

As nanotechnology advances, the deployment of nanomaterials in various applications increases, inevitably leading to their release into the environment, particularly soil. By elucidating the mechanisms underlying microbial responses to nanomaterials, researchers can develop strategies to mitigate potential adverse effects and harness the beneficial roles of nanomaterials in soil environments. Moreover, the chapter explores the concept of a "balancing act" in microbial-nanomaterial interactions, highlighting the delicate equilibrium between nanomaterial-induced stress and microbial adaptation. It discusses how microbial communities may undergo shifts in composition and function in response to nanomaterial exposure, potentially affecting nutrient cycling, soil fertility, and ecosystem resilience. By advancing our understanding of these interactions, we can develop sustainable practices for the safe deployment of nanomaterials in agriculture, remediation, and other soil-related applications, ensuring the preservation of soil ecosystem health and function.

Chapter 6: Harnessing Nano-Enabled Nutrient Delivery Systems for Enhanced Plant Growth

This chapter focuses on nano-enabled nutrient delivery systems as emerging opportunities in agricultural technology that can address some of the current issues of nutrient use in agriculture. Different forms of nanomaterials such as nanofertilizers, nanosensors, nanopesticides, nanoemulsions, and nanoclays are described about their uses and how they work. The chapter focuses on the mode of utilization of nano-nutrient by the plant through root absorption, foliar absorption, and cellular uptake pathway of nano-particles. These systems show better uptake of the nutrients compared with the basic fertilizers hence reducing pollution while increasing the yields of crops. Case studies shows how nano-enabled systems are a better solution than traditional ones. This chapter lays out the opportunities of nano-enabled nutrient delivery systems in agriculture sector, but also the need to further investigate the valorization of these systems.

Chapter 7: Nano-Biochar in Sustainable Agriculture: Applications, Benefits, and Future Perspectives

Innovative soil amendment research is driven by the need for sustainable agriculture to provide global food security and reduce environmental damage. Biochar, made from organic feedstocks that promote soil health and sequester carbon. Bulk characteristics may limit its efficacy. Nanotechnology can modify biochar's functioning. Sub-micrometer particle size, large surface area, and unique physicochemical properties make nano-biochar (NBC) a significant breakthrough. Nano-biochar in sustainable agriculture is covered in this chapter. NBC production and characterization are explained, along with its properties compared to regular biochar. In this chapter, we discuss NBC's many uses as a soil conditioner, remediator, agrochemical transporter, and plant stress reducer. A heavy metal contaminated soil remediation case study shows NBC's practicality and efficacy. Finally, the chapter discusses nano-biochar's role in next-generation sustainable agriculture systems.

Chapter 8: Monitoring Earth's Vital Signs: Utilizing Nanosensors for Precision Soil Monitoring and Analysis

This chapter explores the innovative use of nanosensors for precision monitoring and analysis of soil health and environmental conditions. Nanotechnology offers unprecedented opportunities to develop sensors capable of detecting and quantifying various soil parameters with high sensitivity and specificity. By leveraging nanomaterials and nanotechnology principles, these sensors can provide real-time data on soil moisture, pH levels, nutrient concentrations, contaminant presence, and microbial activity, among other critical factors. This chapter discusses the principles underlying nanosensor design, fabrication, and deployment in soil environments. It explores the diverse applications of nanosensors in precision agriculture, environmental monitoring, and soil remediation efforts. Furthermore, the chapter examines the challenges and future prospects of integrating nanosensor technologies into soil science practices, emphasizing their potential to revolutionize soil management strategies and promote sustainable agriculture and environmental stewardship.

Chapter 9: Global Web of Science-Based Bibliometric Analysis of Heavy Metal Current Research Gaps and Biochar-Nanotechnology Based Solutions for Armenia

Heavy metals are one of the major persistent environmental threats to both human health and environmental quality, so that the determination of baseline and background becomes an indispensable step before reliable risk evaluation. This paper provides a full bibliometric analysis of the Web of Science-indexed literature (2000–2025) to analyzed global and Armenia-contextualized research on heavy metals. Global findings show that the publication landscape is very concentrated, with a few countries contributing most of the output and focussing on baseline environmental characterisation. While, Armenia stands out as a topically specialized and highly productive contributor with most of its publication trends leaning toward contamination sources, ecological risk, and human health impacts.

Acknowledgement

We express our sincere gratitude for the internal research grants from Yerevan State University, Republic of Armenia, awarded to AS in 2024, and Higher education and science committee MESCS Republic of Armenia provided the 24FP-4C013 grant to support KG, which played a crucial role in the preparation and successful development of the edited book, ***Nanotechnology Applications in Soil Science: From Remediation to Precision Agriculture***, published by IGI Global (USA).

VDR and TM also extend their heartfelt thanks to the Strategic Academic Leadership Program of Southern Federal University ("Priority 2030") and the Ministry of Science and Higher Education of the Russian Federation (Grant No. FENW-2023-0008) for their support. Their ongoing research contributions have greatly enhanced the scientific depth and interdisciplinary value of this volume.

RKS is deeply appreciative of the support from the ECONUTRI project, funded by the European Union Horizon Europe Innovation Programme (Grant Agreement No. 101081858), as well as the backing from the Portuguese Foundation for Science and Technology under the projects UID/04033/2025 and LA/P/0126/2020 (https://doi.org/10.54499/LA/P/0126/2020).

We extend our profound gratitude to all supporting institutions for facilitating the integration of innovative research, collaborative expertise, and transformative advancements in soil science, nanotechnology-based remediation, and precision agriculture.

Chapter 1
The Rise of Nanotechnology in Soil Science:
A Bibliometric Meta-Analysis of Global Research Trends (2010-2025)

Abhishek Singh
http://orcid.org/0000-0002-0049-6910
Yerevan State University, Armenia

Karen Ghazaryan
http://orcid.org/0000-0002-8973-2595
Faculty of Biology, Yerevan State University, Armenia

Rupesh Kumar Singh
http://orcid.org/0000-0002-2536-1967
Universidade de Trás-os-Montes e Alto Douro, Vila Real, Portugal

João Ricardo Sousa
http://orcid.org/0000-0003-4365-9086
Universidade de Trás-os-Montes e Alto Douro, Vila Real, Portugal

Tatiana Minkina
Academy of Biology and Biotechnology, Southern Federal University, Rostov-on-Don, Russia

Vishnu D. Rajput
http://orcid.org/0000-0002-6802-4805
Academy of Biology and Biotechnology, Southern Federal University, Rostov-on-Don, Russia

ABSTRACT

Soil is an important part of the Earth's ecosystem because it supports farming, nutrient cycling, and the health of the environment. But practices of modern farming and industry have damaged soil to such an extent that it is no longer as fertile or as stable, imperiling both food security and the stability of ecosystems. Nanotechnology

DOI: 10.4018/979-8-3693-7006-3.ch001

have gained importance in the context of soil science as a novel approach for new avenues to enhance soil, remediate pollution, and manage land in an environment-friendly manner. This study presents an in-depth meta-analysis of 252 publications (2010–2025) extracted from Scopus to evaluate the current state, research trends and future prospects for nanotechnology applications in soil science.

1. INTRODUCTION

While aligning with the 2030 Agenda for Sustainable Development, the U.N has pledged to meet ambitious goals of eradicating hunger and poverty as well as achieving agricultural sustainability. The world population is projected to rise up to 9.8 billion people by the year 2050, mainly in developing countries and ensuring food security requires that agricultural production increases by at least 50% (FAO, 2017). The land surface area is 29% of total the Earth's surface and among them only 11% of the land can be used for cultivation that is not enough for future increasing population in entire world (FAO, 2003, 2017). This not only compromise food security but also raises poverty rates that can also lead to serious economic consequences that is undermine for sustainable development objectives (FAO, 2003, 2017). Soil is a crucial component of land global resources and plays a vital role in terrestrial system functioning (Vogel et al., 2019). Soil is created by the combination of parent material, climate, biota and topography over time. Recent research publications on agricultural soils have, among other topics, focused on the following subjects: (1) microbial communities in soils (Dai et al., 2024), (2) agricultural soil pollution (Baqar et al., 2024), (3) soil pollution and risk assessment under a variety of climatic conditions and soil types (Wang et al., 2024), (4) understanding the challenges that introducing pharmaceuticals into agricultural soils raises (Cela-Dablanca et al., 2024), (5) long-term stocks of soil organic carbon in agricultural soils under bioeconomy (Andrade Díaz et al., 2024), (6) how agricultural soils support human health (Brevik et al., 2018), and (7) functions of soil onsite agro-ecosystems (D'Acunto et al., 2024). Soil environment or soil ecology changes are major global challenges which is indispensable for soil biodiversity conservation (Jansson, 2023; Kudureti et al., 2023; Steffan et al., 2020). Nanotechnology has recently generated great applications in almost all fields of life, ranging from basic sciences to multidisciplinary engineering (Ashraf et al., 2021; A. Singh et al., 2022, 2025). Soil is an important storehouse of global biodiversity, which is mainly regulated by the soil microorganism (Costa et al., 2018). For fulfilling the worldwide requirement NPs have also become a potential approach to mitigating the detrimental impacts of intensive agriculture on soil ecology and environment. Recent research reports that the uses of NPs are enhanced microbial diversity in soil helps to stabilization of environment and sustainable for agriculture management (Dinesh et al., 2012). Interestingly, in the recent decades- NPs have demonstrated great potential in soil remediation for health. The environmentally safe green-synthesized nanoparticles (NPs) have been reported as potential promising agents for soil health management because of their beneficial sustainability (Bandala & Berli, 2018). Nevertheless, although extensive efforts have been made to assess and understand their potential beneficial or harmful effects on agriculture, there are still major gaps in knowledge and further research needs to be carried out to investigate the yet unclear specific behavior features of NPs in soils (El-Ramady et al., 2024). Recent years have witnessed an increasing number of publications on the use of nanotechnology in environmental and agriculture policies based on the special properties of such materials. Nanomaterials have been widely used in soil fertilization and food production with unique physical and chemical properties, high efficiency, cost-effectiveness and environmental friendliness. Only 1 publication utilizing both "nanoparticles" and "soil" was found in 2010, whereas maximum number of publications (more than "252") with the same keywords were published for the last 15 years (Fig 1). In this chapter discussed the recent advances of NPs and their role in changes of global proposition of sustainable management of soil resources dynamics. For current study used bibliometric frequency analysis, inter-active analysis, citation analysis, keywords analysis, evolution analysis, and hotspot identification to identify current research focuses, knowledge gaps, and trends in the field. The findings provide insights for future research prospects and opportunities in NPs in the research area of soil science research.

2. MATERIAL AND METHOD

2.1 Data Collection

All the data are extracted from Scopus databased from 2010 to 2025 timeline by giving some search keywords query for first step data identification: "*Nanotechnology Applications in Soil Science*" with selection of search category are: *Article, Abstract, Keywords* (Luo et al., 2025; Quintarelli et al., 2024; Stopar, 2016). The first identification stage of data extraction provided 252 documents in Scopus database. After data identification process performed the screening of data-based on "*subject area (Environmental Science and Agricultural and Biological Sciences)*"; "*Document type (Articles, Reviews, Book chapters*"; "*Source type (Journals, Book)*"; "*Language (English)*"; "*Publication stage (Final)*". In final stage of inclusion, we selected total 252 more appropriate scientific documents for our further analysis. The collection encompasses 252 documents from 107 sources, exhibiting a 28.5% annual growth rate between the year 2010-2025 (Fig 2). These relatively recent documents, with an average age of 4 years, demonstrate significant impact, garnering 31.36 citations per document and amassing 1990 references in total. The breadth of topics is evident through 751 Author Keywords. Contributions from 1097 authors showcase strong collaborative efforts, with an average of 8.83 co-authors per document and 33.73% international collaborations (Figure 1).

Figure 1. Scopus database summary report on biochar and heavy metals, generated using R-studio

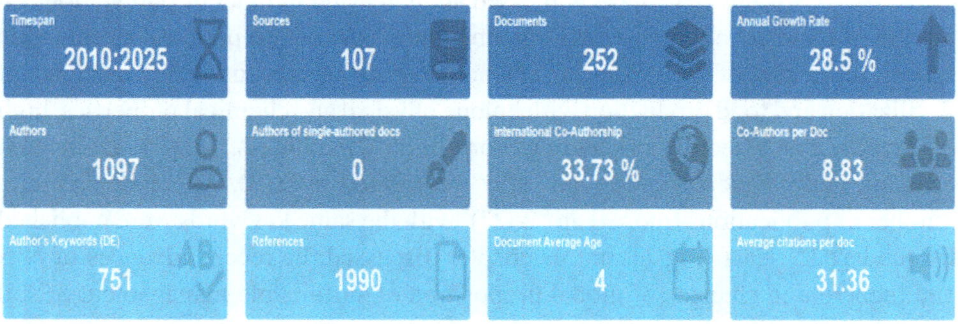

2.2 Data Visualization Tools

We used R-Studio-Biblioshiny (https://www.bibliometrix.org/home/index.php/layout/biblioshiny) (R version 4.4.1 (2024-06-14 ucrt) emerges as the primary bibliometric analytical instrument. The research methodology involves processing all collected data through R-Studio, which then produces bibliographic information essential for the ensuing analyses (Waseem Ahmad Bhat, 2023).

3. RESULTS AND DISCUSSION

3.1 Performance Analysis

(A) Annual Scientific Publication

Figure 2 provided outlines the annual scientific publications trend over time from 2010 to 2025. The number of articles per year steadily increased over the next decade, starting from 1 article in 2010 and reaching up to 252 articles in 2025 which indicating an increase in the interest on the topic. Overall, the data highlights a remarkable expansion in scientific output over the 15-year period (Figure 2).

Figure 2. The figure illustrates the yearly scientific publications concerning the use of nanotechnology in soil science

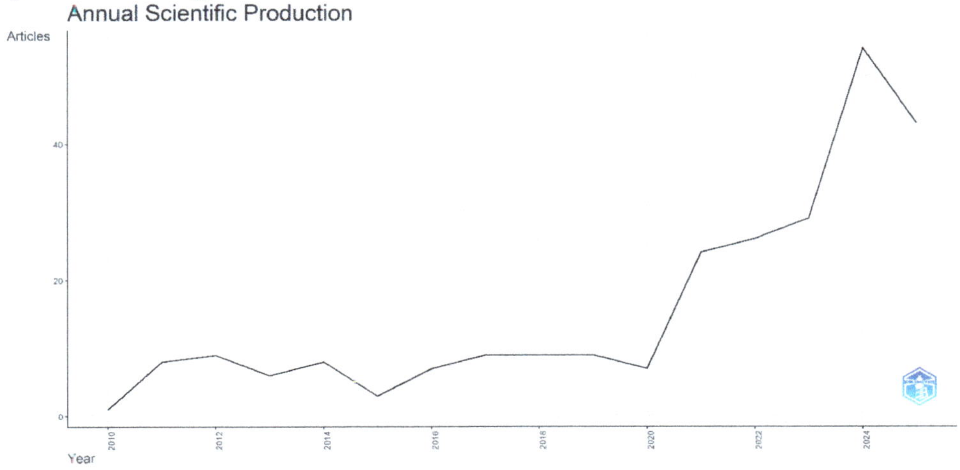

(B) Annual Citation Analysis

Annual citation analysis is a quantitative approach to the assessment of scholarly impact and influence, applied at various levels (articles, authors, journals, institutions). It is a measure of how often particular publications (or data sets etc) have been cited by other publishers or researchers over a given period of time. In current analysis show that citation is increased to 2010 to 2024 and after that show application of nanotechnology increased in soil science field for soil health management (Figure 3) (Singh et al., 2025; Sousa et al., 2025).

Figure 3. The diagram shows the annual average of citations related to the application of nanotechnology in the field of soil science

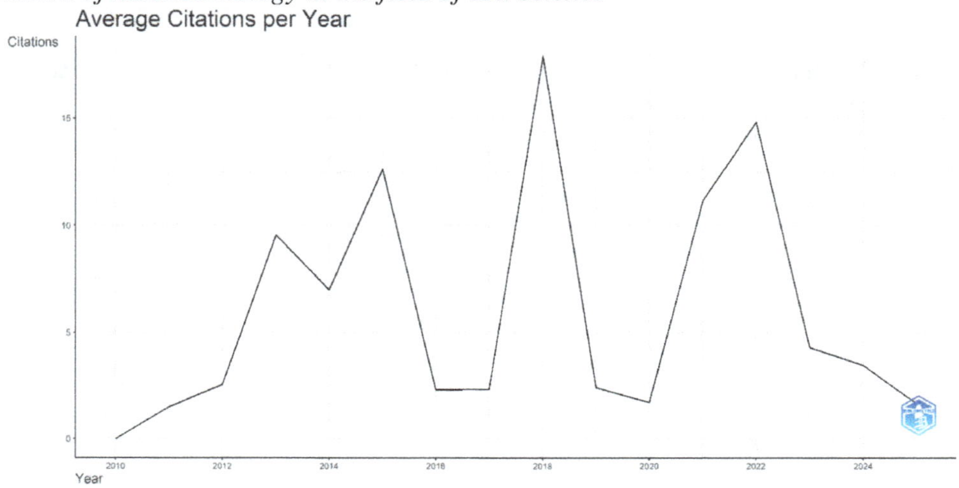

(C) Nations Scientific Production Trend

The recent publication trend indicates the diversity in studies related to use of nanotechnology in soil science and the interest there of, among different countries. India is ranked first with a very high scientific production (420 papers) and suggests the strong country interest to soil nanotechnologies, particularly nano-based remediation, fertilization and stress reduction (Figure 4). China ranks second in terms of an intermediate yield (109 publications), which is testament to its increasing investment in nanomaterials for soil improvement and pollution removal. Medium contributors also come from Mexico (61), Pakistan (56), Egypt (50), the USA (49) and Iran (44) working with such topics as nano-biochar interactions, heavy metal stabilization and nano-fertilizer development. Emerging interest is also evident in countries such as Saudi Arabia (31), Malaysia (15), and Bangladesh (13) which has gained momentum by expanding the research toward sustainable soil management with use of nanotechnology tools (Figure 4). Taken together, these observations reflect an emerging global focus on the use of nanotechnology-based solutions toward sustained soil health, environmental cleanup and crop resilience (Atanda et al., 2025; Futa et al., 2024; A. Singh et al., 2024).

Figure 4. Nation regarding scientific production publication associated with nanotechnology and soil science

Country Scientific Production

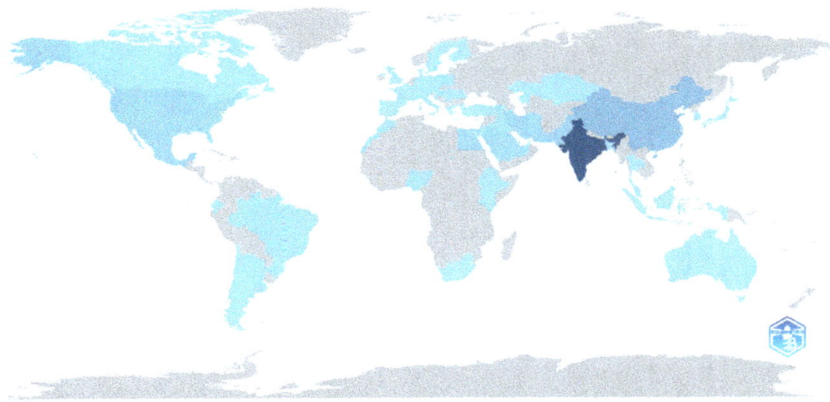

(D) Nation's Citation Trend

Citation analysis indicates the worldwide impact of research in nanotechnology approaches to soil science. India ranks first in citation count (1935) among all other countries indicating its remarkable global influence and high-quality contributions particularly related to nano-enabled soil remediation, nano-fertilizers and soil NPs interactions (Figure 5). Korea is the second (1020 citations) Country of origin which has a high impact in Prospective on engineered nanomaterials and soil ecological risk. China is the second most productive country (909 citations), which reflects its significant work in nano-biochar, heavy-metal immobilization and the nano-facilitated nutrient delivery system. Iran (355), Pakistan (252) and USA (248) constitute the second-tier effective contributors, concentrating on soil pollution reduction, toxicology of nanomaterials in soils and sustainable management of soil system (Figure 5). European nations such as Croatia (159) and Czech Republic (125) also some other nations like Mexico (150), Qatar (138) have indicated presence through streamlines studies on nano-remediation, soil quality improvement and environmental nanotechnology (Fig 5). In general, the trend of citations indicates that although there are many countries who are currently contributing to this scientific research and promoting awareness through collaboration, India followed by Korea and China are most influential in shaping global scientific discourse on nanotechnology-based soil health and sustainable agricultural interventions.

Figure 5. The current image show the citation of a scientific publication concerning nanotechnology and soil science

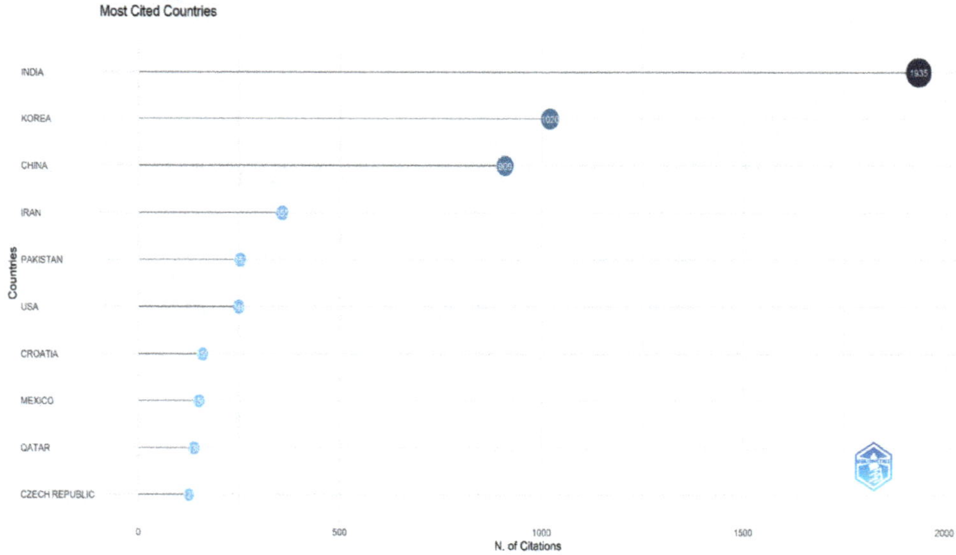

3.2 Science Mapping

(A) Keyword Analysis

I. Most Frequent Word

Keyword co-occurrence analysis reveals hot topics in the nanotechnology applications in soil science. The term with the highest frequency is "nanotechnology" (59), which indicates the prevalence of nano-based methodologies for soil improvement, treatment and sustainable agriculture. Both "nanoparticles" (27) and "nanoparticle" (21) collectively illustrate the intense concentration of papers on engineered nanomaterials, their characteristics, behaviour in environmental media and interactions. "Sustainable development" (27) stands out as one of the strong terms, suggesting that researchers perceive nanotech for enhancing long-term and sustainable conditions in agricultural and environmental sector (Figure 6). Terms including "agriculture" (23) and "nonhuman" (23) indicate plant–soil systems, ecological effects, and other non-human biological responses to nanomaterials. "soil pollution" and "fertilizers" (21, 19) indicate two main application areas at remediation of contaminated soils, and in the development of nano-fertilizer technologies with improved nutrient utilization efficiency. The word "review" (19) shows the increased number of review papers on their advancements in soil nanotechnology. Last, "nanostructured materials" (18) highlights the continued fascination with new nano-engineered materials used for soil modification, contaminant stabilization, and productivity boosting (Figure 6). In summary, the top ten keywords support that research is highly focused in direction of nano-facilitated soil management, environmental safeguard and sustainable agricultural efficiency.

Figure 6. Most frequent keywords linked to the use of nanotechnology in the field of soil science

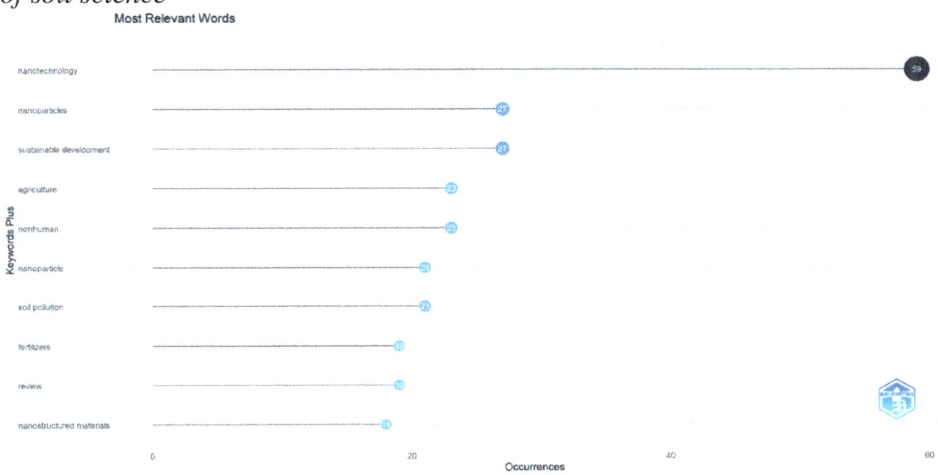

II. Word Cloud

In a bibliometric analysis, a word cloud is a chart representing the frequency of keywords within a data collection of publications; the size of each keyword represents its weight (Figure 7). This technique is employed to detect new trends, examine author-contributed keywords, and extract an immediate profile of the main terms related with a particular research field (Kumar et al., 2023; Mathews et al., 2015; Patil et al., 2023).

Figure 7. Word cloud related to nanotechnology and soil science

III. Tree Map

A treemap is a visualization method widely used in bibliometric analysis for presenting hierarchical data with nested rectangles. It is useful for visual identification and visual differentiation of key themes, keywords, sources, authors or countries by their frequency and significance with in a particular literature body (Chandran et al., 2024; Ismail et al., 2025). The most common term, "soil" (425), reflects a focus of current research on soil health, soil remediation and the interactions of soils with nanomaterials (Figure 8). A close second is "nanotechnology" (422), substantiating the level of emphasis given to nano-enabled solutions for enhancing soil quality, nutrient use efficiency, and environmental sustainability. The high frequency words "nanoparticles" (317) and "agriculture" (316) indicate the double focus of the field: how engineered nanoparticles are used for increasing agricultural yield and learning their behavior in soil systems. Keywords such as "environmental" (283) and "plant" (257) underscore trends of investigation about environmental risk assessment, plant responses, and ecological safety problems of nanomaterials. The word combinations "reserved", "rights" and the noun "bv" (each 252–257) stem mostly from publisher copyright notices ("All rights reserved"; "BV"), which are generated automatically in keyword phrase extraction, so should also be removed during bibliometric filtering, as they do not reflect scientific content (Figure 8). The final section "applications" (246) (Figure 8) puts into perspective the wide range of nano-enabled applications, from soil pollution abatement and nutrient supply to mitigation of crop stress, showing that practical use of nanotechnology in the soil–plant–environment continuum is an expanding concept.

Figure 8. The tree map illustrates connections between nanotechnology and soil science, highlighting their contributions to soil, agriculture, and nanotechnology

IV. Trend Topic

Trend topics chart (Figure 9) presents a decade landscape, of which it provides insights into the general trends and movements in the identity research.

Figure 9. The trending topic reveals the links between nanotechnology and soil science, emphasizing their roles in soil, agriculture, and nanotechnology

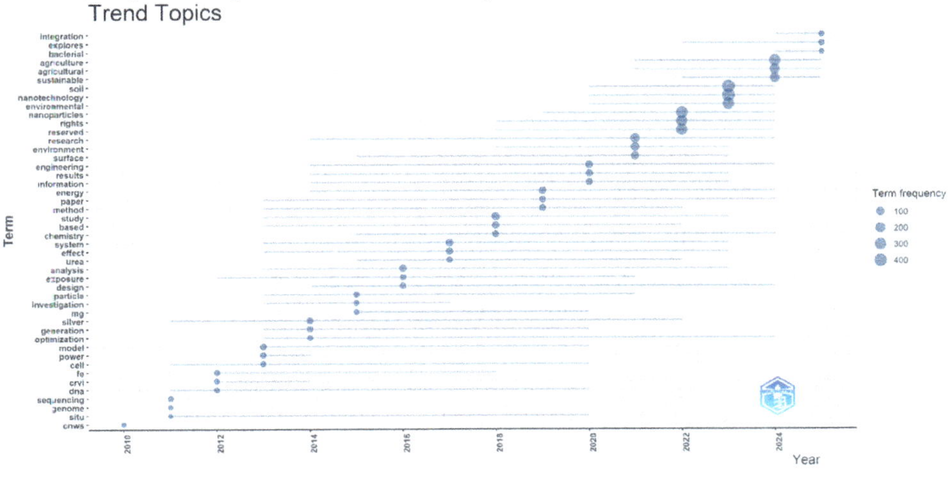

(B) Network Approaches

I. Co-Occurrence Network

In bibliometric analysis, a co-occurrence network is a visualization and analysis technique for mapping research fields in terms of the conceptual structure, emerging trends and broad themes and the specific areas of research conducted within the domain of the research (Figure 10) (Klarin, 2024; Narong et al., 2023). It identifies and visualizes the links between entities (i.e. keywords, terms, authors or organizations) that co-occur within a given document level unit of text e.g. research article, sentence (Ninkov et al., 2024).

Figure 10. The figure illustrates the co-occurrence network based on a keyword analysis, all of which pertain to nanotechnology and soil science

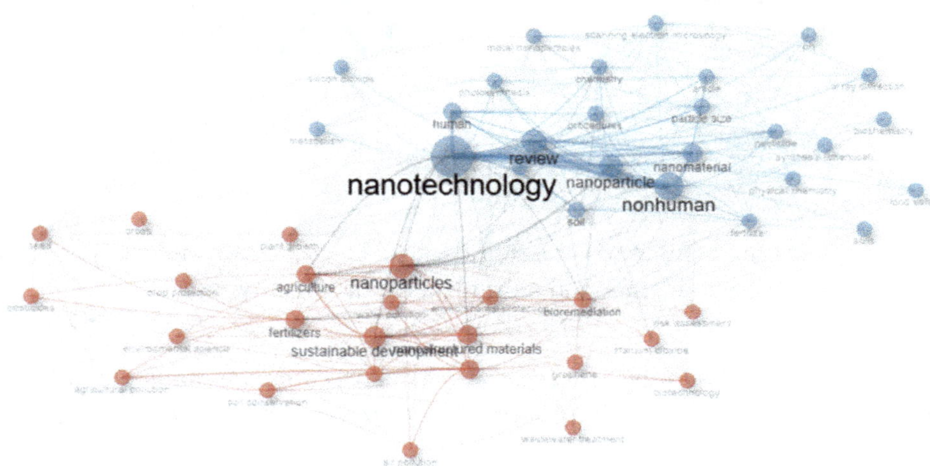

II. Countries Collaboration World Map

The bibliometrix country collaboration world map Visualization in scientific querying is a valuable tool to illustrate the worldwide picture of international cooperation in one specific field (Figure 11) (Akrami et al., 2023). The map visualizes complex data with graphic detail: countries are expressed as nodes which often have color or size coding to illustrate a country's output and volume of work in the field (Aria & Cuccurullo, 2017). Lines linking these nations (often rendered in colors to represent clusters, or weighted by thickness according to frequency) are used to depict current collaborative connections, with thicker lines representing more frequent (or more substantial) co-authored publications among the countries. This visual presentation is efficient in pinpointing leading global research nodes, showcasing established and strong international networks, and showing possible deficiency of collaboration that can be very useful to both researchers as well as funding bodies and policy makers interested in global soil science and nanotechnology dynamics.

Figure 11. The map illustrating international collaboration highlights the connections between nanotechnology and soil science, underscoring their significance in the fields of soil, agriculture, and nanotechnology

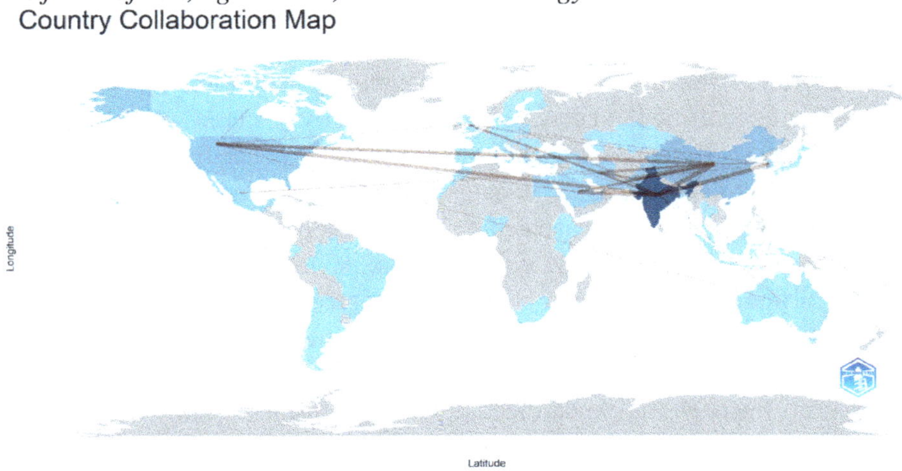

4. FUTURE PROSPECTIVE

The soil structure is an important factor for good healthy soils and it affects such essential physical and biological aspects as water holding capacity, aeration, penetration of roots, activity of microorganism etc. Because such problems as compaction, erosion and poor water absorption hinder plant growth and degrade soil for the future. The opportunity that nanomaterials (e.g. Nano clays, Nano silica etc.) offer for such challenges and improving soil properties (Saadi et al., 2020; Wahab et al., 2024). These nanoparticles act as efficient binding agents that enable soil particles to aggregate into stable aggregates. This clumping action creates pore spaces and water holding capacity for plant roots and soil microorganisms. In addition, higher porosity enhances aeration required for root respiration and microbiological activities. Better soil aeration and water holding capacity in combination with the in-situ fertilizer along with an enhanced till provides for a more diverse and flexible microbial ecology without sacrificing consistent and health plant growth. The physical attributes provided by nanoparticles enhance soil resistance to compaction and erosion, which supports its long-term fertility & utility. These are success stories that illustrate the promise of nanotechnology in soil health management, evidencing new approaches that serve as an alternative to achieving the goals of sustainable agriculture. As the research progresses, it is expected that through application of nanomaterials, soil conservation practices will change, ensuring stability and sustainability of agricultural ecosystems for humanity (Wahab et al., 2024).

The present bibliometric analysis results show the increasing volume and the wide spectrum of this research area of nanotechnology applications in soil science. Even so, several scientific and practical gaps will need to be addressed in future studies. First, future work needs mechanistic insights into nano–soil–plant–microbe interactions under realistic field conditions. The majority of studies conducted have small time span experiments under controlled conditions hence they do not reflect the complexity of soils ecosystems in nature. Thus, long-term studies are required to understand the fate, dynamics and ecological effects of engineered and biogenic nanoparticles in soils with varying compositions, environmental conditions, and climates. A further important avenue is the design of green, environmentally benign nanomaterials. Green synthetic and biodegradable nanoparticles that exhibit low toxicity to soil biota and environmental surrounding-ecosystem must to be emphasized. This fills the need for standardized toxicity assessments and harmonization of evaluation protocols so that nano-enabled agricultural inputs can be applied safely. Moreover, there is a demand for the field-based studies mainly for multi or combined abiotic stresses like salinity, drought, heavy metal, heat interactions to follow the realistic scenario imposed on sustainable agriculture systems. Efforts for future work should also investigate combined strategies where nanotechnology can be integrated with nature-based solutions (e.g., biochar, compost, benign microbes, phytoremediation strategies). Co-benefit systems can increase nutrient use efficiency, improve soil quality, and enhance remediation processes, contributing to sustainability purposes. In parallel, we call for global collaboration and transparent data-sharing platforms that promote scientific transparency, accelerate innovation, and sustainably strengthen emerging research capacity in low- and middle-income countries. Finally, clearer policy guidelines and regulations concerning the application of nanomaterials in agricultural systems are also urgently required. It would be important for governments and institutions to strive for harmonized legislation that minimizes the risk to the environment and human health but at the same time seeks to promote critical innovation. Practical advances resting on scientific ones will also require the development of nano-enabled technologies that are affordable and simple enough for smallholder farmers to deploy. The above directions together will promote the safe, effective and sustainable integration of nanotechnology into sustainable soil health management and sustainable and climate-resistant agricultural systems.

5. CONCLUSION

This bibliometric analysis illustrates the impressive expansion of scientific work that connects nanotechnology and soil science in the last 15 years. These trends have been reflected by the consistent growth in annual publications, citations, and international involvement leading to an acknowledgement of nanotechnology as a valuable instrument for sustainable agriculture, soil health management, and environmental remediation. India, China, and Korea are the top three largest contributes in productivity and citation impact which indicate the active voice from the nations in nano-enabled soil research. We find that the dominating research themes, as confirmed by keyword mapping, co-occurrence network and trend analysis, refers to soil remediation, nanoparticle interactions, sustainable development, nano-fertilizers, and environmental safety. A growing range of research topics indicates the continued diversification of interest in the field, from nano-biochar to nutrient delivery systems to ecological risk and plant–soil responses. In conclusion, the results highlight that nanotechnology represents a promising tool to tackle some of the soil-related big challenges. Nevertheless, further nanostructure-based research needs interdisciplinary studies, validation in the field on long terms, safety and implementation, and strengthening of fair international collaborations to be successful to plug proposed nano-emerging technologies into soil and agriculture management sustainability.

ACKNOWLEDGEMENT

KG is support by internal grant program of Yerevan State University (2023). AS is support by internal grant program of Yerevan State University (2024).

REFERENCES

Andrade Díaz, C., Albers, A., Zamora-Ledezma, E., & Hamelin, L. (2024). The interplay between bioeconomy and the maintenance of long-term soil organic carbon stock in agricultural soils: A systematic review. *Renewable & Sustainable Energy Reviews*, *189*, 113890. DOI: 10.1016/j.rser.2023.113890

Aria, M., & Cuccurullo, C. (2017). bibliometrix: An R-tool for comprehensive science mapping analysis. *Journal of Informetrics*, *11*(4), 959–975. DOI: 10.1016/j.joi.2017.08.007

Ashraf, S. A., Siddiqui, A. J., Elkhalifa, A. E. O., Khan, M. I., Patel, M., Alreshidi, M., Moin, A., Singh, R., Snoussi, M., & Adnan, M. (2021). Innovations in nanoscience for the sustainable development of food and agriculture with implications on health and environment. *The Science of the Total Environment*, *768*, 144990. DOI: 10.1016/j.scitotenv.2021.144990 PMID: 33736303

Atanda, S. A., Shaibu, R. O., & Agunbiade, F. O. (2025). Nanoparticles in agriculture: balancing food security and environmental sustainability. *Discover Agriculture 2025 3:1, 3*(1), 26-. DOI: 10.1007/s44279-025-00159-x

Bandala, E. R., & Berli, M. (2018). Nanomaterials: New Agrotechnology Tools to Improve Soil Quality? *Agricultural Nanobiotechnology*, 127–140. DOI: 10.1007/978-3-319-96719-6_7

Baqar, M., Saleem, R., Zhao, M., Zhao, L., Cheng, Z., Chen, H., Yao, Y., & Sun, H. (2024). Combustion of high-calorific industrial waste in conventional brick kilns: An emerging source of PFAS emissions to agricultural soils. *The Science of the Total Environment*, *906*, 167612. DOI: 10.1016/j.scitotenv.2023.167612 PMID: 37804982

Brevik, E. C., Pereg, L., Steffan, J. J., & Burgess, L. C. (2018). Soil ecosystem services and human health. *Current Opinion in Environmental Science & Health*, *5*, 87–92. DOI: 10.1016/j.coesh.2018.07.003

Cela-Dablanca, R., Míguez-González, A., Barreiro, A., Rodríguez-López, L., Arias-Estévez, M., Núñez-Delgado, A., Fernández-Sanjurjo, M. J., Castillo-Ramos, V., & Álvarez-Rodríguez, E. (2024). Simultaneous adsorption of amoxicillin and ciprofloxacin on agricultural soils and by-products used as bio-adsorbents: Unraveling the interactions in complex systems. *Environmental Research, 240*, 117535. DOI: 10.1016/j.envres.2023.117535 PMID: 39492492

Chandran, R., & Chandran MC, S. (2024). Green finance and sustainability: mapping research development through bibliometric analysis. *Discover Sustainability 2024 5:1, 5*(1), 339-. DOI: 10.1007/s43621-024-00549-z

Costa, J., Oliveira, R. S., Tiago, I., Ma, Y., Galhano, C., Freitas, H., & Castro, P. (2018). Soil Microorganisms. *Advances in Plant Ecophysiology Techniques*, 457–482. DOI: 10.1007/978-3-319-93233-0_27

D'Acunto, L., Iglesias, M. A., Poggio, S. L., & Semmartin, M. (2024). Land cover, plant residue and soil microbes as drivers of soil functioning in temperate agricultural lands. A microcosm study. *Applied Soil Ecology, 193*, 105133. DOI: 10.1016/j.apsoil.2023.105133

Dai, W., Liu, Y., Yao, D., Wang, N., Shao, J., Ye, X., Cui, Z., Zong, H., Tian, L., Chen, X. Z., & Wang, H. (2024). Biogeographic distribution, assembly processes and potential nutrient cycling functions of myxobacteria communities in typical agricultural soils in China. *The Science of the Total Environment, 906*, 167255. DOI: 10.1016/j.scitotenv.2023.167255 PMID: 37741390

Dinesh, R., Anandaraj, M., Srinivasan, V., & Hamza, S. (2012). Engineered nanoparticles in the soil and their potential implications to microbial activity. *Geoderma, 173–174*, 19–27. DOI: 10.1016/j.geoderma.2011.12.018

El Akrami, N., Hanine, M., Flores, E. S., Aray, D. G., & Ashraf, I. (2023). Unleashing the Potential of Blockchain and Machine Learning: Insights and Emerging Trends From Bibliometric Analysis. *IEEE Access : Practical Innovations, Open Solutions, 11*, 78879–78903. DOI: 10.1109/ACCESS.2023.3298371

El-Ramady, H., Prokisch, J., Sári, D., Singh, A., Ghazaryan, K., Rajput, V. D., & Brevik, E. C. (2024). Nanotechnology in the soil system: An ecological approach towards sustainable management. *Applied Soil Ecology, 204*, 105669. Advance online publication. DOI: 10.1016/j.apsoil.2024.105669

FAO. (2003). Prospects for aggregate agriculture and major commodity groups. *World Agriculture: Towards 2015/2030 An FAO Perspective*, 57–123.

FAO. W. I. U. W. (2017). The State of Food Security and Nutrition in the World 2017. *United Nations Human Settlements Programme (UN-Habitat): Addis Ababa, Ethiopia. (2021), 1*, 1–132. Retrieved from https://openknowledge.fao.org/handle/20.500.14283/i7695en

Futa, B., Gmitrowicz-Iwan, J., Skersienė, A., Šlepetienė, A., & Parašotas, I. (2024). Innovative Soil Management Strategies for Sustainable Agriculture. *Sustainability 2024, Vol. 16, Page 9481, 16*(21), 9481. DOI: 10.3390/su16219481

Ismail, A., Munsi, H., Yusuf, A. M., & Hijjang, P. (2025). Mapping One Decade of Identity Studies: A Comprehensive Bibliometric Analysis of Global Trends and Scholarly Impact. *Social Sciences 2025, Vol. 14, Page 92, 14*(2), 92. DOI: 10.3390/socsci14020092

Jansson, J. K. (2023). Soil viruses: Understudied agents of soil ecology. *Environmental Microbiology, 25*(1), 143–146. DOI: 10.1111/1462-2920.16258 PMID: 36271323

Klarin, A. (2024). How to conduct a bibliometric content analysis: Guidelines and contributions of content co-occurrence or co-word literature reviews. *International Journal of Consumer Studies, 48*(2), e13031. DOI: 10.1111/ijcs.13031

Kudureti, A., Zhao, S., Zhakyp, D., & Tian, C. (2023). Responses of soil fauna community under changing environmental conditions. *Journal of Arid Land, 15*(5), 620–636. DOI: 10.1007/s40333-023-0009-4

Kumar, D., Shandilya, A. K., & Srivastava, S. (2023). The journey of F1000Research since inception: through bibliometric analysis. *F1000Research 2023 12:516, 12*, 516. DOI: 10.12688/f1000research.134244.1

Luo, X., Li, J., Guo, S., Yu, H., Zeng, X., Zhou, Z., Shangguan, Y., He, M., Ouyang, Y., Chen, K., Chen, Z., & Qin, Y. (2025). Analysis of research status and trends on nano-agricultural application: A bibliometric study. *Frontiers in Plant Science, 16*, 1530629. DOI: 10.3389/fpls.2025.1530629 PMID: 40034150

Mathews, D., Franzen-Castle, L., Colby, S., Kattelmann, K., Olfert, M., & White, A. (2015). Use of Word Clouds as a Novel Approach for Analysis and Presentation of Qualitative Data for Program Evaluation. *Journal of Nutrition Education and Behavior, 47*(4), S26. DOI: 10.1016/j.jneb.2015.04.071

Narong, D. K., & Hallinger, P. (2023). A Keyword Co-Occurrence Analysis of Research on Service Learning: Conceptual Foci and Emerging Research Trends. *Education Sciences, 13*(4), 339. DOI: 10.3390/educsci13040339

Ninkov, A. B., Frank, J. R., Costello, J. A., Artino, A. R.Jr, & Maggio, L. A. (2024). Bibliometric Networks for Researchers in Health Professions Education. *Academic Medicine, 99*(9), 1048. DOI: 10.1097/ACM.0000000000005700 PMID: 38478805

Patil, R. R., Kumar, S., Rani, R., Agrawal, P., & Pippal, S. K. (2023). A Bibliometric and Word Cloud Analysis on the Role of the Internet of Things in Agricultural Plant Disease Detection. *Applied System Innovation 2023, Vol. 6, Page 27, 6*(1), 27. DOI: 10.3390/asi6010027

Quintarelli, V., Ben Hassine, M., Radicetti, E., Stazi, S. R., Bratti, A., Allevato, E., Mancinelli, R., Jamal, A., Ahsan, M., Mirzaei, M., & Borgatti, D. (2024). Advances in Nanotechnology for Sustainable Agriculture: A Review of Climate Change Mitigation. *Sustainability 2024, Vol. 16, Page 9280, 16*(21), 9280. DOI: 10.3390/su16219280

Saadi, M. A. S. R., Uluutku, B., Parvini, C. H., & Solares, S. D. (2020). Soft sample deformation, damage and induced electromechanical property changes in contact- A nd tapping-mode atomic force microscopy. *Surface Topography : Metrology and Properties, 8*(4), 045004. Advance online publication. DOI: 10.1088/2051-672X/abb888

Singh, A., Bol, R., Lovynska, V., Singh, R. K., Sousa, J. R., & Ghazaryan, K. (2025). Application of nanoparticles for salinity stress management and biofortification in wheat: A review of dual approaches and insights. *Frontiers in Plant Science, 16*, 1592866. Advance online publication. DOI: 10.3389/fpls.2025.1592866 PMID: 40688693

Singh, A., Sengar, R. S., Rajput, V. D., Minkina, T., & Singh, R. K. (2022). Zinc Oxide Nanoparticles Improve Salt Tolerance in Rice Seedlings by Improving Physiological and Biochemical Indices. *Agriculture 2022, Vol. 12, Page 1014, 12*(7), 1014. DOI: 10.3390/agriculture12071014

Singh, A., Sharma, A., Singh, O., Rajput, V. D., Movsesyan, H., Minkina, T., Alexiou, A., Papadakis, M., Singh, R. K., Singh, S., Sousa, J. R., El-Ramady, H. R., Zulfiqar, F., Kumar, R., Al-Ghamdi, A. A., & Ghazaryan, K. (2024). In-depth exploration of nanoparticles for enhanced nutrient use efficiency and abiotic stresses management: Present insights and future horizons. *Plant Stress, 14*, 100576. DOI: 10.1016/j.stress.2024.100576

Singh, S., Singh, A., Chakhmakhchyan, A., Hayrapetyan, A., Darbinyan, N., & Ghazaryan, K. (2025). Global Research Trends on Drought Stress and Nanoparticles (2012–2025): A Bibliometric Analysis. *Journal of Innovative Solutions for Eco-Environmental Sustainability, 18*, 018–018. DOI: 10.46991/JISEES.2025.SI1.018

Sousa, J. R., Ferreira, J., Saraiva, F., Singh, R. K., Singh, A., Ghazaryan, K., & Trindade, H. (2025). Soil Amendments for the Remediation of Cu-Contaminated Vineyard Soil: Effects of Biochar, Nanobiochar and Chitosan on Copper Bioavailability and Ecotoxicity. *Journal of Innovative Solutions for Eco-Environmental Sustainability*, 033–033. DOI: 10.46991/JISEES.2025.SI1.033

Steffan, J. J., Derby, J. A., & Brevik, E. C. (2020). Soil pathogens that may potentially cause pandemics, including severe acute respiratory syndrome (SARS) coronaviruses. *Current Opinion in Environmental Science & Health*, *17*, 35–40. DOI: 10.1016/j.coesh.2020.08.005 PMID: 33521411

Stopar, K. (2016). Presence of nanotechnology in agriculture: Bibliometric approach. *Acta Agriculturae Slovenica*, *107*(2), 497–507. DOI: 10.14720/aas.2016.107.2.20

Vogel, H. J., Wollschläger, U., Helming, K., Heinrich, U., Willms, M., Wiesmeier, M., Russell, D., & Franko, U. (2019). Assessment of Soil Functions Affected by Soil Management. *Atlas of Ecosystem Services: Drivers, Risks, and Societal Responses*, 77–82. DOI: 10.1007/978-3-319-96229-0_13

Wahab, A., Muhammad, M., Ullah, S., Abdi, G., Shah, G. M., Zaman, W., & Ayaz, A. (2024). Agriculture and environmental management through nanotechnology: Eco-friendly nanomaterial synthesis for soil-plant systems, food safety, and sustainability. *The Science of the Total Environment*, *926*, 171862. DOI: 10.1016/j.scitotenv.2024.171862 PMID: 38527538

Wang, C., Jiang, Y., Shao, Y., Chen, Z., Gao, Y., Liang, J., Gao, J., Fang, F., & Guo, J. (2024). The influence and risk assessment of multiple pollutants on the bacterial and archaeal communities in agricultural lands with different climates and soil properties. *Applied Soil Ecology*, *193*, 105130. DOI: 10.1016/j.apsoil.2023.105130

Waseem Ahmad Bhat, N. L. K. A. M. Z. A. D. R. A. Q. (2023). How to Conduct Bibliometric Analysis Using R- Studio: A Practical Guide. [EEL]. *European Economic Letters*, *13*(3), 681–700. DOI: 10.52783/eel.v13i3.350

Chapter 2

Nanotechnology From Soil to Shoot: Nanomaterial Dynamics for Soil Health and Sustainability

Mohit Yadav
http://orcid.org/0009-0005-4511-3935
Chaudhary Charan Singh Haryana Agricultural University, India

Aryan Dhamija
Chaudhary Charan Singh Haryana Agricultural University, India

Reena Sharma
Chaudhary Charan Singh Haryana Agricultural University, India

Manju Verma
Chaudhary Charan Singh Haryana Agricultural University, India

Jyoti Taunk
Chaudhary Charan Singh Haryana Agricultural University, India

Shikha Yashveer
http://orcid.org/0000-0001-6997-141X
Chaudhary Charan Singh Haryana Agricultural University, India

ABSTRACT

Nanotechnology has become an important technological advancement in increasing soil functionality, crop productivity, and the long-term sustainability of agricultural production systems. In this chapter, we highlight how soil physicochemical properties (pH, texture, organic matter and salinity) determine the fate, transport and availability of nanomaterials for uptake by plants, translocated through the plant and influence physiological responses to nanomaterials. A mechanistic insight of how nanomaterials modulate use efficiency of nutrients, redox balance, and plant growth will also be provided. Overall, our chapter highlights the importance of developing a mechanistic, soil-mediated understanding of the nanomaterials to

DOI: 10.4018/979-8-3693-7006-3.ch002

maximize agronomic benefits while minimizing risks to the environment and human health so that nanotechnology may be incorporated into soil-based agricultural systems in a sustainable manner.

INTRODUCTION

Nanoparticles are particles with sizes ranging from 1 to 100 nm which possess specific chemical and physical properties that are quite different from those of the bulk form (Joudeh & Linke, 2022). Despite their small size, nanoparticles also exhibit a high surface area-to-mass ratio. These properties of nanoparticles allow them to control their physical and chemical properties, known as quantum effects (Chanu et al., 2021). In the modern world, nanotechnology is considered an emerging field in sustainable agriculture. The use of nanomaterials has further extended to Genetically Modified Crops (GMC), animal production input, biocides, and precision farming using nano sensors (Chanu et al., 2021). The rapid increase in population and climate change hinder crop productivity. To ensure food security in 21st century, various approaches are used to achieve sustainable agriculture with limited resources. (Fanzo et al., 2017; Watts et al., 2018). Many significant advancements have been made in conventional agriculture; however, there is still a lot of scope for improvement in global agriculture(Kakaei et al. 2022). Conventional fertilizers, pesticides, and soil amendments often results for inefficiencies. They contribute to nutrient imbalances, leaching losses, and environmental pollution. Another major problem are the abiotic stresses affecting soil that contribute majorly for crop losses(Verma et al., 2020; Kumari et al., 2022; Patni et al., 2022; Verma et al., 2023). Among them soil salinity is one of the major stresses affecting agricultural production, approx. 1125 mha of world's total land area is affected by salinity, among which 20% is agricultural land and 33% of which is irrigated land. (Hossain, 2019; Negacz et al., 2022).On an average 1.5mha of agricultural land is lost because of salt water. Due to soil salinity, one of another major abiotic stress emerges in agricultural field that is drought. (Kumari et al., 2022).

Around the globe, 10 million sites of soil contamination are reported, of which 50% are polluted with toxic metal ions. These toxic metal ions are introduced into the soil through various direct and indirect routes, including industrial emissions, coal combustion, wastewater-based irrigation, petrochemical and hydrocarbon spills, animal waste, and sewage sludge application. Prolonged heavy metal contamination results into change into physical and chemical properties of soil which lead to disturbance of soil profile (Boretti and Rosa, 2019; Urra et al., 2019; Rashid et al., 2023; Verma et al., 2023; Singh et al., 2025a).

Soil acts as a central nexus of disturbed agricultural output, emphasizing the necessity for targeted soil management and remediation approaches to support sustainable agricultural systems (Jatav et al., 2025). The capability of soil to function within an ecosystem or land use constraint, at the same time support plant and animal health, and protect the environment, is referred to as soil quality from the perspectives of both sustainable agriculture and environmental sustainability (Anikwe and Ife, 2023; Verma et al., 2022a; Sharma et al., 2023). There are several strategies being utilised to improve soil health for enhancing crop yields and/or production levels including modification of water management, application of nano-enhanced fertilizers, pesticides, plant growth regulators, and soil amendments (Abdul Aziz et al., 2022; Urra et al., 2019; Verma et al., 2022b; Akanmu et al., 2023). In recent years, soil amendments have become a primary method for managing soil profiles to improve plant growth and production in agroecosystems. (Rubin et al., 2023; Urra et al., 2019; Kumari et al., 2023). A number of researchers have indicated that soil amendments are a viable alternative for improving soil profiles, restoring soil health, and increasing crop yields (Urra et al., 2019; Rani et al., 2023).

Chemical, physical, and biological changes in the soil properties are a direct or indirect result from the interaction between nanomaterials and the soil due to the high reactivity, high surface area, surface charge and morphological structure of nanomaterials.(Babu & Joseph, 2016). The pH and the presence of organic matter in addition to the amount of macro (NPK) and micro nutrients are the most significant chemical components of soils that determine soil quality. The soil microbiome, nitrogen mineralization, and soil aeration are the main biological indicators of soil quality while the water retaining capacity, density, and stability of the soil are the principal physical attributes (Nielsen and Winding, 2002).

Another approach to improve soil quality is the combination of plant growth promoters with nanoparticles. Due to their increased surface area and stability, nanoparticles improve the effectiveness of plant growth promoters. (Akhtar et al., 2021; Alharbi et al., 2023; Nayana et al., 2020; Rajput et al., 2023; Verma et al., 2023b; Singh et al., 2025b-c). The objective of this book chapter is to emphasize the contributions of nanoparticles to the improvement of soil quality and remediation.

ROLE OF SOIL AS REGULATORS OF NANO PARTICLE BEHAVIOUR

Soil determines how nanoparticles behave within it due to its capacity to act as a chemical and biological reactor where nanoparticles are regulated by the formation of their aggregates, transformations, absorption and bioavailability, and thus influencing its environmental impact on ecosystems and agriculture (Zhang et al., 2020; Ma et al., 2023).

Entry of nanoparticles into the plant system is achieved via two pathways: apoplastic and symplastic. Extracellular spaces and xylem vessels represent the apoplastic pathways of nanoparticles, while the cytoplasm, plasmodesmata, and sieve plates represent the symplastic pathways (Aslani et al., 2014). The bioavailability of nanoparticles is primarily influenced by their unique characteristics, for example pH, ionic strength, dissolved oxygen, nitrogen and sulphate concentrations and many other geochemical properties such as, groundwater, soil porosity, hydraulic flow, conductivity, and gradient (Figure 1).

Figure 1. Major processes involved in soil-nanomaterial interaction

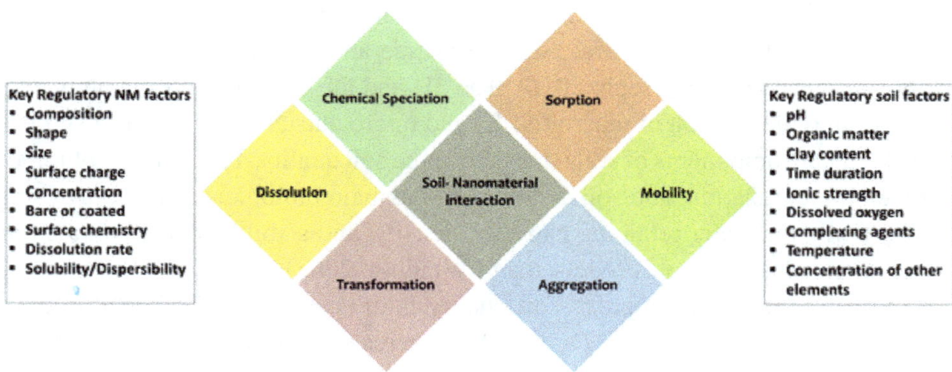

The aforementioned properties regulate nanoparticle behaviour by affecting the surface charge, type, and interaction with other organic substances and nutrients contained within the soil (Ma et al., 2023; Zhang et al., 2020).

Types of Nanomaterials in Soil

A variety of Nanomaterials (NMs) have been deliberately or accidentally introduced into soils including metal/metal oxides, Quantum Dots, Carbon-based NMs and Hybrid Systems specifically developed for enhancing remediation, improving nutrient uptake, and protecting crops (Haydar et al., 2025; Zhang et al., 2020). The core composition and outer surface coatings are two major components of nanomaterials which govern their initial reactions in soil and further transformation in the environment (Zhang et al., 2020; Haydar et al., 2025). Metal/metal oxide nanomaterials such as nano zero-valent iron (nZVI), Ag, CuO, ZnO, TiO_2, and CeO_2, are widely used for soil remediation and as nano-fertilizers or nano-pesticides in modern agriculture. Redox-Active NPs, such as nZVI, have the potential to be immobilized soil contaminants (Haydar et al., 2025), while Metal Oxide NPs, such as ZnO and CuO NPs can be utilized as micronutrient delivery systems in agricultural soils (Haydar et al., 2025) and (Zhang et al., 2020). Nanocarbons (Fullerene Derivatives, Graphene Derivatives, Carbon Nanotubes, Nano-Biochar) are effective carriers/sorbents for Agrochemicals due to their large surface areas and variable surface chemistries. As a result of their ability to strongly interact with both organic contaminants and nutrient compounds, these carbon-based nanomaterials lead to decreases in the bioavailability of organic contaminants and nutrients in the soil (Zhang et al., 2020). Composites/Hybrid Materials are specially developed composite materials which consist of, among others, metal oxides combined with clays/biochar to improve material stability, and specifically target contaminants in soil. Material performance is very dependent upon the chemical/physical properties of the individual components of the composites/hybrid materials (Haydar et al., 2025).

Effects of Soil Properties on ENMs Fate

Physical and chemical characteristics of the soil influence the fate of ENMs through alteration of surface charge, speciation, aggregation, and changes in redox reactions that occur in soil (Waalewijn-Kool et al., 2014; Tourinho et al., 2012; Zhang et al., 2020). They also influence the exposure of soil microorganisms and plant roots to both intact and dissolved ENM (Tourinho et al., 2012; Waalewijn-Kool et al., 2014). Variations in soil pH strongly influence the dissolution and associated dynamics of nanomaterials. For instance, ZnO NP dissolution and toxicity are mainly dependent on porewater pH and free Zn^{2+} ions in soils (Waalewijn-Kool et al., 2014). As the pH decreases, Zn dissolution increases, and while the pH is higher, Zn^{2+} free ions are low despite a higher total NPs dosage (Waalewijn-Kool et al., 2014). Texture and mineralogy of soils is another important factor as the soils with high surface area and reactive minerals are more capable to withhold metal based nanopartciles (Tourinho et al., 2012). Clay rich soils are more efficient than sandy soils in this context, as the sorption and filtration in fine textured soils often immobilize nanoparticle constrain their mobility yet increasing their persistence and contact with soil invertebrates (Tourinho et al., 2012). Soil organic matter affects speciation, toxicity, and mobility of nanoparticles. By complexing free metal ions, it can regulate metal speciation for nanomaterials. Additionally, organic matter alters transport and aggregation behavior of nanoparticles through adsorption and steric sterilization (Zhang et al., 2020). Soil salinity is a major factor in regulating the dissolution and adsorption of nanomaterials. As the ionic strength increases, it compresses the electrical double layer, leading to the aggregation of nanoparticles in the soil (Zhang et al., 2020). This aggregation affects mobility and core behavior of nanomaterials in saline soils.

Soil - Nanomaterial Interactions

The interactions of nanomaterials with the soil environment involve a variety of chemical and physical mechanisms including dissolution, adsorption, aggregation, and degradation that influence their distribution, bioavailability and reactivity in the soil (Zhang et al., 2020; Lowry et al., 2012; Haydar et al., 2025; Singh et al., 2025d). Redox reactions and biotic interactions play a significant role in controlling these mechanisms (Haydar et al., 2025; Lowry et al., 2012).

Aggregation: Aggregation is a common mechanism for nanomaterials in both soil and water due to van der Waal's forces and changes in surface charge causing the formation of larger aggregates (Lowry et al., 2012). In addition, the adsorption of nanomaterials on metal oxides, clays, and organic colloids increases the aggregation rate and causes the permanent immobilization of nanomaterials (Haydar et al., 2025; Lowry et al., 2012).

Adsorption: The adsorption of nanomaterials, particularly metal based nanomaterials, into soils will lead to increased plant nutrient availability (Tourinho et al., 2012). This adsorption can also result in the formation of a corona, in which the nanoparticle surface is coated with organic matter and ions, and regulates their behavior with biota (Lowry et al., 2012; Zhang et al., 2020).

Dissolution: Zhang et al. (2020) highlighted that the dissolution of metal-based ENMs (e.g. ZnO, Ag, and CuO) is strongly pH- and ligand-dependent and often governs plant uptake and toxicity, especially where ions are more bioavailable than particles. In Waalewijn-Kool et al. (2014), free Zn^{2+} in porewater explained ZnO-NP toxicity patterns across soil better than the total added nanoparticle mass, underscoring the central role of dissolution and speciation (Waalewijn-Kool et al., 2014).

Transformation & Bioavailability of Nanomaterial in Soil-Plant Systems

Once ENMs are retained in the soil, they undergo transformation and alteration of their reactivity resulting in reduced accumulation and acute toxicity of nanomaterials (Lowry et al., 2012). The transformation process commonly involves oxidation, reduction, carbonation, and sulfidation, resulting in the regulation of solubility and toxicity (Lowry et al., 2012).

In addition to above mention parameters, persistence and exposure of nanoparticles in soil-plant systems (Haydar et al., 2025; Tourinho et al., 2012; Zhang et al., 2020) are critical to both the performance and evaluation of risks for nano-enabled agriproducts (Zhang et al., 2020; Haydar et al., 2025).

Table 1. Key features regulating the behavior of nanomaterials in soil ecosystems

Feature	Function	Examples	References
Primary particle size	Regulates surface area, dissolution kinetics, and mobility within soil environment	AgNPs, ZnO NPs	(Ferdous & Nemmar, 2020)
Surface charge (zeta potential)	Regulation of electrostatic interactions with clay minerals, soil organic matter, and biological membranes	Engineered Nanomaterials majorly metal NPS	(Alkilany & Murphy, 2010)
Surface Chemistry	Controls nanomaterials affinity with biomolecules, microbial regulation and rhizosphere stability.	Coated carbon nanomaterial	(Vardharajula et al., 2012)
Particle morphology	Controls shape and orientation dependent behaviour of nanomaterial such as transport and penetration in soil environment	Carbon nanotubes, nanorods	(Mahor et al., 2021)
Dispersion	Controls transport and residence time of nanomaterials in soil solution	Zinc oxide nanoparticles (ZnO NPs)	(Vimercati et al., 2020)
Aggregation	Controls retention in soil aggregates	Silicon dioxide nanoparticles (SiO_2 NPs)	(Murugadoss et al., 2017)
Dissolution rate	Controls transformation of nanomaterials into ionic and secondary forms with in soil	Anatase vs rutile TiO_2	(Naz et al., 2019)
Surface Coating	Enhance stability and persistence of nanomaterials	Polymer-coated AuNPs	(Zamora-Justo et al., 2019)

In contrast to the form in which they were introduced, dissolved components from the transformation of engineered nanomaterials exhibit higher mobility, stability and ability to be transported into deeper soil layers (Tourinho et al., 2012). Role of all these features is given in *Table1*.

Notably, the controlled application of nanomaterials to soil leads to the transformation into simpler molecules, resulting in improved nutrient availability and increased crop production (Li et al., 2024).

IMPACT OF NANOPARTICLES ON RHIZOSPHERE

Exposure of ENMs to plants primarily occurs through dissolved ions released in the rhizosphere, whereas direct uptake of nanoparticles is secondary, as it is affected by various factors such as soil pH, organic matter, and redox conditions (Zhang et al., 2020). The rhizosphere refers to the region surrounding plant roots generally extending 2-80 mm from the root, depending on the species, and is rich in enzymes, microbes, and nutrients (Igiehon & Babalola, 2018). It is the richest source of organic matter and the principal zone for microbial proliferation and activity (Reinhold-Hurek et al., 2015). Rhizosphere-associated microbial communities contribute significantly to plant growth by mediating nitrogen fixation, nutrient decomposition, and the synthesis of bioactive compounds (Chauhan et al.,2023). The interaction between rhizosphere secretions and rhizosphere microorganisms is bidirectional, as metabolites secreted by plant roots shape microbial communities, and, in turn, rhizosphere microorganisms maintain a close association with plant growth and health (Trivedi et al., 2020). Nanoparticle applications in the rhizosphere offer great potential for enhancing nutrient acquisition and for mitigating abiotic stresses in plants due to their unique nanoscale characteristics, surface chemistry, and distinct physical properties (Lv et al., 2019). Once nanoparticles are introduced into the soil they may interact with soil components, such as soil organic matter and root microorganisms, and plants, through several mechanisms including adsorption, flocculation and precipitation (Chang et al., 2024). Soil is an ecosystem where various physical, biochemical and genetic mechanisms can be influenced by the presence of nanoparticles (Ahmed et al., 2023). Nanoparticles may also change structure and diversity of microbial communities, therefore changing plant –soil microbial community systems (You et al., 2023). Xin et al. (2022), found that application of CNPs to soil improved water retention and functionally increased microbial activity in the soil which indirectly supported plant growth. Their ability to enter plant roots via stomata or adhere to the epidermis of the roots stimulated the activity of soil microorganisms and enzymes in the soil. Interaction between nanoparticles and plant cells can stimulate metabolite synthesis in the plant cells and stimulate the release of substances in the rhizosphere (Francis et al., 2024). These biomolecules include amino acids, sugars, phenolic compounds and other secondary metabolites, which contain many active functional groups that are able to quickly adsorb to the surface of the nanoparticles, forming a surface ecological corona. This corona may modify some of the physicochemical properties of the nanoparticles and potentially affect their movement, transformations and biological activity in the soil environment creating a very complex ecological network (Wheeler et al., 2021; Kang et al., 2024).

These rhizosphere secretions could help plants to alleviate various stresses and enhance essential nutrient uptake. ZnO NPs adsorbed to the root epidermis of legumes trigger the stress signaling pathway resulting in increased production of terpenoids and amino acids (Liu et al., 2023). Combined foliar and soil application of ZnO NPs alleviated the Cd and phenanthrene stress in lettuce by means of combined metabolic adjustments and restructuring of the rhizosphere microbiome, including: modulation of lipid and amino acid metabolism in leaves, enhancement of phosphorus and carbon cycling in soil, and selective enrichment of plant growth-promoting, stress tolerant, heavy metal resistant and PAH degrading microbial taxa (Cao et al., 2024). Selenium nanoparticles enhance rice growth by up-regulating genes involved in organic acid biosynthesis and transport, increasing malate and citric acid exudation from roots and selectively recruiting beneficial rhizosphere bacteria, such as *Sphingomonas* and *Streptomyces*, thereby enhancing plant-microbe interactions. Root-exuded compounds and rhizosphere microbial metabolites can interact with metal ions to form complexes, which can modulate the mobility and bioavailability of metals in the plant-soil system (Chen et al., 2017). Citrate-encapsulated Fe_3O_4 nanoparticles increase iron solubility and interaction with root exudates, resulting in hormonal modulation that promotes root elongation and plant growth (Sun et al., 2023

Crop yields and plant performance are well known to be greatly influenced by both soil supplied nitrogen and the ability of plants to efficiently take up this nutrient. Nitrogen fixing by microorganisms within the rhizosphere has been shown to be increased through the use of iron nanoparticles in alfalfa resulting in an enhanced nitrogen fixing capability of the root system (Zhang et al., 2024) The current data clearly indicate that nanoparticles have the capability to alter rhizosphere microbial populations and promote beneficial microorganisms; however, it is essential to understand all of the interactions that occur between the root exudate and the rhizosphere microbiome prior to fully utilizing the agronomic potential of these products.

UPTAKE, TRANSLOCATION, AND ACCUMULATION OF NANOMATERIALS IN PLANTS

Plants are continually exposed to airborne and soil-sourced nanoparticles as they are primary producers in an ecosystem. As such, understanding how plants take up and transport internally these particles is crucial to the study of plant-nanoparticle interactions. The means by which aerosol nanoparticles enter plant aerial tissues include stomatal entrance; entry through hydathodes; or either directly penetrating the cuticle through surface wounds; and/or lenticel, tissue rupture, or root hair passage for those entering the plant via soil (Hussain et al., 2019). Root surface adsorption represents the initial step in the nanoparticle's uptake by plants (Figure 2).

Figure 2. Uptake, translocation and accumulation of nanomaterials

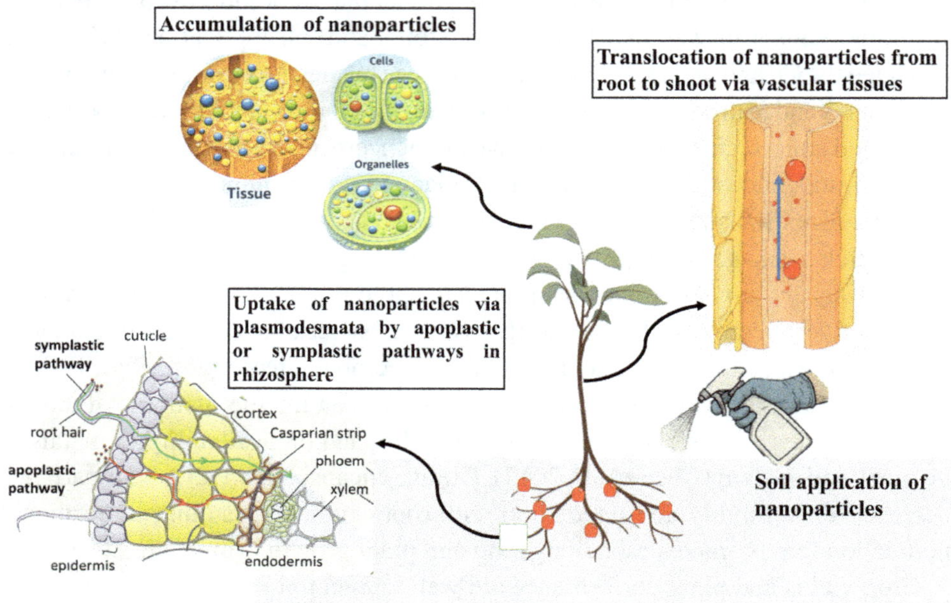

Chemical exudates released from the root hairs generate a negatively charged root surface, enhancing the affinity of a positively charged nanoparticles, whereas the lateral root development increases the available adsorption area and promote nanoparticles penetration into inner root tissues (Lv et al., 2019; Peng et al., 2015). The epidermis at root tip and on root hairs of primary and secondary roots is often incompletely developed and nanoparticle exposure at these sites enables direct contact with the root epidermis and facilitate trans-epidermal entry. Although root cell wall is semi-permeable and contains nanoscale pores that restrict the movement of larger particles, the absence of fully developed exodermis allows nanoparticles into the stele or xylem (Lv et al., 2019). In addition, certain nanoparticles can disrupt plasma membrane integrity and induce the formation of new pores in the epidermal cell walls, thereby permitting the entry of larger particles. Once inside plant tissues, nanoparticles may be internalized through multiple mechanisms including ion transport pathways, endocytosis, membrane protein binding and physical damage (Ali et al., 2021; Rajput et al., 2020). Evidence indicates that plant absorb predominantly through hydrophilic pathways; however small pore size of this route severely restricts nanoparticle entry. Endocytosis therefore represents an important cellular uptake mechanism where plasma membrane invagination internalizes nanoparticles into plant cells (Li et al., 2020).

Endocytosis enables plant protoplast to internalize sub-micron particles, with studies showing uptake of particles smaller that 1um, thereby implying limited size selectivity in nanoparticles absorption (Ma & Yan, 2018). The uptake of nanoparticles by plant roots depends on the size of the nanoparticles, which is an important variable for the uptake. It has been shown that nanoparticles smaller than 10nm could be absorbed by the roots of plants, for example Au nanoparticles (3.5nm) in Vicia faba L. and CeO_2 (8 ± 1nm) in maize (Hu et al., 2022).

Nanoparticles will enter plant roots after entering the root hair cells of the root, then they will enter the cell wall to enter the inner tissue of the root. Transport of the nanoparticles from the epidermis to the endodermis may occur either through an apoplastic pathway (through cell wall spaces), or a symplastic pathway (through cell-to-cell connections), then they will travel up the aerial parts of the plant through the xylem (Tripathi et al., 2017).

Adsorption of the nanoparticles onto the surface of plant roots is the first step in the process of bioaccumulation. For example, CuO nanoparticles demonstrate a strong and consistent adsorption to root surfaces through physical interactions and remain strongly attached to the root surfaces and resistant to desorption even when there are competing agents present (Rajput et al., 2020). Bioaccumulation is the subsequent accumulation of the nanoparticles in the plant tissues, and this can involve the redistribution of the nanoparticles throughout the plant tissues and can also involve the dissolution, biotransformation and other physico-chemical changes of the nanoparticles. These processes collectively highlight that both root and shoot tissue serve as a sink nanoparticle accumulation (Rajput et al., 2018). Studies have confirmed the accumulation of ZnO nanoparticles in the roots and shoots of *Brassica juncea* and *Medicago sativa*. Such nanoparticle buildup has the potential to alter key physiological processes and adversely affect the structural organization of cellular and sub-cellular components (Rao and Shekhawat, 2016; Fedorenko et al., 2021).In addition, the accumulation and distribution of nanoparticles have been shown to suppress plant antioxidative defense system including both enzymatic and non-enzymatic components such as catalase, superoxide dismutase, glutathione reductase, ascorbate and phenolic compounds (Badawy et al., 2021). Overall, the uptake, internal transport, and accumulation of nanoparticles in plants are governed by combination of nanoparticles properties, root and shoot anatomical features, and physiological features, collectively determine their distribution, biological effects, and potential risks within plant systems.

PLANT PHYSIOLOGICAL AND BIOCHEMICAL RESPONSES TO SOIL-APPLIED NANOMATERIALS

Effects of nanoparticle on plant morpho-physiology, overall growth and yield could range from subtle beneficial improvement to phytotoxicity depending upon dosage, size, physio-chemical properties, method of usage and plant species specific responsive mechanism (Mohamed and Kumar, 2016; Siddiqui et al., 2025). Several nanoparticles have been reported to improve seed germination, plant growth by enhancing nutritional content and some are shown to mitigate abiotic stress and their adverse effects on plants (

Table 2. Physiological and biochemical responses of plants to soil-applied engineered nanomaterials

Nanoparticles	Concentration or dosage	Target crop	Key Physiological and Biochemical response	References
Titanium nanoparticle (TiO_2 NPs)	50 mg Ti kg^{-1}	*Oryza sativa*	Significant, increase of Antioxidant enzymes (SOD, CAT, APX) in root and shoot	Machanuru et al. (2024)
Zinc nanoparticles (ZnO NPs)	150 ppm under drought stress	*Triticum aestivum*	Significant increase in total yield and zinc content	Raza et al. (2024)
Iron nanoparticles (FeO NPs)	100 ppm under Cd stress	*Solanum lycopersicum*	Improved growth parameters and mitigation of metal stress	Ahmad et al. (2024)
Silver nanoparticle (Ag NPs)	150 mg/kg of soil	Helianthus annuus	Increment in lipid peroxidation and antioxidant activity	Saleeb et al. (2019)
Zinc nanoparticles (ZnONPs)	20 ppm under fungal pathogen stress	*Vigna radiata*	Boost in disease resistance mechanism and agronomic traits	Mazhar et al. (2024)
Cerium oxide nanoparticles (CeO_2 NPs)	500 mg/kg of soil	*Triticum aestivum*	Improved shoot biomass and grain yield	Rico et al. (2014)
Iron Nanoparticle (Fe_3O_4 NPs)	500 mg/kg of soil	*Zea mays*	Beneficial effects on root development and membrane integrity	Yan et al. (2020)
Silicon Nanoparticle (Si NPs)	75 mg/kg of soil under salinity stress	*Zea mays*	Positive effect on physiological parameters and nutrient availability	Rizwan et al. (2023)
Nano-graphene oxide	100 g kg^{-1} of soil under salinity stress	*Vitis vinifera*	Improved osmolyte and the mineral nutrients balance	Aazami et al. (2022)

For instance, SiO_2 nanoparticle in low dosage shown to be improved seed germination tomato. Similarly, ZnO nanoparticles in the highest conc (100% ZnO) under salinity stress improved germination and early seedling growth by reducing accumulation of toxic Na^+ ions in *Brassica napus* (El-Badri et al., 2021). In another study, positive effect of similar nanoparticles was reported in wheat seeds at a concentration of 10 ppm with 18 hours of priming. They showed that, ZnO nano priming significantly improved seed vigour, fresh Biomass, water uptake ability as well as imbibition rate without any stress condition (Rai-Kalal & Jajoo, 2021). One comprehensive study indicated that Sulphur nanoparticles (200 ppm) effectively improve watermelon growth and quality attributes by coordinated regulation of Sulphur uptake, transport and nutrient homeostasis (Yang et al., 2025). Additionally, CeO_2 and ZnO nanoparticles at lower concentrations has been shown to significantly improved accumulation of essential elements (Peralta-Videa et al.,2014).

Photosynthesis, a key physiological process of plant system that determines growth and final yield, have been shown to improve in presence of nanomaterials at lower dosage. For instance, Addition of SiO_2, TiO_2 nanoparticles and their nanocomposite in soil at conc of 100 ppm on maize seedling positively influenced chlorophyll content, photosynthetic rate, transpiration rate, stomatal conductance. Chlorophyll content increased by 75% in SiO_2 treated plants, 66% in TiO_2 treated plants and 34% in NCs treated plants at 100 ppm concentrations, respectively. Significant stimulation was observed at 200 ppm nanocomposite that promote translocation of Si and Ti to the leaves leading to enhanced photosynthetic capacity of the Z. mays plants (Kumari et al., 2024). Moreover, treatment of ZnO NP in rice plants grown under saline sodic soil, indicated increased amount of chlorophyll and carotenoid by 54 and 22% as compared to control (Dang et al., 2024). Beside photosynthetic pigments, nanomaterial also help to maintain plant water balance under salt stress when osmotic stress limits the water uptake causing reduction in leaf water content, stomatal conductance and growth. One study indicated that application of copper nanoparticles (300 ppm) has been positively maximize water use efficiency (3.78%), stomatal conductance in compare to control treatments, whereas leaf relative water content (%), excised leaf water loss (%), leaf water potential, all parameters were improved in comparison to drought stress followed by controls (Raza et al., 2024). In addition, nanoparticles increase root hydraulic conductivity by affecting aquaporin transduction, channel proteins which result in better water uptake (Rios et al., 2017). For example, application of multiwalled carbon nanotube in broccoli under saline stress suggested increase of PIP (aquaporins) proteins under NaCl + nanotube treatment (Martínez-Ballesta et al., 2016).

A study demonstrated that cerium oxide NPs (CeO_2NPs) (nanoceria) at low Ce3+/Ce4+ ratio (PNC1) exhibit catalytic behavior in scavenging free radicals in the chloroplast of mesophyll cells protecting photosynthetic system (Wu et al.,

2017). Selenium nanoparticles (50 ppm) evaluated under saline (150mM) stress in *Melissa officinalis* and *Solanum lycopersicum* showed a dramatic rise in POX activity followed by CAT, SOD indicating role in antioxidant stress regulation (Ghasemian et al., 2021). Similar efficient modulation of key antioxidant enzyme like SOD, CAT, APX, POD was observed in rice plants under nano TiO_2 suggesting adaptive response to maintain ROS detoxification and redox homeostasis (Machanuru et al., 2024). The exact mechanism carrying out this defense has not been fully understood. Accumulation of nonenzymatic antioxidant due to SiO_2 NP was observed more pronounced in presence of metal toxicity such as aluminum, suggesting alone there is minimal changes in non-enzymatic metabolites level (De Sousa et al., 2019). Notably, a particular study revealed the improved level of nonenzymatic antioxidants compound phenolics and flavonoids under Cd stress when treated with iron nanoparticles (300 ppm) in *Solanum lycopersicum* (Ahmad et al., 2024).

Several metal nanoparticles have been reported to participate in plant hormone-antioxidant crosstalk and modulate hormone biosynthesis. For example, the contents of several plant hormones, such as ABA and auxin, were evaluated in peanut plants grown with EDTA-Fe fertilisers or Fe2O3 NPs. The results demonstrated that Fe_2O_3 NP-treated peanut plants had lower levels of ABA and Auxin in the root and shoot areas than the controls. Lower concentrations of ABA promote normal physiological growth and delay senescence (Rui et al., 2016). Interestingly, in non-stress conditions, nanoparticles are perceived by plants as foreign materials, triggering the plant's internal defense mechanism which involves the biosynthesis of the stress hormone ABA. This behavior has been observed in silver (Ag) nanoparticles, which indirectly increase ABA levels in plants (Abasi et al. 2022).

Despite the vast advancement of nanomaterials in agriculture, their unintended release into environment and bioaccumulation in plants is one major concern. The size and dosage depended characteristics of nanomaterial decide its reactivity and toxicity. Major adverse effects shown by plant system include alter plant morphology, physiology, biochemical function, plant metabolism namely micronutrient uptake, disrupt soil microbiome, and impair soil health.

ENVIRONMENTAL SAFETY, THRESHOLDS, AND RISK CONCERNS

The phytotoxicity thresholds for nanomaterials (NMs) have been found to be significantly different among various types of NMs. The differences in phytotoxicity among NMs are also influenced by their size, surface characteristics, and the type of plant they come into contact with. In several instances, metal-based NMs (e.g. Ag and ZnO) have been shown to cause significant phytotoxic effects to plants growing in soils that contain more than 100 – 500 mg/kg of the NMs (Elsayed et al., 2025). There are concerns about accumulation risks associated with NMs because there is evidence that once NMs accumulate in plant tissue, they can enter food chains and exceed safe intake limits for humans and animals. Cerium oxide NPs have demonstrated a root absorption rate of approximately 0.11% demonstrating an opportunity for uptake of NMs in crops. Therefore, it will be important to monitor crop exposure to NMs in order to reduce the potential for chronic adverse effects caused by reduced photosynthesis and increased oxidative stress. Nanomaterials can alter soil microbial community structure and function even at relatively low concentrations (approximately 50 mg/kg). These changes primarily influence nitrogen cycle processes and soil enzyme activity, leading to decreases in bacterial diversity and earthworm populations (McKee & Filser, 2016).

In agricultural ecosystems, long term exposure modeling of NMs is necessary to account for the fact that NMs can form long-lasting aggregates and interact physically with soil fauna (e.g. nematode gut occlusion), which could disrupt trophic relationships and lead to ecosystem imbalance (Zhang et al., 2020). These findings illustrate the necessity for conducting soil specific risk assessments for NMs as opposed to simply extrapolating results from hydroponic or laboratory studies.

There are concerns regarding the bioaccumulation of NMs and its potential risks to herbivorous animals and humans through dietary consumption. Although the primary route of entry into the plant is through root uptake, smaller NMs (<50 nm) have been shown to be readily translocated to aerial parts of the plant, resulting in higher accumulation factors (Zakaria et al., 2021). Genotoxicity thresholds for Cu-NPs in plant biomass have been identified in controlled experiments with wheat and lettuce within the range of 10-100 mg/kg (Hong et al., 2015).

Although regulatory agencies (e.g. EPA) have proposed provisional exposure limits for NMs, establishing universal thresholds has proven difficult due to the variability in soil chemistry. The variability in soil chemistry (pH, OM content, etc.) creates uncertainty in the prediction of NM behavior and accumulation in soils, complicating safety assessments of nanotechnology enabled agricultural products. These uncertainties reinforce the importance of designing NMs to be persistent and minimize ecological risks.

Safe-by-design (SbD) approaches focus on integrating hazard mitigation during the design and synthesis of NMs. This includes the application of biocompatible coatings to the surfaces of NMs, reducing the dissolution rate of NMs, and selecting NMs with low persistence. SbD approaches are assessed using life cycle assessments and ecotoxicological studies prior to widespread use and reflect the core tenets of green nanoscience (Subramanian et al., 2023).

Presently, there are no universally accepted regulatory standards for NMs; however, empirical benchmarks for model crops (e.g., lettuce, wheat, tomato, A thaliana) have been developed. For example, nano-TiO2 inhibited cress root growth by 4.6 – 13.9% at 10 – 1000 mg/L; the greatest effects were observed at intermediate concentrations. Cucumber root growth was suppressed by >100 mg/L CuO via ion release; in contrast, growth of Arabidopsis was inhibited by Fe3O4 NPs at 100 – 5000 mg/L but promoted at lower concentrations (Fatima et al., 2020).

Phytotoxicity thresholds for carbon-based nanomaterials also differ considerably. Graphene oxide (GO) inhibited root development in tomato and cabbage at 50–100 mg L^{-1} after ten days, but did not affect red spinach until at concentrations ranging from 500 to 2,000 mg L^{-1}.Nano-CuO appears to be less toxic than nano-ZnO, however, it does produce some adverse effects. At 50 mg kg−1 soil in Brassica nigra, a positive effect was seen, which included increased plant height, number of leaves, and leaf area (apical leaf increase of 232%), with no adverse effects on germination. However, in sandy soils toxicity appeared at values greater than 50 mg kg−1, with a 30–39% reduction in primary root length at values of 12.5–50 mg kg−1 and a 60% reduction in shoot growth at 500 mg kg−1. The distribution of nano-CuO in plants follows the sequence: roots > stems > leaves > seeds (Zhang et al., 2024). Overall, these results indicate that while nano-matls can provide enhanced crop yields at low doses, values above soil specific thresholds pose ecological and food chain risk.

FUTURE RESEARCH DIRECTIONS AND GAPS IN KNOWLEDGE

Although there is increasing knowledge regarding nanomaterial-soil-plant interactions, many fundamental questions remain unanswered, hindering the ability to safely scale-up nanotechnology for use in agriculture. Many studies have lacked sufficient detail regarding the physiological process of plants, including nutrient transport, hormone mediated stress signaling, and a complete description of photosynthetic efficiency beyond the measurement of chlorophyll content. There has been little work done to study the dynamic nature of soils, specifically how microbial communities shift in response to nanomaterial introduction and changes in nutrient bioavailability, making it difficult to translate laboratory results to field conditions. Additionally, many environmental safety issues, including persistence of nanomaterials in soil, non-target organism toxicity, and the creation of hazardous by-product materials have not been adequately addressed.

Future research must focus on multidisciplinary approaches utilizing modern analytical techniques, such as LA-ICP-MS to map the spatial distribution of nutrients in plants, X-ray absorption spectroscopy (XAS) to determine the chemical form of introduced nanomaterials, and omics based techniques to understand the molecular mechanisms involved in plant responses to nanomaterials. Also, long-term field studies must utilize soil physical/chemical properties to monitor, microbial metagenomics to assess, and comprehensive risk assessment methodologies to evaluate potential impacts. It would be beneficial to validate biofortification results with bioavailability measurements and to conduct multi-environment studies that relate soil processes to human nutrition.

Implications for Precision and Sustainable Agriculture

Resolving the previously stated issues will create opportunities for the use of nanotechnology in precision and sustainable agricultural systems. Nanotechnology is potentially useful in developing nanosensors for continuous monitoring of nutrient levels and plant stress. These sensors could be used with nano-enabled slow-release fertilizers and with data driven decision-support systems to allow for targeted input applications, improved resource utilization, and reduced production costs. These integrated systems will allow for soil fertility remediation and plant resilience thereby providing a means to achieve "smart" farming practices that meet long-term sustainability objectives (Ibrahim et al., 2024).

CONCLUSION

Nanotechnology for agricultural uses requires the incorporation of plant physiology, soil science, and the characterization of materials. Through the use of new analytical tools and long term field studies, the limitations of previous knowledge can be overcome and maximum benefits of increased micronutrient uptake and biofortification can be realized while minimizing environmental and health risks. As long as there is a mechanistic understanding of how nano-matls interact with soils, soil-specific risk assessments are performed, and nano-materials are delivered with precision, nanotechnology has the potential to contribute significantly to the development of more sustainable and resilient agricultural systems. Collaboration between policy makers and interdisciplinary researchers will be necessary to develop and deploy nano-materials based solutions in future agro-ecosystems.

ACKNOWLEDGMENTS

The authors would like to thank the world-wide scientific community for their research and scholarly works on which this chapter is based; the editors of this book and IGI Global Scientific Publishing are gratefully thanked for the chance to participate in this work as well as for their editorial assistance and for providing institutional/academic resources supporting manuscript development.

REFERENCES

Aazami, M. A., Mehrabani, L. V., Hashemi, T., Hassanpouraghdam, M. B., & Rasouli, F. (2022). Soil-based nano-graphene oxide and foliar selenium and nano-Fe influence physiological responses of' Sultana' grape under salinity. *Scientific Reports*, *12*(1), 4234. DOI: 10.1038/s41598-022-08251-8 PMID: 35273327

Abasi, F., Raja, N. I., Mashwani, Z. U. R., Amjad, M. S., Ehsan, M., Mustafa, N., Haroon, M., & Proćków, J. (2022). Biogenic silver nanoparticles as a stress alleviator in plants: A mechanistic overview. *Molecules (Basel, Switzerland)*, *27*(11), 3378. DOI: 10.3390/molecules27113378 PMID: 35684312

Abdul Aziz, M., Brini, F., Rouached, H., & Masmoudi, K. (2022). Genetically engineered crops for sustainably enhanced food production systems. *Frontiers in Plant Science*, *13*, 1027828. DOI: 10.3389/fpls.2022.1027828 PMID: 36426158

Ahmad, A., Javad, S., Iqbal, S., Shahid, T., Naz, S., Shah, A. A., Shaffique, S., & Gatasheh, M. K. (2024). Efficacy of soil drench and foliar application of iron nanoparticles on tomato exposed to cadmium stress. *Scientific Reports*, *14*, 27920. DOI: 10.1038/s41598-024-79270-w PMID: 39538010

Ahmed, T., Noman, M., Gardea-Torresdey, J. L., White, J. C., & Li, B. (2023). Dynamic interplay between nano-enabled agrochemicals and the plant-associated microbiome. *Trends in Plant Science*, *28*(11), 1310–1325. DOI: 10.1016/j.tplants.2023.06.001 PMID: 37453924

Akanmu, A. O., Akol, A. M., Ndolo, D. O., Kutu, F. R., & Babalola, O. O. (2023). Agroecological techniques: Adoption of safe and sustainable agricultural practices among smallholder farmers in Africa. *Frontiers in Sustainable Food Systems*, *7*, 1143061. DOI: 10.3389/fsufs.2023.1143061

Akhtar, N., Ilyas, N., Mashwari, Z., Hayat, R., Yasmin, H., Noureldeen, A., & Ahmad, P. (2021). Synergistic effects of plant growth-promoting rhizobacteria and silicon dioxide nanoparticles to ameliorate drought stress in wheat. *Plant Physiology and Biochemistry*, *166*, 160–176. DOI: 10.1016/j.plaphy.2021.05.039 PMID: 34116336

Alharbi, K., Hafez, E. M., Omara, A. E. D., Rashwan, E., & Alshaal, T. (2023). Zinc oxide nanoparticles and PGPR strengthen salinity tolerance and productivity of wheat irrigated with saline water in sodic–saline soil. *Plant and Soil*, *493*(1-2), 475–495. DOI: 10.1007/s11104-023-06245-7

Ali, S., Mehmood, A., & Khan, N. (2021). Uptake, translocation, and consequences of nanomaterials on plant growth and stress adaptation. *Journal of Nanomaterials*, *6677616*, 1–17. Advance online publication. DOI: 10.1155/2021/6677616

Alkilany, A. M., & Murphy, C. J. (2010). Toxicity and cellular uptake of gold nanoparticles: What we have learned so far? *Journal of Nanoparticle Research*, *12*(7), 2313–2333. DOI: 10.1007/s11051-010-9911-8 PMID: 21170131

Anikwe, M. A., & Ife, K. (2023). The role of soil ecosystem services in the circular bioeconomy. *Frontiers in Soil Science*, *3*, 1209100. DOI: 10.3389/fsoil.2023.1209100

Aslani, F., Bagheri, S., Muhd Julkapli, N., Juraimi, A. S., Hashemi, F. S. G., & Baghdadi, A. (2014). Effects of engineered nanomaterials on plant growth: An overview. *TheScientificWorldJournal*, *2014*, 641759. Advance online publication. DOI: 10.1155/2014/641759 PMID: 25202734

Babu, S., & Joseph, S. (2016). Effect of nanomaterials on properties of soft soil. *International Journal of Scientific Research*, *5*, 634–637.

Badawy, A. A., Abdelfattah, N. A., Salem, S. S., Awad, M. F., & Fouda, A. (2021). Efficacy assessment of biosynthesized copper oxide nanoparticles on stored grain insects and their impacts on wheat. *Biology (Basel)*, *10*(3), 233. DOI: 10.3390/biology10030233 PMID: 33802973

Boretti, A., & Rosa, L. (2019). Reassessing the projections of the World Water Development Report. *npj. NPJ Clean Water*, *2*(1), 1–6. DOI: 10.1038/s41545-019-0039-9

Cao, M., Lv, W., Wang, F., Ma, S., Geng, H., Li, J., Gao, Z., Xu, Q., Guo, J., Leng, W., Chen, K., Tan, Z., Zhang, P., Sun, K., & Xing, B. (2024). Foliar application of zinc oxide nanoparticles alleviates phenanthrene and cadmium-induced phytotoxicity in lettuce. *Environmental Science & Technology*, *59*(1), 730–743. DOI: 10.1021/acs.est.4c07881 PMID: 39704184

Chang, M., Liu, Y., Xu, M., Li, H., & Li, S. W. (2024). Particle morphology and soil properties affect the retention of copper oxide nanoparticles in agricultural soils. *Environmental Geochemistry and Health, 46*(8), 281. DOI: 10.1007/s10653-024-02057-5 PMID: 38963650

Chanu, N. B., Alice, A. K., & Thokchom, A.. (2021). Engineered nanomaterials and their interactions with the plant–soil system. *Nanotechnology for Environmental Engineering, 6*, 36. Advance online publication. DOI: 10.1007/s41204-021-00130-3

Chauhan, P., Sharma, N., Tapwal, A., Kumar, A., Verma, G. S., Meena, M., Seth, C. S., & Swapnil, P. (2023). Soil microbiome: Diversity, benefits, and interactions with plants. *Sustainability (Basel), 15*(19), 14643. DOI: 10.3390/su151914643

Chen, Y. T., Wang, Y., & Yeh, K. C. (2017). Role of root exudates in metal acquisition and tolerance. *Current Opinion in Plant Biology, 39*, 66–72. DOI: 10.1016/j.pbi.2017.06.004 PMID: 28654805

Dang, K., Wang, Y., Tian, H., Bai, J., Cheng, X., Guo, L., & Shao, X. (2024). Impact of ZnO nanoparticles on rice photosynthesis in saline–sodic soil. *Scientific Reports, 14*, 16233. DOI: 10.1038/s41598-024-66935-9 PMID: 39004658

De Sousa, A., Saleh, A. M., Habeeb, T. H., Hassan, Y. M., Zrieq, R., & Wadaan, M. A. AbdElgawad, H. (. (2019). Silicon dioxide nanoparticles ameliorate aluminum phytotoxicity in maize. *The Science of the Total Environment, 693*, 133636. DOI: 10.1016/j.scitotenv.2019.133636 PMID: 31377375

El-Badri, A. M., Batool, M., Mohamed, I. A., Khatab, A., Sherif, A., Wang, Z., & Zhou, G. (2021). Nano-priming with ZnO mitigates salinity stress in rapeseed. *Plant Physiology and Biochemistry, 166*, 376–392. DOI: 10.1016/j.plaphy.2021.05.040 PMID: 34153882

Elsayed, M. E., Ayoub, H. A., Helal, M. I., Sang, W., Shen, Z., & Abdelhafeez, I. A. (2025). Nanotechnology-enabled soil management for sustainable agriculture: Interactions, challenges, and prospects. *Environmental Science. Nano, 12*(4), 2128–2153. DOI: 10.1039/D4EN00943F

Fanzo, J., McLaren, R., Davis, C., & Choufani, J. (2017). *Climate change and variability: What are the risks for nutrition, diets, and food systems?* International Food Policy Research Institute.

Fatima, A., Singh, S., & Prasad, S. M. (2020). Interaction between copper oxide nanoparticles and plants: Uptake, accumulation and phytotoxicity. In *Sustainable Agriculture Reviews 41: Nanotechnology for Plant Growth and Development* (pp. 143–161). Springer International Publishing. DOI: 10.1007/978-3-030-33996-8_7

Fedorenko, A. G., Minkina, T. M., Chernikova, N. P., Fedorenko, G. M., Mandzhieva, S. S., Rajput, V. D., & Soldatov, A. V. (2021). Toxic effects of CuO nanoparticles on barley cells. *Environmental Geochemistry and Health, 43*(4), 1673–1687. DOI: 10.1007/s10653-020-00530-5 PMID: 32026274

Ferdous, Z., & Nemmar, A. (2020). Health impact of silver nanoparticles: A review of the biodistribution and toxicity following various routes of exposure. *International Journal of Molecular Sciences, 21*(7), 2375. DOI: 10.3390/ijms21072375 PMID: 32235542

Francis, D. V., Abdalla, A. K., Mahakham, W., Sarmah, A. K., & Ahmed, Z. F. (2024). Interaction of plants and metal nanoparticles. *Environment International, 190*, 108859. DOI: 10.1016/j.envint.2024.108859 PMID: 38970982

Ghasemian, S., Masoudian, N., Nematpour, F. S., & Afshar, A. S. (2021). Selenium nanoparticles alleviate salt stress in *Melissa officinalis*. *Biologia, 76*(10), 2879–2888. DOI: 10.1007/s11756-021-00854-2

Haydar, M. S., Kaur, S., Singh, P., & Talukdar, D. (2025). From synthesis to soil: Transformations and impacts of nanoparticles on soil ecosystems. *The Science of the Total Environment, 944*, 173926. DOI: 10.1016/j.scitotenv.2025.180374 PMID: 40914116

Hong, J., Rico, C. M., Zhao, L., Adeleye, A. S., Keller, A. A., Peralta-Videa, J. R., & Gardea-Torresdey, J. L. (2015). Toxic effects of copper-based nanoparticles or compounds to lettuce (*Lactuca sativa*) and alfalfa (*Medicago sativa*). *Environmental Science. Processes & Impacts, 17*(1), 177–185. DOI: 10.1039/C4EM00551A PMID: 25474419

Hossain, M. S. (2019). Present scenario of global salt-affected soils, its management and importance of salinity research. *International Research Journal of Biological Sciences Perspectives, 1*(1), 1–3.

Hu, D., Bai, Y., Li, L., Ai, M., Jin, J., Song, K., & Liu, D. (2022). Phytotoxicity assessment of gold nanoclusters in broad bean (*Vicia faba* L.). *Environmental Pollutants and Bioavailability, 34*(1), 284–296. DOI: 10.1080/26395940.2022.2101543

Hussain, I., Singh, A., Singh, N. B., Singh, P., & Singh, P. (2019). Plant–nanoceria interaction. *South African Journal of Botany, 121*, 239–247. DOI: 10.1016/j.sajb.2018.11.013

Ibrahim, A. A., Abd-Ellatif, S., Abdel Razik, E. S. S., Hamedo, H. A., & Salem, K. F. (2024). Nanosensors for enhancing plant growth and productivity. In *Innovative Methods in Horticultural Crop Improvement: Biosensors and Nanosensors* (pp. 25–50). Springer Nature Switzerland. https://doi.org/DOI: 10.1007/978-3-031-61095-0_2

Igiehon, N. O., & Babalola, O. O. (2018). Below-ground–above-ground plant–microbial interactions in soybean. *The Open Microbiology Journal*, *12*, 261–279. DOI: 10.2174/1874285801812010261 PMID: 30197700

Jatav, H. S., Rajput, V. D., Raza, T., & Maurya, P. (2025). Mitigation strategies to reduce the climate change impact in agriculture. *Journal of Innovative Solutions for Eco-Environmental Sustainability*, *1*(1), 13493. DOI: 10.46991/JISEES.2025.01.1.13493

Joudeh, N., & Linke, D. (2022). Nanoparticle classification, physicochemical properties, characterization, and applications: A comprehensive review for biologists. *Journal of Nanobiotechnology*, *20*(1), 262. Advance online publication. DOI: 10.1186/s12951-022-01477-8 PMID: 35672712

Kakaei, H., Nourmoradi, H., Bakhtiyari, S., Jalilian, M., & Mirzaei, A. (2022). Effect of COVID-19 on food security, hunger, and food crisis. In *COVID-19 and the Sustainable Development Goals* (pp. 3–29). Elsevier., DOI: 10.1016/B978-0-323-91307-2.00005-5

Kang, M., Bai, X., Liu, Y., Weng, Y., Wang, H., & Ye, Z. (2024). Zinc oxide nanoparticle size and hydrophobicity regulate eco-corona formation. *Environmental Science & Technology*, *58*(22), 9875–9886. DOI: 10.1021/acs.est.4c01731 PMID: 38722770

Kumari, A., Lakshmi, G. A., Krishna, G. K., Patni, B., Prakash, S., Bhattacharyya, M., Singh, S. K., & Verma, K. K. (2022). Climate change and its impact on crops: A comprehensive investigation for sustainable agriculture. *Agronomy (Basel)*, *12*(12), 3008. DOI: 10.3390/agronomy12123008

Kumari, A., Singh, S. K., Mathpal, B., Verma, K. K., Garg, V. K., Bhattacharyya, M., & Bhatt, R. (2023). The biosynthesis, mechanism of action, and physiological functions of melatonin in horticultural plants: A review. *Horticulturae*, *9*(8), 913. DOI: 10.3390/horticulturae9080913

Kumari, K., Rani, N., & Hooda, V. (2024). Effects of nano-SiO_2, nano-TiO_2 and nanocomposites on maize growth and soil health. *Scientific Reports*, *14*, 13996. DOI: 10.1038/s41598-024-61456-x PMID: 38886471

Li, L., Luo, Y., Li, R., Zhou, Q., Peijnenburg, W. J. G. M., Yin, N., Yang, J., Tu, C., & Zhang, Y. (2020). Uptake of submicrometre plastics by crop plants. *Nature Sustainability*, *3*(11), 929–937. DOI: 10.1038/s41893-020-0567-9

Li, M.. (2024). Dynamic transformation of nano-MoS_2 in a soil–plant system and its implications for biological nitrogen fixation and crop yield. *Environmental Science & Technology*, *58*(2), 1234–1246. DOI: 10.1021/acs.est.3c09004

Liu, L., Tsyusko, O. V., Unrine, J. M., Liu, S., Liu, Y., Guo, L., & Chen, C. (2023). Pristine and sulfidized zinc oxide nanoparticles promote organic carbon release and decomposition in the legume rhizosphere. *Environmental Science & Technology*, *57*(24), 8943–8953. DOI: 10.1021/acs.est.3c02071 PMID: 37285309

Lowry, G. V., Gregory, K. B., Apte, S. C., & Lead, J. R. (2012). Transformations of nanomaterials in the environment. *Environmental Science & Technology*, *46*(13), 6893–6899. DOI: 10.1021/es300839e PMID: 22582927

Lv, J., Christie, P., & Zhang, S. (2019). Uptake, translocation, and transformation of metal-based nanoparticles in plants. *Environmental Science. Nano*, *6*(1), 41–59. DOI: 10.1039/C8EN00645H

Ma, C., Han, L., Shang, H., Hao, Y., Xu, X., White, J. C., Wang, Z., & Xing, B. (2023). Nanomaterials in agricultural soils: Ecotoxicity and application. *Current Opinion in Environmental Science & Health*, *31*, 100432. DOI: 10.1016/j.coesh.2022.100432

Ma, C., Han, L., Shang, H., Hao, Y., Xu, X., White, J. C., Wang, Z., & Xing, B. (2023). Nanomaterials in agricultural soils: Ecotoxicity and application. *Current Opinion in Environmental Science & Health*, *31*, 100432. DOI: 10.1016/j.coesh.2022.100432

Ma, X., & Yan, J. (2018). Plant uptake of engineered metallic nanoparticles. *Current Opinion in Environmental Science & Health*, *6*, 16–20. DOI: 10.1016/j.coesh.2018.07.008

Machanuru, R., Shrivastava, M., Singh, R., Singh, B., Chakraborty, D., Ramalingappa, P. L., & Narayan, M. (2024). Plant enzymatic activity as an indicator of nano-TiO2 exposure in rice ecosystems. *Plant Nano Biology*, *10*, 100117. DOI: 10.1016/j.plana.2024.100117

Mahor, A., Singh, P., Bharadwaj, P., Sharma, N., Yadav, S., Rosenholm, J. M., & Bansal, K. K. (2021). Carbon-based nanomaterials for delivery of biologicals and therapeutics: A cutting-edge technology. *C*, *7*(1), 19. DOI: 10.3390/c7010019

Martínez-Ballesta, M. C., Zapata, L., Chalbi, N., & Carvajal, M. (2016). Multi-walled carbon nanotubes enter broccoli cells enhancing growth and water uptake of plants exposed to salinity. *Journal of Nanobiotechnology*, *14*(1), 42. DOI: 10.1186/s12951-016-0199-4 PMID: 27278384

Mazhar, M. W., Ishtiaq, M., Maqbool, M., Mazher, M., Amai, S., Siddiqui, M. H., & Bhatt, R. (2024). Management of yield losses in Vigna radiata (L.) R. Wilczek crop caused by charcoal-rot disease through synergistic application of biochar and zinc oxide nanoparticles as boosting fertilizers and nanofungicides. *BMC Plant Biology*, *24*(1), 1099. DOI: 10.1186/s12870-024-05813-y PMID: 39563266

McKee, M. S., & Filser, J. (2016). Impacts of metal-based engineered nanomaterials on soil communities. *Environmental Science. Nano*, *3*(3), 506–533. DOI: 10.1039/C6EN00007J

Murugadoss, S., Lison, D., Godderis, L., Van Den Brule, S., Mast, J., Brassinne, F., Sebaihi, N., & Hoet, P. H. (2017). Toxicology of silica nanoparticles: An update. *Archives of Toxicology*, *91*(9), 2967–3010. DOI: 10.1007/s00204-017-1993-y PMID: 28573455

Nayana, A. R., Joseph, B. J., Jose, A., & Radhakrishnan, E. K. (2020). Nanotechnological advances with PGPR applications. In Hayat, S., Pichtel, J., Faizan, M., & Fariduddin, Q. (Eds.), *Sustainable Agriculture Reviews* (Vol. 41). Springer., DOI: 10.1007/978-3-030-33996-8_9

Naz, S., Gul, A., & Zia, M. (2019). Toxicity of copper oxide nanoparticles: A review study. *IET Nanobiotechnology / IET*, *14*(1), 1–13. DOI: 10.1049/iet-nbt.2019.0176 PMID: 31935671

Negacz, K., Malek, Z., De Vos, A., & Vellinga, P. (2022). Saline soils worldwide: Identifying the most promising areas for saline agriculture. *Journal of Arid Environments*, *203*, 104775. DOI: 10.1016/j.jaridenv.2022.104775

Nielsen, M. N., & Winding, A. (2002). *Microorganisms as indicators of soil health* (Technical Report No. 388). National Environmental Research Institute, Denmark.

Patni, B., Bhattacharyya, M., Kumari, A., & Purohit, V. K. (2022). Alarming influence of climate change and compromising quality of medicinal plants. *Plant Physiology Reports*, *27*(1), 1–10. DOI: 10.1007/s40502-021-00616-x

Peng, C., Duan, D., Xu, C., Chen, Y., Sun, L., Zhang, H., Yuan, X., Zheng, L., Yang, Y., Yang, J., Zhen, X., Chen, Y., & Shi, J. (2015). Translocation and biotransformation of CuO nanoparticles in rice. *Environmental Pollution*, *197*, 99–107. DOI: 10.1016/j.envpol.2014.12.008 PMID: 25521412

Peralta-Videa, J. R., Hernandez-Viezcas, J. A., Zhao, L., Diaz, B. C., Ge, Y., Priester, J. H., & Gardea-Torresdey, J. L. (2014). CeO_2 and ZnO nanoparticles alter soybean nutrition. *Plant Physiology and Biochemistry*, *80*, 128–135. DOI: 10.1016/j.plaphy.2014.03.028 PMID: 24751400

Rai-Kalal, P., & Jajoo, A. (2021). ZnO nanoparticle priming improves wheat photosynthesis. *Plant Physiology and Biochemistry*, *160*, 341–351. DOI: 10.1016/j.plaphy.2021.01.032 PMID: 33548801

Rajput, V. D., Minkina, T., Mazarji, M., Shende, S., Sushkova, S., Mandzhieva, S., Burachevskaya, M., Chaplygin, V., Singh, A., & Jatav, H. (2020). Accumulation of nanoparticles in soil–plant systems and their effects on human health. *Annals of Agricultural Science*, *65*(2), 137–143. DOI: 10.1016/j.aoas.2020.08.001

Rajput, V. D., Minkina, T., Sushkova, S., Tsitsuashvili, V., Mandzhieva, S., Gorovtsov, A., Nevidomskyaya, D., & Gromakova, N. (2018). Effects of nanoparticles on crops and soil microbial communities. *Journal of Soils and Sediments*, *18*(6), 2179–2187. DOI: 10.1007/s11368-017-1793-2

Rajput, V. D., Singh, A., Ghazaryan, K., Alexiou, A., & Minkina, T. M. (2025). Biogenic nanoparticles. In *Elsevier eBooks*. https://doi.org/DOI: 10.1016/C2024-0-00214-7

Rajput, V. D., Singh, A., Minkina, T. M., Verma, K. K., & Singh, A. K. (2023). *Nanotechnology for sustainable agriculture: An innovative and eco-friendly approach* (1st ed.). Apple Academic Press., DOI: 10.1201/9781003333128

Rani, M., Kaushik, P., Bhayana, S., & Kapoor, S. (2023). Impact of organic farming on soil health and nutritional quality of crops. *Journal of the Saudi Society of Agricultural Sciences*, *22*(8), 560–569. DOI: 10.1016/j.jssas.2023.07.002

Rao, S., & Shekhawat, G. S. (2016). Phytotoxicity and oxidative stress of nanoparticles in *Brassica juncea. 3 Biotech, 6*(2), 244. https://doi.org/DOI: 10.1007/s13205-016-0565-0

Rashid, A., Schutte, B. J., Ulery, A., Deyholos, M. K., Sanogo, S., Lehnhoff, E. A., & Beck, L. (2023). Heavy metal contamination in agricultural soil: Environmental pollutants affecting crop health. *Agronomy (Basel)*, *13*(6), 1521. DOI: 10.3390/agronomy13061521

Raza, M. A. S., Amin, J., Valipour, M., Iqbal, R., Aslam, M. U., Zulfiqar, B., & Elsalahy, H. H. (2024). Copper nanoparticles enhance drought tolerance in wheat. *Scientific Reports*, *14*, 14254. DOI: 10.1038/s41598-024-62680-1 PMID: 38902296

Reinhold-Hurek, B., Bünger, W., Burbano, C. S., Sabale, M., & Hurek, T. (2015). Roots shaping their microbiome. *Annual Review of Phytopathology*, *53*(1), 403–424. DOI: 10.1146/annurev-phyto-082712-102342 PMID: 26243728

Rico, C. M., Lee, S. C., Rubenecia, R., Mukherjee, A., Hong, J., Peralta-Videa, J. R., & Gardea-Torresdey, J. L. (2014). Cerium oxide nanoparticles impact yield and modify nutritional parameters in wheat (Triticum aestivum L.). *Journal of Agricultural and Food Chemistry*, *62*(40), 9669–9675. DOI: 10.1021/jf503526r PMID: 25220448

Rios, J. J., Martínez-Ballesta, M. C., Ruiz, J. M., Blasco, B., & Carvajal, M. (2017). Silicon-mediated salinity tolerance. *Frontiers in Plant Science*, *8*, 948. DOI: 10.3389/fpls.2017.00948 PMID: 28642767

Rizwan, A., Zia-ur-Rehman, M., Rizwan, M., Usman, M., Anayatullah, S., Areej, , Alharby, H. F., Bamagoos, A. A., Alharbi, B. M., & Ali, S. (2023). Effects of silicon nanoparticles and conventional Si amendments on growth and nutrient accumulation by maize (Zea mays L.) grown in saline-sodic soil. *Environmental Research*, *227*, 115740. DOI: 10.1016/j.envres.2023.115740 PMID: 36997044

Rubin, R., Oldfield, E., Lavallee, J., Griffin, T., Mayers, B., & Sanderman, J. (2023). Climate mitigation through soil amendments: Quantification, evidence, and uncertainty. *Carbon Management*, *14*(1), 1. DOI: 10.1080/17583004.2023.2217785

Rui, M., Ma, C., Hao, Y., Guo, J., Rui, Y., Tang, X., & Zhu, S. (2016). Iron oxide nanoparticles as iron fertilizer for peanut. *Frontiers in Plant Science*, *7*, 815. DOI: 10.3389/fpls.2016.00815 PMID: 27375665

Saleeb, N., Robinson, B., Cavanagh, J., Hossain, A. K. M., & Gooneratne, R. (2019). Biochemical changes in sunflower plant exposed to silver nanoparticles/silver ions.

Sharma, A., Singh, R. N., Song, X. P., Singh, R. K., Guo, D. J., Singh, P., Verma, K. K., & Li, Y.-R. (2023). Genome analysis of a halophilic *Virgibacillus halodenitrificans* ASH15. *Frontiers in Microbiology*, *14*, 1229955. DOI: 10.3389/fmicb.2023.1229955 PMID: 37808307

Siddiqui, H., Ahmed, K. B. M., Faizan, M., Baran, M. F., Alam, P., & Hayat, S. (2025). Pre-Sowing Seed Treatment with Glucose Boosts the Photosynthetic Efficiency, Metabolic Enzymes and Antioxidant Defense System to increase the Growth of <i>Abelmoschus esculentus</i> (L.) Moench. *Journal of Innovative Solutions for Eco-Environmental Sustainability*, *1*(1), 13854. DOI: 10.46991/JISEES.2025.01.1.13854

Singh, A., Bol, R., Lovynska, V., Singh, R. K., Sousa, J. R., & Ghazaryan, K. (2025b). Application of nanoparticles for salinity stress management and biofortification in wheat: A review of dual approaches and insights. *Frontiers in Plant Science*, *16*, 1592866. DOI: 10.3389/fpls.2025.1592866 PMID: 40688693

Singh, A., Bol, R., Lovynska, V., Singh, S., Singh, R. K., Sousa, J. R., Elshikh, M. S., Khachatryan, H., & Ghazaryan, K. (2025c). Nanoparticles: Dual role in alleviating abiotic stresses and boosting nutrient efficiency for sustainable agriculture. In *Advances in agronomy* (pp. 117–180). https://doi.org/DOI: 10.1016/bs.agron.2025.08.001

Singh, O., Pandey, D., Parmar, S. S., Minkina, T., & Rajput, P. (2025d). Emerging role of nanotechnology in biofortification, plant growth, and crop protection. *Journal of Innovative Solutions for Eco-Environmental Sustainability*, *1*(1), 13490. DOI: 10.46991/JISEES.2025.01.1.13490

Subramanian, V., Peijnenburg, W. J., Vijver, M. G., Blanco, C. F., Cucurachi, S., & Guinée, J. B. (2023). Approaches to implement safe by design in early product design through combining risk assessment and life cycle assessment. *Chemosphere*, *311*, 137080. DOI: 10.1016/j.chemosphere.2022.137080 PMID: 36328317

Sun, X. D., Ma, J. Y., Feng, L. J., Duan, J. L., Xie, X. M., Zhang, X. H., Kong, X., Ding, Z., & Yuan, X. Z. (2023). Magnetite nanoparticle coating chemistry regulates root uptake. *Proceedings of the National Academy of Sciences of the United States of America*, *120*(27), e2304306120. DOI: 10.1073/pnas.2304306120 PMID: 37364127

Tourinho, P. S., van Gestel, C. A. M., Lofts, S., Svendsen, C., Soares, A. M. V. M., & Loureiro, S. (2012). Metal-based nanoparticles in soil: Fate, behavior, and effects on soil invertebrates. *Environmental Toxicology and Chemistry*, *31*(8), 1679–1692. DOI: 10.1002/etc.1880 PMID: 22573562

Tripathi, D. K., Singh, S., Singh, S., Pandey, R., Singh, V. P., Sharma, N. C., & Chauhan, D. K. (2017). Manufactured nanoparticles in plants. *Plant Physiology and Biochemistry*, *110*, 2–12. DOI: 10.1016/j.plaphy.2016.07.030 PMID: 27601425

Trivedi, P., Leach, J. E., Tringe, S. G., Sa, T., & Singh, B. K. (2020). Plant–microbiome interactions: From community assembly to plant health. *Nature Reviews. Microbiology*, *18*(11), 607–621. DOI: 10.1038/s41579-020-0412-1 PMID: 32788714

Urra, J., Alkorta, I., & Garbisu, C. (2019). Potential benefits and risks for soil health derived from organic amendments. *Agronomy (Basel)*, *9*(9), 542. DOI: 10.3390/agronomy9090542

Vardharajula, S., Ali, S. Z., Tiwari, P. M., Eroğlu, E., Vig, K., & Dennis, V. A.. (2012). Functionalized carbon nanotubes: Biomedical applications. *International Journal of Nanomedicine*, *7*, 5361–5374. DOI: 10.2147/IJN.S35832 PMID: 23091380

Verma, K. K., Li, D. M., Singh, M., Rajput, V. D., Malviya, M. K., & Minkina, T.. (2020). Interactive role of silicon and plant–rhizobacteria mitigating abiotic stresses. *Plants*, *9*(9), 1055. DOI: 10.3390/plants9091055 PMID: 32824916

Verma, K. K., Song, X. P., Joshi, A., Rajput, V. D., Singh, M., Sharma, A., Singh, R. K., Li, D.-M., Arora, J., Minkina, T., & Li, Y.-R. (2022a). Nanofertilizer possibilities for healthy soil, water, and food in the future. *Frontiers in Plant Science*, *13*, 865048. DOI: 10.3389/fpls.2022.865048 PMID: 35677230

Verma, K. K., Song, X. P., Joshi, A., Tian, D. D., Rajput, V. D., & Singh, M. (2022b). Recent trends in nanofertilizers for advancing sustainable agriculture. *Nanomaterials (Basel, Switzerland)*, *12*, 173. DOI: 10.3390/nano12010173 PMID: 35010126

Verma, K. K., Song, X. P., Singh, M., Tian, D. D., Rajput, V. D., & Minkina, T.. (2023). Association of silicon and soil microorganisms induces stress mitigation. In de Mello Prado, R. (Ed.), *Benefits of silicon in the nutrition of plants* (pp. 299–328). Springer., DOI: 10.1007/978-3-031-26673-7_17

Vimercati, L., Cavone, D., Caputi, A., De Maria, L., Tria, M., Prato, E., & Ferri, G. M. (2020). Nanoparticles: An experimental study of zinc nanoparticles toxicity on marine crustaceans and a general overview on the health implications in humans. *Frontiers in Public Health*, *8*, 192. DOI: 10.3389/fpubh.2020.00192 PMID: 32509719

Waalewijn Kool, P. L., Diez Ortiz, M., & van Gestel, C. A. M. (2014). Effect of soil organic matter content and pH on ZnO nanoparticle toxicity to *Folsomia candida. Ecotoxicology and Environmental Safety*, *108*, 9–15. DOI: 10.1016/j.ecoenv.2014.06.031 PMID: 25038266

Watts, N., Amann, M., Ayeb-Karlsson, S., Belesova, K., Bouley, T., Boykoff, M., Byass, P., Cai, W., Campbell-Lendrum, D., Chambers, J., Cox, P. M., Daly, M., Dasandi, N., Davies, M., Depledge, M., Depoux, A., Dominguez-Salas, P., Drummond, P., Ekins, P., & Costello, A. (2018). The Lancet Countdown on health and climate change. *Lancet*, *391*(10120), 581–630. DOI: 10.1016/S0140-6736(17)32464-9 PMID: 29096948

Wheeler, K. E., Chetwynd, A. J., Fahy, K. M., Hong, B. S., Tochihuitl, J. A., Foster, L. A., & Lynch, I. (2021). Environmental dimensions of the protein corona. *Nature Nanotechnology*, *16*(6), 617–629. DOI: 10.1038/s41565-021-00924-1 PMID: 34117462

Wu, H., Tito, N., & Giraldo, J. P. (2017). Cerium oxide nanoparticles protect photosynthesis. *ACS Nano*, *11*(11), 11283–11297. DOI: 10.1021/acsnano.7b05723 PMID: 29099581

Xin, X., Nepal, J., Wright, A. L., Yang, X., & He, Z. (2022). Carbon nanoparticles improve corn (Zea mays L.) growth and soil quality: Comparison of foliar spray and soil drench application. *Journal of Cleaner Production*, *363*, 132630. DOI: 10.1016/j.jclepro.2022.132630

Xin, X., Nepal, J., Wright, A. L., Yang, X., & He, Z. (2022). Carbon nanoparticles improve corn (Zea mays L.) growth and soil quality: Comparison of foliar spray and soil drench application. *Journal of Cleaner Production*, *363*, 132630. DOI: 10.1016/j.jclepro.2022.132630

Yan, L., Li, P., Zhao, X., Ji, R., & Zhao, L. (2020). Physiological and metabolic responses of maize (Zea mays) plants to Fe3O4 nanoparticles. *The Science of the Total Environment*, *718*, 137400. DOI: 10.1016/j.scitotenv.2020.137400 PMID: 32105936

Yang, R., Gao, L., Liu, Z., Li, C., & Shen, Y. (2025). Sulfur nanoparticles enhance watermelon growth. *Plant Physiology and Biochemistry*, *229*, 110359. DOI: 10.1016/j.plaphy.2025.110359 PMID: 40811908

You, Y., Kerner, P., Shanmugam, S., & Khodakovskaya, M. V. (2023). Differential effects of carbon nanotubes and graphene on the tomato rhizosphere microbiome. *Environmental Science. Nano*, *10*(6), 1570–1584. DOI: 10.1039/D2EN01026G

Zakaria, Z., Zulkafflee, N. S., Mohd Redzuan, N. A., Selamat, J., Ismail, M. R., Praveena, S. M., & Abdull Razis, A. F. (2021). Understanding potential heavy metal contamination, absorption, translocation and accumulation in rice and human health risks. *Plants*, *10*(6), 1070. DOI: 10.3390/plants10061070 PMID: 34073642

Zamora-Justo, J. A., Abrica-González, P., Vázquez-Martínez, G. R., Muñoz-Diosdado, A., Balderas-López, J. A., & Ibáñez-Hernández, M. (2019). Polyethylene glycol-coated gold nanoparticles as DNA and atorvastatin delivery systems and cytotoxicity evaluation. *Journal of Nanomaterials*, *5982047*, 1–11. Advance online publication. DOI: 10.1155/2019/5982047

Zhang, M., Tao, S., & Wang, X. (2020). Interactions between organic pollutants and carbon nanomaterials and the associated impact on microbial availability and degradation in soil: A review. *Environmental Science. Nano*, *7*(9), 2486–2508. DOI: 10.1039/D0EN00515K

Zhang, M. X., Zhao, L. Y., He, Y. Y., Hu, J. P., Hu, G. W., Zhu, Y., & Zhang, J. L. (2024). Iron nanomaterials enhance growth and nitrogen fixation in alfalfa. *Bioresource Technology*, *391*, 129987. DOI: 10.1016/j.biortech.2023.129987 PMID: 37951551

Chapter 3
Introduction to Biochar: Composition, Production, and Applications

Omkar Singh
http://orcid.org/0000-0003-1971-185X
Keral Verma Subharti College of Science, Swami Vivekanand Subharti University, Meerut, India

Uday Pratap Shahi
College of Agriculture, Sardar Vallabhbhai Patel University of Agriculture and Technology, India

Vaishali Singh
College of Agriculture, Sardar Vallabhbhai Patel University of Agriculture and Technology, India

Sakshi Singh
Faculty of Biology, Yerevan State University, Armenia

Tatevik Derdzyan
Faculty of Biology, Yerevan State University, Armenia

ABSTRACT

Biochar is a substance high in carbon that is produced by thermochemically converting biomass in areas with limited oxygen. It has garnered significant attention because it can address climate change, soil degradation, and waste management simultaneously. The chapter begins with the history of biochar in Amazonian Dark Earths (Terra Preta) and then proceeds to discuss its complex physicochemical properties, including surface area, porosity, elemental composition, and functional groups. We examine various methods for producing biochar, including gasification,

DOI: 10.4018/979-8-3693-7006-3.ch003

hydrothermal carbonization, slow pyrolysis, and fast pyrolysis. Different methods produce biochar with varying qualities that are best suited for different purposes. It also shows how the choice of feedstock changes the properties of biochar. This technology can be applied in various ways, including cleaning up polluted soils, storing carbon, treating and purifying water, enhancing agricultural productivity, and even storing energy.

1. OVERVIEW

Biochar is a carbon-rich material produced from biomass through a process known as pyrolysis. It has garnered significant attention due to its potential to address multiple global challenges simultaneously. This chapter provides a comprehensive overview of biochar, detailing its composition, production methods, and diverse applications. The definition and historical context of biochar are first established, followed by an examination of current research interests and potential applications across various fields.

2. DEFINITION AND HISTORICAL CONTEXT OF BIOCHAR

Definition of Biochar

In the absence of oxygen and low temperatures (<700°C), organic matter undergoes thermochemical conversion to form biochar, a porous, carbonaceous substance (Lehmann & Joseph, 2015). The pyrolysis process yields a durable carbon form that can endure in the environment for extended periods, ranging from hundreds to thousands of years (Woolf et al., 2014). The International Biochar Initiative (IBI) defines biochar as: "A solid material obtained from the thermochemical conversion of biomass in an oxygen-limited environment. Biochar can be used for a range of applications as an agent for soil improvement, improved resource use efficiency, remediation, and/or protection against particular environmental pollution, and as an avenue for greenhouse gas (GHG) mitigation" (IBI, 2015). It is essential to distinguish biochar from charcoal, which is primarily produced for fuel purposes. While both materials are derived from the pyrolysis of biomass, biochar is specifically produced with the intention of application to soil and for environmental management (Lehmann & Joseph, 2015).

Historical Context and Traditional Uses

Application of burned organic materials to enhance soil fertility is not a novel idea. Actually, the origin of contemporary biochar research and applications can be attributed to the identification of Terra Preta de úndio, also known as Amazonian Dark Earths (ADEs). European archaeologists initially documented Terra Preta soils in the Amazon Basin during the late 19th century (Woods & Denevan, 2009). These anthropogenic soils, formed by indigenous populations before the arrival of Europeans, are distinguished by their black hue, superior fertility, and higher carbon content compared to the soils in the vicinity (Glaser & Birk, 2012). Extensive archaeological and soil science investigations have shown that Terra Preta soils are rich in charred organic matter, as well as various organic and inorganic substances, including animal bones, fish bones, and pottery fragments **(Glaser et al., 2001).** It is widely believed that the deliberate incorporation of burned organic material into these soils has been crucial in maintaining their fertility for generations, even in the face of challenging tropical conditions that typically lead to rapid soil deterioration (Lehmann et al., 2003). Traditional char use in agriculture has been documented in many civilisations and historical eras, extending beyond the Amazon Basin. The Japanese practice of "Haigoe" involves combining human feces with rice husk charcoal and applying it to rice paddies (Ogawa & Okimori, 2010). Furthermore, in the Mediterranean region, historical records exist of combining char with animal manure for the purpose of producing fertilizers (Wilson, 2014). These traditional practices demonstrate an intuitive understanding of the benefits of charred organic matter in soil management, long before the scientific understanding of biochar's properties and mechanisms was developed.

3. BIOCHAR COMPOSITION

The chemical composition of biochar plays a crucial role in defining its characteristics and possible uses. The next section examines the physical and chemical properties of biochar, along with the variables that influence its composition. A comprehensive understanding of these factors is essential for maximizing biochar production and tailoring its characteristics to specific applications.

Physical Attributes

The physical properties of biochar are crucial in determining its usefulness and interactions with the environment, encompassing surface area, porosity, particle size distribution, and density. A key physical characteristic of biochar is its substantial surface area, primarily attributed to its porous composition, which can range from 1 to more than 1000 square meters per gram, depending on the type of raw material and the synthesis conditions (Lehmann & Joseph, 2015; Singh et al., 2024). This extensive surface area provides numerous locations for chemical reactions and sorption technologies. Biochar porosity is characterized by a network of pores of various sizes, including macropores (>50 nm), mesopores (2-50 nm), and micropores (<2 nm), with this arrangement having a substantial impact on the biochar's ability to retain water, adsorb nutrients, and provide a habitat for microorganisms (Brewer et al., 2014; Ghazaryan et al., 2024). Micropores play a vital role in the adsorption of small molecules, such as gases and solvents, while macro- and mesopores are essential for sequestering soil water and facilitating microbial colonization. The particle size distribution of biochar exhibits significant variation, including a spectrum from small fragments to pieces measuring several centimetres in diameter, with the distribution being affected by both the initial composition of the raw materials and the specific manufacturing procedure, and particle size influencing the interaction of biochar with soil, its vulnerability to erosion, and its impact on soil physical characteristics such as bulk density and water holding capacity (Sohi et al., 2010; Shahi et al., 2024). Finally, biochar generally exhibits a lower density in comparison to mineral soil particles, thereby resulting in a reduction in soil bulk density upon its addition to the soil, with the skeletal density of biochar being approximately 1.5-1.7 g/cm^3 while the bulk density, which includes pore spaces, can be as low as 0.1-0.5 g/cm^3 (Brewer et al., 2014; Singh et al., 2024), and by virtue of its low density, biochar enhances soil aeration and water retention.

Chemical Properties

The chemical composition of biochar is intricate and subject to variation based on the source material and prevailing manufacturing circumstances, with fundamental chemical characteristics encompassing elemental composition, functional groups, pH, and cation exchange capability. A typical carbon content of biochar ranges from 50% to over 90% of the dry mass (Enders et al., 2012), with the residual mass comprising a variety of additional constituents, including oxygen, hydrogen, nitrogen, sulfur, and ash (inorganic minerals). The proportions of these elements, specifically the hydrogen-to-carbon and oxygen-to-carbon ratios, are frequently used as diagnostic markers for the stability and aromaticity of biochar, with lower H/C and O/C ratios typically indicating increased aromaticity and stability (Spokas, 2010). The surface of biochar is characterised by a diverse array of functional groups that significantly influence its reactivity and sorption activities, including carboxyl (-COOH), hydroxyl (-OH), phenolic (-C6H5OH), carbonyl (C=O), quinone, and lactone groups, with the presence of these functional groups enhancing the capacity of biochar to engage with soil components, impact the availability of nutrients, and adsorb pollutants (Tan et al., 2015). As biochar matures in the soil, the composition and quantity of functional groups can undergo changes through oxidative ageing. An alkaline pH, ranging from 7.5 to 10.5, is characteristic of biochar, although it can vary based on the feedstock and production temperature (Lehmann et al., 2011), with biochar's alkaline properties imparting a liming effect to soil that is potentially advantageous for acidic soils but necessitating cautious handling in neutral or alkaline soils. The cation exchange capacity of biochar experiences a progressive rise as it matures in the soil, primarily attributed to oxidation and the development of carboxyl groups on its surface, with the initial cation exchange capacity (CEC) of fresh biochar frequently being somewhat low but capable of rapidly rising over a period of several months to years in the soil (Cheng et al., 2008), and high cation exchange capacity enhancing biochar's capacity to retain nutrients and improve soil fertility.

Factors Affecting Biochar Composition

Biochar composition is largely determined by the selection of feedstock material and the specific production conditions.

Feedstock Material: The choice of biomass as feedstock greatly influences the composition of the arising biochar. Typical feedstock materials consist of:

- Wood and forestry residues
- Agricultural crop residues
- Animal manures
- Municipal solid waste
- Sewage sludge

Each feedstock imparts specific properties to the resulting biochar. For example, woody biomass typically produces biochar with higher carbon concentration and lower ash content than agricultural residues or manures. In contrast, nutrient-rich feedstocks such as manures yield biochar with increased concentrations of plant nutrients, including phosphorus and potassium.

Production Temperature: The production temperature of biochar is a critically important determinant of its composition and characteristics. With an increase in pyrolysis temperature:

- Carbon content generally increases
- Hydrocarbon and oxygen saturation ratios decline
- Ash content rises

- Surface area typically increases, although it may decrease at very high temperatures due to pore collapse
- the pH level tends to rise
- the prevalence of functional groups varies, characterised by a general inclination towards a reduced number of oxygen-containing functional groups.

For instance, biochar generated at a temperature of 300 °C may exhibit an oxygen-to-carbon (O/C) ratio of 0.4, whereas biochar created from the same raw material at a temperature of 700 °C may have an O/C ratio of 0.05 (Lehmann & Joseph, 2015).

Residence Time: Residence time, which refers to the duration of the pyrolysis process, also influences the characteristics of biochar. Extended residence times typically result in:

- Increased carbon concentration
- Decreased volatile species concentration
- Expanded surface area
- Modified distribution of pore sizes

Nevertheless, the impact of the duration of residence is generally less significant compared to that of temperature (Manyà, 2012).

Heating Rate: The temperature at which the feedstock is heated during pyrolysis might impact the physical structure of the biochar. Elevated heating rates typically lead to the formation of biochar with a more vesicular structure and larger pores, whereas reduced heating rates commonly yield a finer pore structure (Cetin et al., 2004).

Post-Production Treatments: A range of post-production processes can be used to alter the characteristics of biochar:

- Activation: This process can greatly enhance the surface area and porosity by physical or chemical means.
- Impregnation: The functionality of biochar can be enhanced by including nutrients or other chemicals throughout its structure.
- Oxidation: The controlled process of oxidation can enhance the presence of oxygen-containing functional groups upon the surface of biochar.
- Particle size reduction: Grinding or milling can modify the distribution of particle sizes and enhance the open surface area accessible for reactions.

Various procedures enable the customisation of biochar characteristics to suit particular uses (Rajapaksha et al., 2016).

Analytical Techniques for Biochar Characterization: A diverse array of analytical methods is used to characterise the content and characteristics of biochar:

- The elemental composition of biochar is determined using CHNS analysers dedicated to elemental analysis.
- Proximate Analysis: Quantifies the levels of moisture, volatile matter, fixed carbon, and ash content.
- Scanning Electron Microscopy (SEM): Offers detailed visual representations of the surface structure of biochar.
- Brunauer-Emmett-Teller (BET) Analysis quantifies the surface area and distribution of pore sizes.
- Fourier Transform Infrared Spectroscopy (FTIR) is a technique used to detect and characterise functional groups present on the surface of biochar.
- X-ray Diffraction (XRD) is a technique used to examine the crystalline structure of biochar and its mineral constituents.
- Nuclear Magnetic Resonance (NMR) Spectroscopy: Offers comprehensive insights into the carbon structures present in biochar.
- X-ray Photoelectron Spectroscopy (XPS) is a technique used to analyse the surface chemistry of biochar.

These methodologies, frequently employed in conjunction, offer a thorough comprehension of the composition and characteristics of biochar (Chia et al., 2012).

Table 1. Typical elemental composition of biochar from different feedstocks

Feedstock	C (%)	H (%)	O (%)	N (%)	S (%)	Ash (%)	H/C	O/C	Reference
Pine wood	84.8	3.09	11.3	0.42	0.03	0.36	0.44	0.10	Kloss et al. (2012)
Oak wood	77.8	3.37	18.5	0.30	0.01	0.02	0.52	0.18	Sun et al. (2014)
Corn stover	60.4	2.43	20.8	1.24	0.15	15.0	0.48	0.26	Spokas et al. (2011)
Rice husk	48.7	1.97	12.7	1.01	0.08	35.5	0.49	0.20	Wu et al. (2012)
Wheat straw	63.4	2.51	25.1	1.45	0.11	7.43	0.47	0.30	Zhao et al. (2013)
Switchgrass	84.4	2.69	10.9	0.91	0.10	1.00	0.38	0.10	Novak et al. (2009)
Peanut hull	75.2	3.11	16.4	1.74	0.08	3.47	0.50	0.16	Novak et al. (2009)
Poultry litter	38.5	1.64	9.86	3.32	0.70	46.0	0.51	0.19	Chan et al. (2008)
Dairy manure	69.6	2.78	15.9	2.56	0.54	8.62	0.48	0.17	Cao & Harris (2010)
Pig manure	56.9	2.89	16.4	3.54	0.43	19.8	0.61	0.22	Tsai et al. (2012)
Sewage sludge	33.2	1.97	10.3	4.21	1.52	48.8	0.71	0.23	Hossain et al. (2011)
Olive mill waste	73.5	3.04	19.3	1.91	0.09	2.16	0.50	0.20	Al-Wabel et al. (2013)

continued on following page

Table 1. Continued

Feedstock	C (%)	H (%)	O (%)	N (%)	S (%)	Ash (%)	H/C	O/C	Reference
Coconut shell	89.6	2.44	6.92	0.41	0.04	0.59	0.33	0.06	Ronsse et al. (2013)
Bamboo	91.5	2.02	5.73	0.51	0.03	0.21	0.26	0.05	Chen et al. (2014)
Miscanthus	79.8	2.71	15.4	0.76	0.09	1.24	0.41	0.14	Crombie et al. (2013)
Sugarcane bagasse	72.5	2.83	22.8	0.71	0.10	1.06	0.47	0.24	Inyang et al. (2010)

Note: All biochars were synthesised within the temperature range of 450-550 °C. The reporting of elemental compositions is based on dry weight. H/C and O/C are fundamental atomic ratios.

4. PRODUCTION METHODS OF BIOCHAR

Biochar is produced by subjecting organic matter to heat decomposition in an environment with limited oxygen availability. The process, referred to as thermo-chemical conversion, can be accomplished using different techniques, each with its own specified parameters and consequent biochar properties. This section examines the primary techniques applied in the creation of biochar, which include pyrolysis, gasification, and hydrothermal carbonisation.

Figure 1. Flowchart of biochar production (Singh et al., 2023)

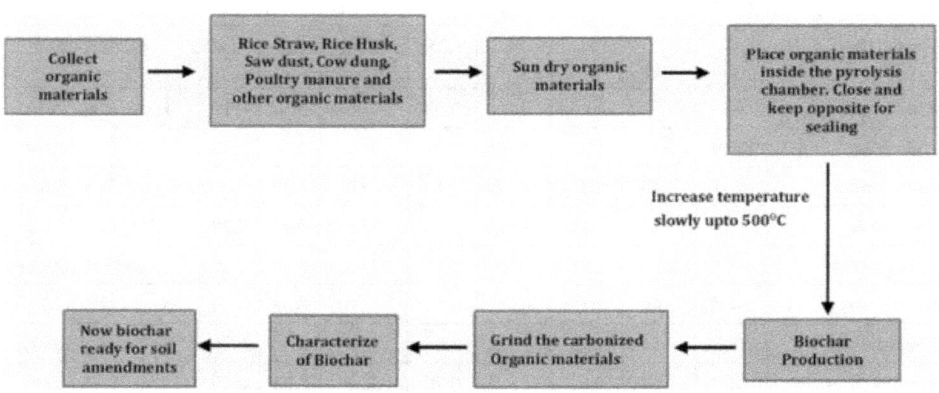

Fig.1: Flowchart of biochar production (Source: Singh et al., 2023)

Pyrolysis: Pyrolysis is the primary method used for synthesizing biochar. Biomass thermal decomposition refers to the breakdown of organic matter in the absence of oxygen, typically occurring at temperatures between 300 °C and 700 °C (Bridgwater, 2012). Pyrolysis can be classified based on heating rate, residence time, and temperature.

Slow Pyrolysis: Slow pyrolysis is characterized by low heating rates of 0.1-1 °C/s, extended residence times ranging from hours to days, and moderate temperatures between 300 and 550 °C (Bridgwater, 2012). This method typically produces the highest proportion of biochar among pyrolysis techniques, with yields ranging from 25% to 35% of the original feedstock mass. Vapour residence times are 5-30 minutes, while solids remain for significantly longer periods. Compared to rapid pyrolysis, slow pyrolysis generates biochar with higher carbon content and greater surface area, due to prolonged residence times that enable more complete carbonization and the development of a more organized biochar structure (Manyà, 2012).

Fast Pyrolysis: Fast pyrolysis involves rapidly heating biomass at rates of 10 to 200 °C per second to temperatures between 400 and 550 °C, with vapour residence times under 2 seconds (Mohan et al., 2006). This process is designed to maximise bio-oil production and also generates biochar as a secondary product. Key characteristics of fast pyrolysis include biochar yields of 10 to 20 percent of the initial biomass mass, heating rates of 10 to 200 °C per second, vapor residence times of less than 2 seconds, solid residence times of under 30 minutes, and operational temperatures ranging from 400 to 550 °C. Compared to slow pyrolysis, fast pyrolysis produces biochar with a higher volatile matter content and lower fixed carbon content, due to the rapid heating and short exposure time, resulting in a less ordered carbon structure (Brewer et al., 2009).

Flash Pyrolysis: Fast pyrolysis involves rapidly heating biomass at rates of 10 to 200 °C per second to temperatures between 400 and 550 °C, with vapour residence times under 2 seconds (Mohan et al., 2006). This process is optimized to maximize bio-oil production and also generates biochar as a secondary product. Key characteristics include biochar yields of 10 to 20 percent of the initial biomass, heating rates of 10 to 200 °C per second, vapor residence times of less than 2 seconds, solid residence times of under 30 minutes, and operational temperatures of 400 to 550 °C. Compared to slow pyrolysis, fast pyrolysis produces biochar with higher volatile matter and lower fixed carbon content due to the rapid heating and short exposure time, which results in a less ordered carbon structure (Brewer et al., 2009).

Gasification: Gasification is a thermal energy conversion process that operates at temperatures above 700 °C, transforming biomass into a combustible gas mixture known as syngas through the controlled addition of oxygen or steam (Basu, 2010). In addition to syngas production, gasification yields a solid carbonaceous residue classified as a type of biochar. The process typically results in a biochar output of 5-

10% of the feedstock mass, with operational temperatures ranging from 700 to 1500 °C and the use of air, oxygen, or steam as oxidizing agents. Compared to pyrolysis biochar, gasification biochar generally contains a higher ash fraction, lower carbon content, and increased pH levels (Hansen et al., 2015). The elevated temperatures and partial oxidation conditions during gasification produce a more resistant carbon structure, which may also contain higher concentrations of polycyclic aromatic hydrocarbons (PAHs) (Wiedner et al., 2013).

Hydrothermal Carbonization (HTC): Hydrothermal carbonisation (HTC) is a wet pyrolysis process that converts biomass into hydrochar, a coal-like material, by applying high temperatures (180-250 °C) and pressure in the presence of water (Libra et al., 2011). Although hydrochar is sometimes considered distinct from biochar due to its unique properties, current research increasingly investigates hydrochar as a biochar variant for targeted applications. Key characteristics of HTC include hydrochar yields of 30-60% of the original feedstock mass, operational temperatures of 180-250 °C, pressures of 2-6 megapascals, and residence times of 1-12 hours. Compared to biochar produced by dry pyrolysis, hydrochar typically contains less carbon and more oxygen, and exhibits a greater abundance of functional groups (Kambo & Dutta, 2015). Hydrochar also generally exhibits a lower surface area and pH. A significant advantage of HTC is its ability to process wet biomass without pre-drying, making it suitable for high-moisture feedstocks such as sewage sludge and marine algae.

Reactor Types and Configurations

The selection of reactor type and configuration is critical in biochar synthesis. Several reactor designs are commonly employed. Fixed-bed reactors are suitable for slow pyrolysis at small to medium scales. In these systems, raw biomass is introduced and externally heated, although temperature gradients and product inhomogeneity may occur. Fluidized bed reactors utilize a heated, inert material, such as sand, that is fluidized by a gas flow to ensure efficient heat transfer to biomass particles. Their high heating rates and precise temperature control make them optimal for rapid pyrolysis. Rotary kiln reactors consist of a rotating cylindrical tube that moves biomass through a heated zone, providing effective mixing and temperature regulation for continuous operation. Auger reactors utilize a screw conveyor to transport biomass through a heated tube, enabling precise control over residence time and facilitating small to medium-scale continuous production. Microwave-assisted pyrolysis utilizes microwave radiation to heat biomass internally, allowing for rapid and uniform heating that can yield unique biochar properties.

Factors Influencing Biochar Synthesis

Multiple variables influence both the biochar production process and the resulting material properties. The type and composition of biomass feedstock, including cellulose, hemicellulose, lignin, ash content, and elemental makeup, play a critical role in determining biochar quality. Pyrolysis temperature is another key parameter; higher temperatures generally increase carbon content and aromaticity, decrease yield and volatile matter, enhance surface area up to a threshold, raise pH, and alter functional group structures. The heating rate also affects product distribution and the characteristics of biochar. Rapid heating rates tend to increase the accumulation of volatile materials and reduce char formation, whereas slower rates favor char development. Longer residence times promote more complete carbonization, resulting in higher fixed carbon and lower volatile matter. Increased pressure during pyrolysis can enhance char yield and alter biochar properties by facilitating secondary char-forming reactions. Finally, smaller biomass particle sizes enhance heat transfer and reaction rates, resulting in more uniform heating during pyrolysis.

Emerging Technologies in Biochar Production

Conventional pyrolysis methods remain the primary technique for biochar synthesis. However, several emerging technologies demonstrate the capacity to enhance production efficiency and process adaptability.

Microwave-Assisted Pyrolysis: Microwave-assisted thermal pyrolysis offers several advantages over conventional heating methods, such as rapid and uniform heating, selective heating of specific biomass components, and the potential to produce biochar with unique properties. Preliminary studies indicate that microwave-assisted pyrolysis generates biochar with higher surface area and pore volume than conventional pyrolysis. Despite these benefits, scaling this technology for industrial production presents significant challenges.

Solar-Assisted Pyrolysis: The use of concentrated solar energy for biomass pyrolysis represents a promising method to reduce the environmental impact of biochar synthesis. Solar-assisted pyrolysis devices utilize mirrors or lenses to focus sunlight and achieve the high temperatures necessary for pyrolysis (Zeng et al., 2017). While this technology remains in the early stages of development, it offers a sustainable pathway for biochar production in regions with abundant solar resources.

Plasma Pyrolysis: Plasma pyrolysis rapidly heats biomass to temperatures exceeding 1000 °C using electrically generated plasma. This process treats diverse feedstocks, including waste materials, under severe thermal conditions and can produce biochar with unique physicochemical properties (Tang et al., 2013). However, the substantial energy requirements currently limit the widespread adoption of this technology.

Torrefaction-Pyrolysis Systems: Torrefaction is a mild thermal pretreatment conducted at temperatures between 200 and 300°C. This process, when combined with pyrolysis, produces biochar with improved properties. Torrefaction increases the grindability of biomass and enables more consistent production of biochar products (Basu et al., 2013). Currently, integrated torrefaction-pyrolysis systems are being developed to improve the efficiency of biochar production from a variety of feedstocks.

Environmental Considerations in Biochar Production: Given the expanding scale of biochar production, a systematic assessment of the associated environmental impacts of the manufacturing process is required.

Emissions: Pyrolysis and gasification processes can emit various gases and particulate matter. Effective emission control systems are essential to mitigate air pollution. Key considerations include the collection and analysis of pyrolysis gases, management of particulate emissions, and monitoring and reduction of hazardous compounds such as polycyclic aromatic hydrocarbons.

Energy Balance: The amount of energy associated with biochar production should be thoroughly evaluated in life cycle studies. Ideally, energy intake should be reduced to the lowest possible level, and, if feasible, renewable energy sources should be employed. Integrating co-products, such as bio-oil and syngas, into the process energy can significantly enhance the overall energy balance (Woolf et al., 2014).

Water Use and Management: Dry pyrolysis applications typically require minimal water, whereas hydrothermal carbonization and specific feedstock pre-treatment processes are often highly water-intensive. Therefore, sustainable biochar production necessitates the adoption of robust water management and recycling strategies.

Economic Aspects of Biochar Production: Economic feasibility of biochar manufacturing is contingent upon several factors:

Feedstock Costs: Feedstock costs represent a significant portion of total biochar production expenses. Utilizing waste biomass or agricultural residues can reduce these costs. However, the expenses related to collection and transportation must also be considered.

Capital and Operating Costs: Capital and operating costs are strongly affected by the choice of manufacturing technology. Simple slow pyrolysis systems require lower initial investments. In contrast, advanced systems offer improved process control and yield higher-value products (Brown et al., 2011).

Product Diversification: The economic feasibility of biochar production can often be enhanced by maximising the value of byproducts such as bio-oil and syngas. The successful establishment of markets for these products is essential for the general economic viability of biochar systems (Shabangu et al., 2014).

Carbon Credits: With the development of carbon markets, the capacity of biochar to capture carbon should offer supplementary sources of income. Nevertheless, the development of reliable carbon accounting methods for biochar systems is crucial for this (Lehmann & Joseph, 2015).

Table 2. Comparison of biochar yield and properties from different production methods

Production Method	Biochar Yield (wt%)	Fixed Carbon (%)	Surface Area (m^2/g)	pH	H/C Ratio	O/C Ratio	Ash Content (%)	Key Characteristics	References
Slow Pyrolysis	25-35	50-90	50-400	7.5-10.5	0.4-0.7	0.1-0.4	5-20	High carbon content, well-developed pore structure	Lehmann & Joseph (2015); Manyà (2012)
Fast Pyrolysis	10-20	60-70	50-300	6.5-7.5	0.5-0.7	0.2-0.4	10-25	Higher volatile matter, less ordered carbon structure	Brewer et al. (2009); Mohan et al. (2006)
Flash Pyrolysis	5-10	70-80	100-500	8.0-11.0	0.2-0.4	0.1-0.3	15-30	High surface area, more ash content	Onay (2007); Onay & Kockar (2003)
Gasification	5-10	80-90	100-400	9.0-12.0	0.1-0.3	0.05-0.2	20-40	High pH, high ash content, potential for PAH formation	Hansen et al. (2015); Wiedner et al. (2013)
Hydrothermal Carbonization	30-60	40-60	10-50	5.0-7.0	0.8-1.5	0.2-0.6	5-15	Higher O/C ratio, more functional groups, lower pH	Kambo & Dutta (2015); Libra et al. (2011)
Microwave-Assisted Pyrolysis	20-30	60-80	200-600	7.0-9.0	0.3-0.6	0.1-0.3	10-25	High surface area, rapid production	Lin et al. (2012); Zhao et al. (2017)

Note: The ranges given are standard values and may differ based on the kind of feedstock and the particular process circumstances. The given values for fixed carbon, ash content, and elemental ratios are based on dry weight criteria.

Feedstock Materials

The careful selection of the suitable raw materials is a crucial determinant in the manufacture of biochar, exerting a substantial impact on the characteristics and possible uses of the developed biochar. This section examines four primary classifications of feedstock materials: agricultural residues, forestry biomass, animal excrement, and municipal solid waste. Each group has distinct attributes and difficulties in the manufacturing of biochar.

Agricultural Residues: Agricultural residues constitute an extensive and easily accessible reservoir of biomass for the synthesis of biochar. These materials are commonly residual results of cultivation and processing of crops, which makes them a desirable choice for the sustainable manufacture of biochar.

Crop Residues: Maize stover, wheat straw, and rice husks are examples of crop wastes that are highly plentiful agricultural feedstocks. These materials contain a high concentration of cellulose and hemicellulose, together with different quantities of lignin **(Tripathi et al., 2016).** The characteristics of the resulting biochar can be markedly affected by the content of crop leftovers. Specifically, biochar produced from rice husks usually has a significant amount of silica, which can improve its ability to retain water in the soil and adsorb heavy metals (Zhao et al., 2017). In contrast, corn stover biochar frequently exhibits desirable cation exchange capacity, thereby conferring advantages in terms of soil nutrient retention (Novak et al., 2009).

Nut Shells and Fruit Stones: Another useful type of agricultural wastes is comprised of nut shells (such as walnut, almond, and peanut) and fruit stones (such as olive pits and peach stones). These materials are distinguished by their significant lignin concentration, which enhances the formation of biochar with a high concentration of fixed carbon and a well-developed pore structure (González et al., 2009). Biochar produced from these substrates frequently demonstrates exceptional adsorption characteristics, rendering it appropriate for both soil amendment and the generation of activated carbon. One study conducted by Ahmadvand et al. (2018) provided evidence that biochar produced from walnut shells had better efficacy in eliminating organic contaminants from water solutions when compared to biochars derived from other agricultural wastes.

Forestry Waste: A substantial source of feedstock for biochar production is derived from forestry waste, which encompasses materials generated by forest management operations and wood processing businesses.

Wood Chips and Sawdust: Wood chips and sawdust derived from different tree species are frequently utilised in the manufacturing of biochar. These materials are distinguished by their significant lignin concentration, which enhances the formation of biochar with a high concentration of fixed carbon and stable aromatic structures (Kloss et al., 2012). Wood-based biochars typically show reduced ash concentration in comparison to biochars produced from herbaceous biomass, rendering them appropriate for uses that demand maximum purity. Furthermore, the abundant carbon content and well-formed pore structure of biochars derived from wood make them very efficient in capturing carbon and retaining water in soil (Brewer et al., 2014).

Bark and Branches: Often regarded as waste materials in forestry operations, tree bark and branches can also be used for the creation of biochar. Typically, these materials include more ash than wood chips and may generate biochar with distinct surface chemistry because of the presence of different extractives (Zhao et al., 2013). Biochar produced from bark has demonstrated potential in selective uses, such as the elimination of heavy metals from water-based systems. For example, Mohan et al. (2007) conclusively showed that oak bark biochar had exceptional efficacy in the removal of lead and cadmium from water.

Animal Manure: Animal manures are a type of feedstock that contains a high concentration of nutrients and can generate biochar with distinct characteristics that are especially advantageous for enhancing soil organic matter and recycling nutrients.

Cattle Manure: The abundant availability of cattle manure enables the production of biochar with a significant ash content and increased concentrations of nutrients, namely phosphorus and potassium. An investigation conducted by Cantrell et al. (2012) revealed that biochar produced from bovine dung displayed superior pH and electrical conductivity in comparison to biochars derived from other sources. This characteristic renders it especially well-suited for the improvement of acidic soils. The substantial nutrient concentration of bovine manure biochar has the potential to enhance soil fertility and increase crop productivity. Nevertheless, the presence of possible problems related to infections and antibiotic residues requires meticulous evaluation of the manufacturing circumstances and approach to application (Lehmann & Joseph, 2015).

Poultry Litter: Another prevalent feedstock in this category is poultry litter, which comprises manure and bedding debris. Typically, biochar derived from poultry litter has a high ash content, raised pH, and increased concentrations of nutrients such as nitrogen, phosphorus, and potassium (Chan et al., 2008). The specific characteristics of poultry litter biochar render it highly appropriate for soil amendment, particularly in soils with high acidity. Nevertheless, the abundant nutrient content also gives rise to apprehensions about possible nutrient leakage, thereby requiring meticulous application rates and monitoring (Novak et al., 2009).

Pig Manure: Biochar produced from pig dung often contains a significant amount of nitrogen and exhibits great potential in enhancing soil fertility. Pig dung biochar was shown by **Ro et al. (2010)** to greatly improve soil cation exchange capacity and serve as a slow-release nitrogen source. Nevertheless, it is important to note that pig manure biochar may have elevated concentrations of heavy metals in comparison to other biochars derived from manure. Therefore, it is crucial to thoroughly evaluate the possible environmental consequences and optimum application rates (Kumari et al., 2014).

Municipal Solid Waste: Utilising municipal solid waste (MSW) for biochar production has a chance to tackle waste management issues while generating a valuable commodity. Nevertheless, the diverse composition of municipal solid waste (MSW) poses distinct difficulties in the manufacture and utilisation of biochar.

Green Waste: Yard clippings and garden debris, which are considered green waste, have the potential to generate biochar that exhibits characteristics comparable to those acquired from agricultural residues. This fuel is especially appealing in metropolitan regions where alternative biomass sources may be scarce **(Waqas**

et al., 2018). Green waste-derived biochar has demonstrated potential in the rehabilitation of urban soil and as a constituent in horticultural growing media. The study conducted by Dispenza et al. (2016) provided evidence that the use of green waste biochar can significantly enhance water retention and nutrient availability in urban soils.

Food Waste: Food waste constitutes a substantial proportion of municipal solid waste (MSW) and is an underutilised asset in certain regions. Although biochar derived from food waste can be rich in nutrients, it may also include a significant amount of ash and exhibit less carbon stability when compared to lignocellulosic feedstocks **(Peng et al., 2016).** Variability in biochar characteristics might arise from the varied composition of food waste, thereby requiring meticulous quality control procedures. Nevertheless, the biochar derived from food waste has demonstrated promise in the realm of soil amendment and as an integral element in composting systems (Lim et al., 2016).

Sewage Sludge: Although sewage sludge is a subject of controversy because of its possible pollutants, it can be utilised to generate biochar with a significant phosphorus concentration. Precision processing and rigorous testing are necessary to guarantee the safety of biochar produced from sewage sludge. However, it can serve as a very efficient method for recycling nutrients from wastewater treatment systems (Hossain et al., 2011). Biochar derived from sewage sludge has shown effectiveness in decontaminating polluted soils, especially in immobilising heavy metals (Lu et al., 2012). Nevertheless, the presence of organic contaminants and viruses requires strict quality control measures and could restrict its utility in food production systems.

Table 3. Common feedstocks and their biochar characteristics

Feedstock Category	Specific Feedstock	Carbon Content (%)	Ash Content (%)	pH	Surface Area (m^2/g)	Key Nutrients	Potential Applications	Reference
Agricultural Residues	Rice Husks	40-50	30-40	8-10	200-400	Si, K	Soil water retention, Heavy metal adsorption	Zhao et al., 2017
Agricultural Residues	Corn Stover	60-70	10-20	9-10	300-450	K, Ca, Mg	Soil amendment, Nutrient retention	Novak et al., 2009

continued on following page

Table 3. Continued

Feedstock Category	Specific Feedstock	Carbon Content (%)	Ash Content (%)	pH	Surface Area (m^2/g)	Key Nutrients	Potential Applications	Reference
Agricultural Residues	Walnut Shells	75-85	1-5	7-8	400-600	Ca, K	Activated carbon, Organic pollutant removal	Ahmadvand et al., 2018
Forestry Waste	Pine Wood Chips	80-90	1-3	7-8	300-500	Ca, K	Carbon sequestration, Soil water retention	Brewer et al., 2014
Forestry Waste	Oak Bark	70-80	5-10	7-8	200-400	Ca, Mg	Heavy metal removal from water	Mohan et al., 2007
Animal Manure	Cattle Manure	50-60	30-40	9-11	50-200	P, K, Ca	Soil pH amendment, Nutrient source	Cantrell et al., 2012
Animal Manure	Poultry Litter	40-50	30-50	9-11	50-150	N, P, K	Soil fertility improvement, pH amendment	Chan et al., 2008
Animal Manure	Pig Manure	50-60	30-40	9-10	50-200	N, P, K	Soil fertility improvement, Nutrient source	Ro et al., 2010
Municipal Solid Waste	Green Waste	60-70	10-20	8-9	200-400	K, Ca	Urban soil amendment, Growing media component	Dispenza et al., 2016
Municipal Solid Waste	Food Waste	50-60	20-30	8-10	100-300	N, P, K	Soil amendment, Composting additive	Lim et al., 2016
Municipal Solid Waste	Sewage Sludge	30-40	40-60	7-8	50-200	P, N	Contaminated soil remediation, Heavy metal immobilization	Lu et al., 2012

Note: The ranges given are standard values and may differ based on the particular properties of the feedstock and the conditions of production.

5. APPLICATIONS OF BIOCHAR

The distinctive physicochemical characteristics of biochar have shown promise in a diverse array of applications. The following section examines the many applications of biochar, ranging from soil amendment and environmental remediation to agricultural uses and energy storage (Singh et al., 2024).

Amendment of Soil and Sequestration of Carbon: A key use of biochar is as a soil supplement, therefore offering the simultaneous advantages of enhancing soil quality and capturing carbon.

Figure 2. Major application area of biochar

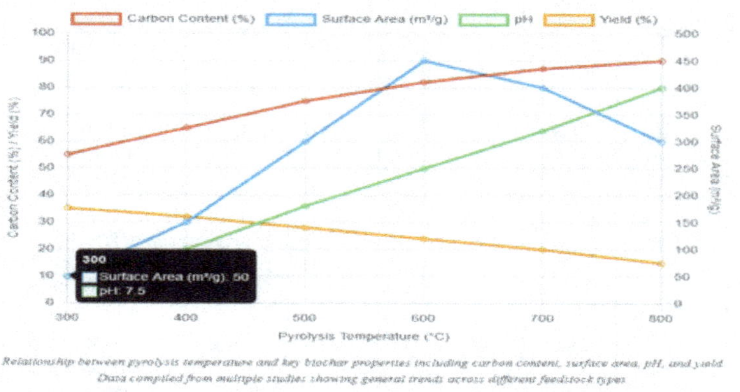

Fig.2: Effect of pyrolysis temperature and Biochar properties

Enhancement of Soil Quality: The application of biochar can greatly improve the physical, chemical, and biological characteristics of soil. The porous structure of the material facilitates enhanced soil water retention capacity and aeration, especially in sandy soils (Basso et al., 2013). The application of biochar has been demonstrated to augment the content of soil organic matter, improve the capacity to exchange cations, and alter the pH of the soil, often resulting in enhanced soil fertility (Lehmann et al., 2011; Selvaganapathi et al., 2025).

Carbon Sequestration: In the realm of long-term carbon sequestration, biochar emerges as a durable carbonaceous material capable of enduring in soil for extended periods ranging from hundreds to thousands of years (Woolf et al., 2016; Dutta et al., 2024). The inherent resistance of biochar carbon, especially resulting from high-temperature pyrolysis, enhances its enduring presence in soil ecosystems. According to Woolf et al. (2016), the adoption of sustainable biochar has the potential to counterbalance as much as 12% of the present anthropogenic CO_2-C emissions.

Water Treatment and Purification: Biochar has demonstrated potential in water treatment applications by utilising its large surface area and remarkable adsorption capabilities.

Removal of Organic Contaminants: The effectiveness of biochar in eliminating an array of organic pollutants from water, such as pesticides, medicines, and industrial chemicals, has been well established. For example, Ahmad et al. (2014) established that biochar produced from rice husk efficiently eliminated atrazine and imidacloprid from water-based solutions.

Heavy Metal Adsorption: Multiple studies have demonstrated the ability of biochar to adsorb hazardous heavy metals from water. A study conducted by Inyang et al. (2016) examined the efficacy of biochar in eliminating lead, cadmium, and copper, therefore emphasising its potential application in the treatment of water and wastewater.

Agricultural Uses: Biochar has several ecological uses in agriculture, enhancing crop yield, nutrient regulation, and pest mitigation.

Crop Productivity: Investigations have shown that the use of biochar is linked to higher agricultural productivity, especially in soils that are acidic and lacking in nutrients. A meta-analysis conducted by Biederman and Harpole (2013) revealed a general beneficial impact of biochar on agricultural output, particularly in tropical areas and soils with high acidity. Yield enhancements are driven by several methods, including optimised nutrient availability, enhanced soil physical characteristics, and heightened soil biological activity. Nevertheless, the impacts can differ according on the type of soil, characteristics of biochar, and choice of crop species (Jeffery et al., 2011).

Nutrient Retention: In soils, biochar can augment nutrient retention, therefore mitigate leaching and enhance the efficiency of fertiliser use. Its propensity to exchange cations enables it to absorb and retain nutrients, hence increasing their availability to plants over time (Clough *et al.*, 2013). The study conducted by Yao et al. (2012) provided evidence that biochar has the potential to effectively mitigate the leaching of nitrate, ammonium, and phosphate in agricultural waste. Furthermore, this capacity to retain nutrients not only enhances the nutritional value of crops but also reduces the possible environmental consequences linked to nutrient runoff.

Pest and Disease Management: Recently published studies indicate that biochar has the potential to be effective in controlling agricultural pests and illnesses. A study conducted by Elad et al. (2010) found that the use of biochar decreased the occurrence of foliar fungal infections in pepper and tomato plants. Potential mechanisms include the development of systemic resistance in plants and changes in the soil microbial community. Furthermore, biochar has demonstrated promise in inhibiting soil-borne infectious agents. A study conducted by Jaiswal et al. (2015) demonstrated that the addition of biochar decreased the occurrence of Fusarium wilt in tomato plants. Nevertheless, the impacts can lack consistency and are contingent upon the particular biochar-pathogen-plant interaction.

Environmental Remediation: Biochar has proven to be effective in several environmental remediation applications, especially in the cleanup of polluted soils and sediments.

Soil Contaminant Immobilization: The immobilisation of several soil pollutants by biochar has been shown to be highly efficient in lowering their bioavailability and associated environmental hazards. The efficacy of biochar in immobilising heavy metals and organic contaminants for the treatment of contaminated soils was examined by Beesley et al. (2011). Jiang et al. (2012) showed that biochar produced from rice straw effectively decreased the amount of cadmium and lead that could be absorbed by tissues in polluted soils. The processes encompass surface adsorption, precipitation, and intricate complexation reactions.

Restoration of Degraded Lands: Biochar has demonstrated potential in the rehabilitation of soils that have been deteriorated, such as mine sites and salty sites. In 2015, Anawar and colleagues conducted a comprehensive assessment of the capacity of biochar to restore soils damaged by mines. They observed enhancements in soil physical characteristics, increased availability of nutrients, and decreased metal toxicity. The application of biochar in saline soils can effectively alleviate salt stress on plants. A study by Akhtar et al. (2015) found that the addition of biochar enhanced the growth and productivity of wheat in salty environments. The researchers attributed these improvements to the enhancement of soil physical characteristics and the decrease in Na^+ absorption by plants.

Energy Production and Storage: Although biochar is commonly regarded as a secondary product of bioenergy generation, it also possesses promising possible uses in energy storage and conversion systems.

Biochar in Fuel Cells: Exploration of biochar as a catalyst support and electrode material in fuel cells has been initiated. By way of illustration, Huggins et al. (2014) showcased the application of biochar as an inexpensive catalytic support in microbial fuel cells, therefore augmenting power generation in comparison to conventional carbon materials.

Super capacitor Applications: The substantial surface area and permeable structure of biochar render it an promising option for electrodes in supercapacitors. A study conducted by Gu et al. (2015) documented the advancement of high-performance supercapacitors utilising KOH-activated biochar obtained from rice husk. The observed material had a high specific capacitance and demonstrated excellent cycling stability.

Biochar in Solar Energy Systems: Furthermore, biochar has been utilised in solar energy systems. The study conducted by Wang et al. (2016) examined the utilisation of biochar as a photothermal material in the production of solar steam. The results showed a high level of efficiency in converting solar energy into vapour, suitable for possible desalination purposes.

6. ENVIRONMENTAL AND ECONOMIC CONSIDERATIONS

The manufacturing and utilisation of biochar have substantial environmental and economic consequences. This section analyses these factors using life cycle assessment, carbon footprint analysis, and cost-benefit evaluation models.

Life Cycle Assessment: Life Cycle Assessment (LCA) is an essential analysis method used to assess the environmental effects of biochar systems across their entire life cycle. Carbon Cycle Assessment (LCA) offers a thorough structure for evaluating the possible environmental impacts linked to every phase of biochar synthesis and utilisation, encompassing the extraction of raw materials, processing, manufacturing, distribution, usage, and disposal or recycling.

System Boundaries and Functional Units: Establishing system boundaries is an essential stage in life cycle assessment (LCA) investigations of biochar. These boundaries usually include the processes of feedstock production or collection, transportation, biochar formation, application, and end-of-life situations. The selection of the functional unit plays an equally significant role and is contingent upon the specific emphasis of the investigation. Typical functional units seen in biochar life cycle assessments (LCAs) are:

- One tonne of biochar manufactured
- One hectare of land subjected to biochar treatment
- One tonne of carbon dioxide equivalent sequestered

The choice of functional unit can have a substantial impact on the interpretation of findings and the feasibility of comparing different studies (Matuštík et al., 2020).

Impact Categories: Ecological impact assessment (LCA) studies on biochar systems commonly evaluate several environmental impact categories, one of which is:

- Global Warming Potential (GWP)
- Potential of Eutrophication
- Potential of acidification
- Human Toxicity
- Ecotoxicology

- Land Utilisation
- Depletion of Water

A thorough life cycle assessment (LCA) of biochar production from waste biomass in the UK was carried out by Hammond et al. (2011). The study revealed significant decreases in greenhouse gas (GWP) emissions across several feedstock and production scenarios. Nevertheless, they acknowledged possible compromises with other impact categories, such as heightened acidification potential in certain instances.

Variability and Uncertainty: The life cycle assessment (LCA) outcomes for biochar systems might vary greatly based on factors such as:

- The kind and source of the feedstock
- Technology and volume of production
- Distances of transportation
- Rates and techniques of application
- Soil and climatic factors

Lorenz and Lal (2014) emphasised the requirement of region-specific Life Cycle Assessments (LCAs) that consider the unique local conditions and practices. They stressed that the environmental advantages of biochar systems are extremely contingent on the specific circumstances and can significantly differ among various geographical and agricultural environments.

7. CHALLENGES IN BIOCHAR LCA: PERSISTENT OBSTACLES IN DOING THOROUGH LIFE CYCLE ASSESSMENTS (LCAS) FOR BIOCHAR SYSTEMS INCLUDE

- The enduring stability of biochar in soil and its impact on soil processes over extended periods of time, ranging from decades to centuries, pose challenges in conventional life cycle assessment (LCA) frameworks.
- Accurate modelling of the intricate interactions of biochar, soil, plants, and microbes in life cycle assessments (LCAs) poses serious challenges.
- Data availability: There is sometimes a dearth of thorough, geographically specific data on the production of biochar and its impact on soil systems.
- Impact allocation: In systems where biochar is a byproduct (such as with bio-oil or syngas), the distribution of environmental impacts among products can have a substantial impact on the outcomes.

Mitigating these issues is essential for enhancing the dependability and comparability of biochar life cycle assessments **(Muñoz et al., 2017).**

Carbon Footprint: Due to biochar's potential contribution to climate change mitigation, the carbon footprint of biochar systems is a crucial factor to consider. A comprehensive investigation of biochar usually encompasses both its capacity to sequester carbon and the greenhouse gas emissions linked to its manufacturing and use.

Carbon Sequestration Potential: The carbon sequestration capacity of biochar arises from its capacity to retain carbon in a highly stable state for extended durations, with Woolf et al. (2016) calculating that the deployment of sustainable biochar could counterbalance as much as 12% of the present anthropogenic CO2-C emissions. The potential in question emerges from three primary mechanisms: soil-based direct carbon sequestration, reduction of soil emissions of other greenhouse gases, namely nitrogen dioxide (N2O), and preventable emissions resulting from the breakdown of biomass. Nevertheless, the true capacity for sequestration is contingent upon variables such as the type of feedstock, the conditions of production, and the rates of application, with the study conducted by Singh et al. (2015) and Singh et al. (2024) revealing that the carbon storage capacity of biochar ranged from 40% to 80% of the initial biomass carbon, contingent upon the pyrolysis parameters and the kind of feedstock.

Emissions from Production and Application: While biochar has the ability to retain carbon, its manufacturing and use also result in the release of emissions, with the primary contributors to emissions being energy use during the collection and transportation of feedstock, hydrogen emissions resulting from the pyrolysis process, mechanical equipment and procedures for application, and potential priming effects on the breakdown of soil organic matter. An examination of the life cycle carbon footprint of biochar systems by Roberts et al. (2010) revealed that switchgrass-derived biochar has the capacity to sequester a maximum of 864 kg CO2e per tonne of dry feedstock, nevertheless, they observed that the overall carbon advantage was influenced by assumptions regarding the durability of biochar and its effects on soil emissions.

Indirect Effects on Soil Greenhouse Gas Emissions: The application of biochar has the ability to impact soil greenhouse gas emissions, so potentially augmenting the overall capacity for climate mitigation (Subash et al.,2023). In 2014, **Cayuela et al.** investigated the impact of biochar on soil N_2O emissions by a meta-analysis. Their findings revealed an average decrease of 54% ± 6% (95% CI). Nevertheless, the impacts exhibited significant variation contingent upon the characteristics of the soil and biochar. Research on the influence of biochar on soil CH_4 emissions is inconclusive, as some research indicate decreases while others demonstrate no substantial effects or even rises (Jeffery et al., 2016).

Life Cycle Carbon Balance: The carbon footprint of biochar systems is determined by the equilibrium between carbon sequestration and emissions across their entire life cycle, with key determinants impacting this equilibrium being the type and origin of feedstock (e.g., waste biomass vs. purpose-grown crops), efficiency of production technology, distances of transportation, rates and techniques of application, soil and climatic factors, and consideration of time horizon. A study conducted by Hammond et al. (2011) revealed that biochar systems have the potential to deliver a net carbon abatement ranging from 0.7 to 1.3 tonnes CO2 per tonne of dry feedstock, depending on the particular circumstances, nevertheless, they stressed the need of attaining sustainable feedstock and implementing efficient production methods in order to optimise the advantages for the environment.

Cost-Benefit Analysis: Economics play a vital role in facilitating the extensive implementation of biochar technologies. Using cost-benefit studies, the economic feasibility of biochar systems can be assessed under several scenarios.

Production Costs: The costs of biochar production differ significantly based on variables such as the type of feedstock, the volume of production, and the method used, with primary cost elements comprising acquisition of feedstock, costs associated with the acquisition of pyrolysis equipment, expenses related to maintenance, energy, and labour, and logistics and operational management. The production costs of biochar in the UK were estimated by Shackley et al. (2011) to range from $51 to $386 per tonne under different scenarios, with their analysis revealing that the expenses related to feedstock and the capital expenditure in pyrolysis equipment were the primary factors influencing the total production costs.

Potential Revenue Streams: Potential revenue sources for biochar systems may encompass the sale of biochar specifically for soil amendment purposes, carbon credits for carbon that has been sequestered, sustainable energy derived from pyrolysis gases, and reduced expenses associated with trash disposal (for feedstocks obtained from waste). Biochar systems' economic feasibility generally relies on their capacity to generate revenue from several value streams, with a study by Clare et al. (2015) examining the economic aspects of biochar production in North America and concluding that biochar systems produced from trash were more likely to be financially viable compared to those utilizing purpose-grown feedstocks.

Agricultural Benefits: Application of biochar in agriculture can yield several economic advantages, such as augmented agricultural productivity, decreased need for fertilizers, enhanced water use control, and improved ability to withstand changes in climate. In their study, Dickinson et al. (2015) examined the cost-benefit ratio of applying biochar in cereal crop production, with their findings indicating that the advantages of biochar may surpass the costs in specific circumstances, especially in soils with lacking nutrients or regions susceptible to drought (Singh et al., 2023).

Non-Market Benefits: There are several possible advantages of biochar that are difficult to measure in monetary terms, including preservation of soil health over an extended period, decreasing water contamination caused by agricultural runoff, increased diversification in soil vegetation, and climate change mitigation, and it is important to include these non-market benefits into economic calculations in order to fully evaluate the value of biochar (Galinato et al., 2011).

Sensitivity Analysis: Financial feasibility of biochar systems is influenced by several aspects, such as analysis of feedstock availability and pricing, energy pricing, prices in the carbon market, regulatory landscape, and prices of agricultural commodities, with comprehensive sensitivity studies being essential for evaluating the resilience of economic evaluations in various situations and pinpointing critical factors that impact profitability (Brown et al., 2011).

8. CHALLENGES AND FUTURE DIRECTIONS

Biochar research and implementation encounters numerous problems that must be resolved to fully harness its potential as a means for sustainable agriculture, environmental management, and climate change mitigation. An inherent obstacle in the biochar industry is the absence of uniformity in the manufacturing and evaluation of its quality, with the varied composition of biochar arising from a variety of feedstocks and manufacturing settings making it challenging to develop globally valid standards, despite efforts by the International Biochar Initiative (IBI, 2015) and the European Biochar Certificate (EBC, 2012) to provide standardised definitions and testing criteria. The task of quality control in biochar manufacturing becomes increasingly critical as the business expands its operations, with Camps-Arbestain et al. (2015) highlighting the importance of implementing in-line monitoring algorithms for pyrolysis operations and developing industry-wide standards. Upgrading biochar production from a small laboratory or pilot scale to a large commercial scale poses many challenges, including assuring a reliable and long-lasting source of raw materials, competition from other biomass applications, variations in the availability of agricultural wastes during different seasons, and logistical complexities in the collecting and shipping processes, with Qian et al. (2015) proposing that future studies should priorities the optimization of biomass supply chains and the investigation of new feedstock sources. Additionally, scaling up pyrolysis technologies while optimizing product quality and process efficiency poses a notable technical challenge, successfully establishing strong markets for biochar and its byproducts is crucial for ensuring commercial feasibility, quantifying the long-term impacts of biochar in soil conditions is essential for evaluating its capacity to sequester carbon and enhance soil quality, and understanding the environmental fate and possible ecological effects of biochar becomes essential as its application grows, with future research needing to focus on comprehensive life cycle assessments and techno-economic evaluations to accurately evaluate the sustainability and economic feasibility of biochar systems.

9. CONCLUSION

Biochar is a promising multipurpose carbonaceous material that has a great deal of potential to address current agricultural and environmental issues through its various uses in environmental remediation, water treatment, soil amendment, and carbon sequestration. The effectiveness of biochar is shown in this chapter to be largely dependent on the careful selection of feedstock materials, the optimization of production parameters (specifically pyrolysis temperature, residence time, and heating rate), and the use of application techniques that are suitable for a given soil and climate. Even though biochar has a lot of potential to increase soil fertility, boost crop yields, and mitigate climate change by storing carbon for a long time, there are still a number of significant obstacles to overcome. These include the requirement for standardized production and quality assessment procedures, the transition from lab to commercial applications, the establishment of reliable supply chains, and the completion of extensive long-term research to better understand ecological impacts and environmental fate. To fully realize its potential as a sustainable solution for environmental management and agricultural productivity enhancement, future research efforts must concentrate on creating biochar systems that are specific to a given region, carrying out comprehensive life cycle assessments and techno-economic evaluations, and filling in knowledge gaps regarding the interactions between biochar, soil, and plants. Overcoming these obstacles while preserving financial sustainability and guaranteeing environmental safety in a variety of applications and geographic settings will ultimately determine the successful deployment of biochar technologies.

REFERENCES

Ahmad, M., Rajapaksha, A. U., Lim, J. E., Zhang, M., Bolan, N., Mohan, D., Vithanage, M., Lee, S. S., & Ok, Y. S. (2014). Biochar as a sorbent for contaminant management in soil and water: A review. *Chemosphere, 99*, 19–33. DOI: 10.1016/j.chemosphere.2013.10.071 PMID: 24289982

Ahmadvand, M., Soltani, J., Hashemi, S. A. H., & Nabipour, M. (2018). The relationship between the characteristics of biochar produced at different temperatures and its impact on the uptake of NO3−. *Environmental Science and Pollution Research International, 25*(4), 3388–3400.

Akhtar, S. S., Andersen, M. N., & Liu, F. (2015). Residual effects of biochar on improving growth, physiology and yield of wheat under salt stress. *Agricultural Water Management, 158*, 61–68. DOI: 10.1016/j.agwat.2015.04.010

Al-Wabel, M. I., Al-Omran, A., El-Naggar, A. H., Nadeem, M., & Usman, A. R. (2013). Pyrolysis temperature induced changes in characteristics and chemical composition of biochar produced from conocarpus wastes. *Bioresource Technology, 131*, 374–379. DOI: 10.1016/j.biortech.2012.12.165 PMID: 23376202

Anawar, H. M., Akter, F., Solaiman, Z. M., & Strezov, V. (2015). Biochar: An emerging panacea for remediation of soil contaminants from mining, industry and sewage wastes. *Pedosphere, 25*(5), 654–665. DOI: 10.1016/S1002-0160(15)30046-1

Basso, A. S., Miguez, F. E., Laird, D. A., Horton, R., & Westgate, M. (2013). Assessing potential of biochar for increasing water-holding capacity of sandy soils. *Global Change Biology. Bioenergy, 5*(2), 132–143. DOI: 10.1111/gcbb.12026

Basu, P. (2010). *Biomass gasification and pyrolysis: practical design and theory*. Academic press.

Basu, P., Sadhukhan, A. K., Gupta, P., Rao, S., Dhungana, A., & Acharya, B. (2013). An experimental and theoretical investigation on torrefaction of a large wet wood particle. *Bioresource Technology, 127*, 449–455. PMID: 24657751

Beesley, L., Moreno-Jiménez, E., Gomez-Eyles, J. L., Harris, E., Robinson, B., & Sizmur, T. (2011). A review of biochars' potential role in the remediation, revegetation and restoration of contaminated soils. *Environmental Pollution, 159*(12), 3269–3282. DOI: 10.1016/j.envpol.2011.07.023 PMID: 21855187

Biederman, L. A., & Harpole, W. S. (2013). Biochar and its effects on plant productivity and nutrient cycling: A meta-analysis. *Global Change Biology. Bioenergy, 5*(2), 202–214. DOI: 10.1111/gcbb.12037

Brewer, C. E., Chuang, V. J., Masiello, C. A., Gonnermann, H., Gao, X., Dugan, B., Driver, L. E., Panzacchi, P., Zygourakis, K., & Davies, C. A. (2014). New approaches to measuring biochar density and porosity. *Biomass and Bioenergy*, *66*, 176–185. DOI: 10.1016/j.biombioe.2014.03.059

Brewer, C. E., Schmidt-Rohr, K., Satrio, J. A., & Brown, R. C. (2009). Characterization of biochar from fast pyrolysis and gasification systems. *Environmental Progress & Sustainable Energy*, *28*(3), 386–396. DOI: 10.1002/ep.10378

Bridgwater, A. V. (2012). Review of fast pyrolysis of biomass and product upgrading. *Biomass and Bioenergy*, *38*, 68–94. DOI: 10.1016/j.biombioe.2011.01.048

Brown, T. R., Wright, M. M., & Brown, R. C. (2011). Estimating profitability of two biochar production scenarios: Slow pyrolysis vs fast pyrolysis. *Biofuels, Bioproducts & Biorefining*, *5*(1), 54–68. DOI: 10.1002/bbb.254

Camps-Arbestain, M., Amonette, J. E., Singh, B., Wang, T., & Schmidt, H. P. (2015). A biochar classification system and associated test methods. Biochar for environmental management: science, technology and implementation, 165-193.

Cantrell, K. B., Hunt, P. G., Uchimiya, M., Novak, J. M., & Ro, K. S. (2012). Impact of pyrolysis temperature and manure source on physicochemical characteristics of biochar. *Bioresource Technology*, *107*, 419–428. DOI: 10.1016/j.biortech.2011.11.084 PMID: 22237173

Cao, X., & Harris, W. (2010). Properties of dairy-manure-derived biochar pertinent to its potential use in remediation. *Bioresource Technology*, *101*(14), 5222–5228. DOI: 10.1016/j.biortech.2010.02.052 PMID: 20206509

Cayuela, M. L., Van Zwieten, L., Singh, B. P., Jeffery, S., Roig, A., & Sánchez-Monedero, M. A. (2014). Biochar's role in mitigating soil nitrous oxide emissions: A review and meta-analysis. *Agriculture, Ecosystems & Environment*, *191*, 5–16. DOI: 10.1016/j.agee.2013.10.009

Cetin, E., Moghtaderi, B., Gupta, R., & Wall, T. F. (2004). Influence of pyrolysis conditions on the structure and gasification reactivity of biomass chars. *Fuel*, *83*(16), 2139–2150. DOI: 10.1016/j.fuel.2004.05.008

Chan, K. Y., Van Zwieten, L., Meszaros, I., Downie, A., & Joseph, S. (2008). Using poultry litter biochars as soil amendments. *Soil Research (Collingwood, Vic.)*, *46*(5), 437–444. DOI: 10.1071/SR08036

Chen, T., Zhang, Y., Wang, H., Lu, W., Zhou, Z., Zhang, Y., & Ren, L. (2014). Influence of pyrolysis temperature on characteristics and heavy metal adsorptive performance of biochar derived from municipal sewage sludge. *Bioresource Technology*, *164*, 47–54. DOI: 10.1016/j.biortech.2014.04.048 PMID: 24835918

Cheng, C. H., Lehmann, J., & Engelhard, M. H. (2008). Natural oxidation of black carbon in soils: Changes in molecular form and surface charge along a climosequence. *Geochimica et Cosmochimica Acta*, *72*(6), 1598–1610. DOI: 10.1016/j.gca.2008.01.010

Chia, C. H., Gong, B., Joseph, S. D., Marjo, C. E., Munroe, P., & Rich, A. M. (2012). Imaging of mineral-enriched biochar by FTIR, Raman and SEM–EDX. *Vibrational Spectroscopy*, *62*, 248–257. DOI: 10.1016/j.vibspec.2012.06.006

Clare, A., Shackley, S., Joseph, S., Hammond, J., Pan, G., & Bloom, A. (2015). Competing uses for China's straw: The economic and carbon abatement potential of biochar. *Global Change Biology. Bioenergy*, *7*(6), 1272–1282. DOI: 10.1111/gcbb.12220

Clough, T. J., Condron, L. M., Kammann, C., & Müller, C. (2013). A review of biochar and soil nitrogen dynamics. *Agronomy (Basel)*, *3*(2), 275–293. DOI: 10.3390/agronomy3020275

Crombie, K., Mašek, O., Sohi, S. P., Brownsort, P., & Cross, A. (2013). The effect of pyrolysis conditions on biochar stability as determined by three methods. *Global Change Biology. Bioenergy*, *5*(2), 122–131. DOI: 10.1111/gcbb.12030

Dickinson, D., Balduccio, L., Buysse, J., Ronsse, F., van Huylenbroeck, G., & Prins, W. (2015). Cost-benefit analysis of using biochar to improve cereals agriculture. *Global Change Biology. Bioenergy*, *7*(4), 850–864. DOI: 10.1111/gcbb.12180

Dispenza, V., De Pasquale, C., Fascella, G., Mammano, M. M., & Alonzo, G. (2016). Use of biochar as peat substitute for growing substrates of Euphorbia × lomi potted plants. *Spanish Journal of Agricultural Research*, *14*(4), e0908. DOI: 10.5424/sjar/2016144-9082

Dutta, D., Meena, A. L., Bhanu, C., Ghasal, P. C., Choudhary, J., Kumar, S., Mishra, R. P., Ansari, M. A., Raghavendra, K. J., Prusty, A. K., Jat, P. C., Kashyap, P., Punia, P., Dixit, M., Singh, O., Rai, A. K., Meena, A. K., Rathi, S., & Yadav, P. (2024). Sustainable soil management for climate resilience: Long- term management effects on soil carbon sequestration and nitrogen dynamics in a semi-arid tropical Inceptisol of India. *Journal of Soil Science and Plant Nutrition*, *24*(3), 4407–4426. DOI: 10.1007/s42729-024-01844-4

Elad, Y., David, D. R., Harel, Y. M., Borenshtein, M., Kalifa, H. B., Silber, A., & Graber, E. R. (2010). Induction of systemic resistance in plants by biochar, a soil-applied carbon sequestering agent. *Phytopathology*, *100*(9), 913–921. DOI: 10.1094/PHYTO-100-9-0913 PMID: 20701489

Enders, A., Hanley, K., Whitman, T., Joseph, S., & Lehmann, J. (2012). Characterization of biochars to evaluate recalcitrance and agronomic performance. *Bioresource Technology*, *114*, 644–653. DOI: 10.1016/j.biortech.2012.03.022 PMID: 22483559

European Biochar Certificate (EBC). (2012). *Guidelines for a Sustainable Production of Biochar. European Biochar Foundation*. EBC.

Galinato, S. P., Yoder, J. K., & Granatstein, D. (2011). The economic value of biochar in crop production and carbon sequestration. *Energy Policy*, *39*(10), 6344–6350. DOI: 10.1016/j.enpol.2011.07.035

Ghazaryan, K., Pandey, D., Singh, S., Varagyan, V., Alexiou, A., Petropoulos, D., Kriemadis, A., Rajput, V., Minkina, T., Singh, R., Sousa, J., Kumar, S., El-Ramady, H., Singh, O., & Singh, A. (2024). Enhancing Crop Production: Unveiling the Role of Nanofertilizers in Sustainable Agriculture and Precision Nutrient Management. *Egyptian Journal of Soil Science*, *64*(3), 981–1007. DOI: 10.21608/ejss.2024.282778.1748

Glaser, B., & Birk, J. J. (2012). State of the scientific knowledge on properties and genesis of Anthropogenic Dark Earths in Central Amazonia (terra preta de Índio). *Geochimica et Cosmochimica Acta*, *82*, 39–51. DOI: 10.1016/j.gca.2010.11.029

González, J. F., Román, S., Encinar, J. M., & Martínez, G. (2009). Pyrolysis of various biomass residues and char utilization for the production of activated carbons. *Journal of Analytical and Applied Pyrolysis*, *85*(1-2), 134–141. DOI: 10.1016/j.jaap.2008.11.035

Gu, X., Wang, Y., Lai, C., Qiu, J., Li, S., Hou, Y., Martens, W., Mahmood, N., & Zhang, S. (2015). Microporous bamboo biochar for lithium-sulfur batteries. *Nano Research*, *8*(1), 129–139. DOI: 10.1007/s12274-014-0601-1

Hammond, J., Shackley, S., Sohi, S., & Brownsort, P. (2011). Prospective life cycle carbon abatement for pyrolysis biochar systems in the UK. *Energy Policy*, *39*(5), 2646–2655. DOI: 10.1016/j.enpol.2011.02.033

Hansen, V., Müller-Stöver, D., Ahrenfeldt, J., Holm, J. K., Henriksen, U. B., & Hauggaard-Nielsen, H. (2015). Gasification biochar as a valuable by-product for carbon sequestration and soil amendment. *Biomass and Bioenergy*, *72*, 300–308. DOI: 10.1016/j.biombioe.2014.10.013

Hossain, M. K., Strezov, V., Chan, K. Y., Ziolkowski, A., & Nelson, P. F. (2011). Influence of pyrolysis temperature on production and nutrient properties of wastewater sludge biochar. *Journal of Environmental Management*, *92*(1), 223–228. DOI: 10.1016/j.jenvman.2010.09.008 PMID: 20870338

Huggins, T., Wang, H., Kearns, J., Jenkins, P., & Ren, Z. J. (2014). Biochar as a sustainable electrode material for electricity production in microbial fuel cells. *Bioresource Technology*, *157*, 114–119. DOI: 10.1016/j.biortech.2014.01.058 PMID: 24534792

International Biochar Initiative (IBI). (2015). Standardized Product Definition and Product Testing Guidelines for Biochar That Is Used in Soil. Accessed from: https://biochar-international.org/

Inyang, M., Gao, B., Pullammanappallil, P., Ding, W., & Zimmerman, A. R. (2010). Biochar from anaerobically digested sugarcane bagasse. *Bioresource Technology*, *101*(22), 8868–8872. DOI: 10.1016/j.biortech.2010.06.088 PMID: 20634061

Inyang, M. I., Gao, B., Yao, Y., Xue, Y., Zimmerman, A., Mosa, A., Pullammanappallil, P., Ok, Y. S., & Cao, X. (2016). A review of biochar as a low-cost adsorbent for aqueous heavy metal removal. *Critical Reviews in Environmental Science and Technology*, *46*(4), 406–433. DOI: 10.1080/10643389.2015.1096880

Jaiswal, A. K., Elad, Y., Graber, E. R., & Frenkel, O. (2015). Contribution of biochar to suppression of Fusarium oxysporum f. sp. lycopersici in tomato plants. *Israel Journal of Plant Sciences*, *62*(1-2), 76–85.

Jeffery, S., Verheijen, F. G., Kammann, C., & Abalos, D. (2016). Biochar effects on methane emissions from soils: A meta-analysis. *Soil Biology & Biochemistry*, *101*, 251–258. DOI: 10.1016/j.soilbio.2016.07.021

Jeffery, S., Verheijen, F. G., Van Der Velde, M., & Bastos, A. C. (2011). A quantitative review of the effects of biochar application to soils on crop productivity using meta-analysis. *Agriculture, Ecosystems & Environment*, *144*(1), 175–187. DOI: 10.1016/j.agee.2011.08.015

Jiang, J., Xu, R. K., Jiang, T. Y., & Li, Z. (2012). Immobilization of Cu(II), Pb(II) and Cd(II) by the addition of rice straw derived biochar to a simulated polluted Ultisol. *Journal of Hazardous Materials*, *229*, 145–150. DOI: 10.1016/j.jhazmat.2012.05.086 PMID: 22704774

Kambo, H. S., & Dutta, A. (2015). A comparative review of biochar and hydrochar in terms of production, physico-chemical properties and applications. *Renewable & Sustainable Energy Reviews*, *45*, 359–378. DOI: 10.1016/j.rser.2015.01.050

Kloss, S., Zehetner, F., Dellantonio, A., Hamid, R., Ottner, F., Liedtke, V., Schwanninger, M., Gerzabek, M. H., & Soja, G. (2012). Characterization of slow pyrolysis biochars: Effects of feedstocks and pyrolysis temperature on biochar properties. *Journal of Environmental Quality*, *41*(4), 990–1000. DOI: 10.2134/jeq2011.0070 PMID: 22751041

Kumari, K. G. I. D., Moldrup, P., Paradelo, M., Elsgaard, L., Hauggaard-Nielsen, H., & de Jonge, L. W. (2014). Effects of biochar on air and water permeability and colloid and phosphorus leaching in soils from a natural calcium carbonate gradient. *Journal of Environmental Quality*, *43*(2), 647–657. DOI: 10.2134/jeq2013.08.0334 PMID: 25602666

Lehmann, J., & Joseph, S. (Eds.). (2015). *Biochar for environmental management: science, technology and implementation*. Routledge. DOI: 10.4324/9780203762264

Lehmann, J., Kern, D. C., Glaser, B., & Woods, W. I. (2003). *Amazonian dark earths: origin properties management*. Springer Science & Business Media. DOI: 10.1007/1-4020-2597-1

Lehmann, J., Rillig, M. C., Thies, J., Masiello, C. A., Hockaday, W. C., & Crowley, D. (2011). Biochar effects on soil biota–a review. *Soil Biology & Biochemistry*, *43*(9), 1812–1836. DOI: 10.1016/j.soilbio.2011.04.022

Libra, J. A., Ro, K. S., Kammann, C., Funke, A., Berge, N. D., Neubauer, Y., Titirici, M.-M., Fühner, C., Bens, O., Kern, J., & Emmerich, K. H. (2011). Hydrothermal carbonization of biomass residuals: A comparative review of the chemistry, processes and applications of wet and dry pyrolysis. *Biofuels*, *2*(1), 71–106. DOI: 10.4155/bfs.10.81

Lim, S. L., Lee, L. H., & Wu, T. Y. (2016). Sustainability of using composting and vermicomposting technologies for organic solid waste biotransformation: Recent overview, greenhouse gases emissions and economic analysis. *Journal of Cleaner Production*, *111*, 262–278. DOI: 10.1016/j.jclepro.2015.08.083

Lin, Y., Chen, Z., Dai, M., Fang, S., Liao, Y., Yu, Z., & Ma, X. (2012). Rapid microwave-assisted pyrolysis of biomass using different heating modes. *Bioresource Technology*, *125*, 308–314.

Lorenz, K., & Lal, R. (2014). Biochar application to soil for climate change mitigation by soil organic carbon sequestration. *Journal of Plant Nutrition and Soil Science*, *177*(5), 651–670. DOI: 10.1002/jpln.201400058

Lu, H., Zhang, W., Yang, Y., Huang, X., Wang, S., & Qiu, R. (2012). Relative distribution of Pb2+ sorption mechanisms by sludge-derived biochar. *Water Research*, *46*(3), 854–862. DOI: 10.1016/j.watres.2011.11.058 PMID: 22189294

Manyà, J. J. (2012). Pyrolysis for biochar purposes: A review to establish current knowledge gaps and research needs. *Environmental Science & Technology*, *46*(15), 7939–7954. DOI: 10.1021/es301029g PMID: 22775244

Matuštík, J., Hnátková, T., & Kočí, V. (2020). Life cycle assessment of biochar-to-soil systems: A review. *Journal of Cleaner Production*, *259*, 120998. DOI: 10.1016/j.jclepro.2020.120998

Mohan, D., Pittman, C. U.Jr, Bricka, M., Smith, F., Yancey, B., Mohammad, J., Steele, P. H., Alexandre-Franco, M. F., Gómez-Serrano, V., & Gong, H. (2007). Sorption of arsenic, cadmium, and lead by chars produced from fast pyrolysis of wood and bark during bio-oil production. *Journal of Colloid and Interface Science*, *310*(1), 57–73. DOI: 10.1016/j.jcis.2007.01.020 PMID: 17331527

Mohan, D., Pittman, C. U.Jr, & Steele, P. H. (2006). Pyrolysis of wood/biomass for bio-oil: A critical review. *Energy & Fuels*, *20*(3), 848–889. DOI: 10.1021/ef0502397

Novak, J. M., Lima, I., Xing, B., Gaskin, J. W., Steiner, C., Das, K. C., & Schomberg, H. (2009). Characterization of designer biochar produced at different temperatures and their effects on a loamy sand. *Annals of Environmental Science (Boston, Mass.)*, *3*(1), 2.

Ogawa, M., & Okimori, Y. (2010). Pioneering works in biochar research, Japan. *Soil Research (Collingwood, Vic.)*, *48*(7), 489–500. DOI: 10.1071/SR10006

Onay, O. (2007). Influence of pyrolysis temperature and heating rate on the production of bio-oil and char from safflower seed by pyrolysis, using a well-swept fixed-bed reactor. *Fuel Processing Technology*, *88*(5), 523–531. DOI: 10.1016/j.fuproc.2007.01.001

Qian, K., Kumar, A., Zhang, H., Bellmer, D., & Huhnke, R. (2015). Recent advances in utilization of biochar. *Renewable & Sustainable Energy Reviews*, *42*, 1055–1064 DOI: 10.1016/j.rser.2014.10.074

Rajapaksha, A. U., Chen, S. S., Tsang, D. C., Zhang, M., Vithanage, M., Mandal, S., Gao, B., Bolan, N. S., & Ok, Y. S. (2016). Engineered/designer biochar for contaminant removal/immobilization from soil and water: Potential and implication of biochar modification. *Chemosphere*, *148*, 276–291. DOI: 10.1016/j.chemosphere.2016.01.043 PMID: 26820777

Roberts, K. G., Gloy, B. A., Joseph, S., Scott, N. R., & Lehmann, J. (2010). Life cycle assessment of biochar systems: Estimating the energetic, economic, and climate change potential. *Environmental Science & Technology*, *44*(2), 827–833. DOI: 10.1021/es902266r PMID: 20030368

Ronsse, F., Van Hecke, S., Dickinson, D., & Prins, W. (2013). Production and characterization of slow pyrolysis biochar: Influence of feedstock type and pyrolysis conditions. *Global Change Biology. Bioenergy*, *5*(2), 104–115. DOI: 10.1111/gcbb.12018

Selvaganapathi, K., Kumar, K. A., Bhople, B. S., Singh, V. K., Jayanthi, J., Singh, O., Bárek, V., Alshehri, M. A., Gaber, A., & Hossain, A. (2025, September). **Singh, O.**, Bárek, V., Alshehri, M. A., Gaber, A., & Hossain, A. (2025). Boosting soil nutrients, carbon storage, and wheat yield with co-compost organic fertilizers and growth-promoting microorganisms in Typic Haplustept soils. *Journal of Soil Science and Plant Nutrition*, *25*(3), 6592–6607. Advance online publication. DOI: 10.1007/s42729-025-02551-4

Shabangu, S., Woolf, D., Fisher, E. M., Angenent, L. T., & Lehmann, J. (2014). Techno-economic assessment of biomass slow pyrolysis into different biochar and methanol concepts. *Fuel*, *117*, 742–748. DOI: 10.1016/j.fuel.2013.08.053

Shackley, S., Hammond, J., Gaunt, J., & Ibarrola, R. (2011). The feasibility and costs of biochar deployment in the UK. *Carbon Management*, *2*(3), 335–356. DOI: 10.4155/cmt.11.22

Shahi, U. P., Singh, O., Singh, V. K., Shivangi, P. K., Singh, R., & Singh, V. K. Anand Singh, Rajput, V. D., Rajput, V. D., Singh, A., Ghazaryan, K. A., & Al Tawaha, A. R. M. (2024). Nanotechnology in rice farming: Optimizing nutrient management with nanofertilizers. In: Sustainable Agriculture: Nanotechnology and Biotechnology for Crop Production and Protection. Walter de Gruyter GmbH. DOI: DOI: 10.1515/9783111234694-003

Singh, B. P., Cowie, A. L., & Smernik, R. J. (2015). Biochar carbon stability in a clayey soil as a function of feedstock and pyrolysis temperature. *Environmental Science & Technology*, *46*(21), 11770–11778. DOI: 10.1021/es302545b PMID: 23013285

Singh, O., Shahi, U.P., Dhyani, B.P., Kumar, S., Vivek, S., & Sengar, R.S., Shivangi, and Singh, A. (2023). Effect of zinc oxide nanoparticles application on growth and yield of basmati rice (Oryza sativa) in alkaline soil. *Indian Journal of Agronomy*, *68*(3), 1–6.

Singh, O., Singh, S., Shahi, U. P., Singh, R., Singh, P. K., Grigoryan, K. K., Khachatryan, H., Rajput, P., Perelomov, L., Kozmenko, S., & Burachevskaya, M. (2024). Nanotechnological Innovations for Soil Pollution Remediation and Environmental Conservation. In Rajput, V., Singh, A., Ghazaryan, K., Alexiou, A., & Said Al-Tawaha, A. (Eds.), *Harnessing NanoOmics and Nanozymes for Sustainable Agriculture* (pp. 237–255). IGI Global., DOI: 10.4018/979-8-3693-1890-4.ch012

Singh, O., Singh, S., Singh, V. K., & Singh, A. (2023). Biochar: An Organic Amendment for Sustainable Soil Health. In Baskar, C., Ramakrishna, S., & Daniela La Rosa, A. (Eds.), *Encyclopedia of Green Materials*. Springer., DOI: 10.1007/978-981-16-4921-9_265-1

Sohi, S. P., Krull, E., Lopez-Capel, E., & Bol, R. (2010). A review of biochar and its use and function in soil. *Advances in Agronomy*, *105*, 47–82. DOI: 10.1016/S0065-2113(10)05002-9

Spokas, K. A. (2010). Review of the stability of biochar in soils: predictability of O: C molar ratios. *Carbon Management*, *1*(2), 289–303. DOI: 10.4155/cmt.10.32

Spokas, K. A., Novak, J. M., Stewart, C. E., Cantrell, K. B., Uchimiya, M., DuSaire, M. G., & Ro, K. S. (2011). Qualitative analysis of volatile organic compounds on biochar. *Chemosphere*, *85*(5), 869–882. DOI: 10.1016/j.chemosphere.2011.06.108 PMID: 21788060

Subash, N., Dutta, D., Ghasal, P., Ravisankar, N., Chaudhary, V. P., Kumar, S., Meena, L. R., & Singh, O. (2023). Brahmdutt, Singh, S., & Sunita, K. (2023). A Composite Index to Assess the Climate-Carbon-Yield-Sustainability of Cereal Based Cropping System. *International Journal of Plant Production*, *17*(4), 729–755. DOI: 10.1007/s42106-023-00268-x

Sun, Y., Gao, B., Yao, Y., Fang, J., Zhang, M., Zhou, Y., Chen, H., & Yang, L. (2014). Effects of feedstock type, production method, and pyrolysis temperature on biochar and hydrochar properties. *Chemical Engineering Journal*, *240*, 574–578. DOI: 10.1016/j.cej.2013.10.081

Tan, X., Liu, Y., Zeng, G., Wang, X., Hu, X., Gu, Y., & Yang, Z. (2015). Application of biochar for the removal of pollutants from aqueous solutions. *Chemosphere*, *125*, 70–85. DOI: 10.1016/j.chemosphere.2014.12.058 PMID: 25618190

Tang, L., Huang, H., Hao, H., & Zhao, K. (2013). Development of plasma pyrolysis/gasification systems for energy efficient and environmentally sound waste disposal. *Journal of Electrostatics*, *71*(5), 839–847. DOI: 10.1016/j.elstat.2013.06.007

Tsai, W. T., Liu, S. C., Chen, H. R., Chang, Y. M., & Tsai, Y. L. (2012). Textural and chemical properties of swine-manure-derived biochar pertinent to its potential use as a soil amendment. *Chemosphere*, *89*(2), 198–203. DOI: 10.1016/j.chemosphere.2012.05.085 PMID: 22743180

Wang, X., Ou, G., Wang, N., & Wu, H. (2016). Graphene-based recyclable photo-absorbers for high-efficiency seawater desalination. *ACS Applied Materials & Interfaces*, *8*(14), 9194–9199. DOI: 10.1021/acsami.6b02071 PMID: 27019007

Wiedner, K., Rumpel, C., Steiner, C., Pozzi, A., Maas, R., & Glaser, B. (2013). Chemical evaluation of chars produced by thermochemical conversion (gasification, pyrolysis and hydrothermal carbonization) of agro-industrial biomass on a commercial scale. *Biomass and Bioenergy*, *59*, 264–278. DOI: 10.1016/j.biombioe.2013.08.026

Wilson, K. (2014). How biochar works in soil. *Biochar Journal*, *32*, 25–33.

Woods, W. I., & Denevan, W. M. (2009). Amazonian dark earths: the first century of reports. In *Amazonian dark earths: Wim Sombroek's vision* (pp. 1–14). Springer. DOI: 10.1007/978-1-4020-9031-8_1

Woolf, D., Lehmann, J., & Lee, D. R. (2014). *Biochar for environmental management: science, technology and implementation*. Routledge.

Woolf, D., Lehmann, J., & Lee, D. R. (2016). Optimal bioenergy power generation for climate change mitigation with or without carbon sequestration. *Nature Communications*, *7*(1), 1–11. DOI: 10.1038/ncomms13160 PMID: 27767177

Wu, W., Yang, M., Feng, Q., McGrouther, K., Wang, H., Lu, H., & Chen, Y. (2012). Chemical characterization of rice straw-derived biochar for soil amendment. *Biomass and Bioenergy*, *47*, 268–276. DOI: 10.1016/j.biombioe.2012.09.034

Yao, Y., Gao, B., Zhang, M., Inyang, M., & Zimmerman, A. R. (2012). Effect of biochar amendment on sorption and leaching of nitrate, ammonium, and phosphate in a sandy soil. *Chemosphere*, *89*(11), 1467–1471. DOI: 10.1016/j.chemosphere.2012.06.002 PMID: 22763330

Zeng, K., Gauthier, D., Soria, J., Mazza, G., & Flamant, G. (2017). Solar pyrolysis of carbonaceous feedstocks: A review. *Solar Energy*, *156*, 73–92. DOI: 10.1016/j.solener.2017.05.033

Zhao, L., Cao, X., Mašek, O., & Zimmerman, A. (2013). Heterogeneity of biochar properties as a function of feedstock sources and production temperatures. *Journal of Hazardous Materials*, *256*, 1–9. DOI: 10.1016/j.jhazmat.2013.04.015 PMID: 23669784

Zhao, X., Zhang, J., Song, Z., Liu, H., Li, L., & Ma, C. (2017). Microwave pyrolysis of straw bale and energy balance analysis. *Journal of Analytical and Applied Pyrolysis*, *127*, 284–292.

Chapter 4
Biochar as a Soil Amendment:
Chemical Properties and Fertility Enhancement

Prabhat Kumar Singh
Acharya Narendra Deva University of Agriculture and Technology, Ayodhya, India

Omkar Singh
http://orcid.org/0000-0003-1971-185X
Swami Vivekanand Subharti University, Meerut, India

Shivangi Singh
Sardar Vallabhbhai Patel University of Agriculture and Technology, Meerut, India

Uday Pratap Shahi
Sardar Vallabhbhai Patel University of Agriculture and Technology, Meerut, India

Vijay Kant Singh
Punjab Agricultural University, India

Krishna Kumar Singh
Sardar Vallabhbhai Patel University of Agriculture and Technology, Meerut, India

Vaishali Singh
Sardar Vallabhbhai Patel University of Agriculture and Technology, Meerut, India

ABSTRACT

Biochar is a carbon-rich substance made by pyrolyzing organic biomass in an environment with no oxygen. It has become a viable soil amendment that might greatly improve the chemical characteristics and fertility of soil. This chapter provides a comprehensive analysis of biochar's diverse effects on soil chemistry, encompassing changes to pH, cation exchange capacity, nutrient retention, and overall soil fertility measurements. Through examination of current research findings, we analysed the

DOI: 10.4018/979-8-3693-7006-3.ch004

mechanisms by which biochar alters soil chemical dynamics, its interaction with native soil components, and the ensuing consequences for agricultural production and environmental sustainability. The conversation explores the role of biochar in carbon sequestration, nutrient cycling, and improving soil conditions. It also discusses how the impacts of biochar can vary depending on the feedstock sources, production settings, and soil types.

1. INTRODUCTION

The application of biochar in agricultural soils exemplifies a synthesis of traditional knowledge and contemporary scientific advancements aimed at sustainable soil management practices. Biochar, a carbon-rich product produced by the thermal decomposition of organic materials in an oxygen-restricted environment via pyrolysis, has attracted significant scientific attention in the last two decades as a soil amendment that can concurrently tackle various agricultural and environmental issues (Lehmann and Joseph, 2015; Singh et al., 2023). Modern biochar research is primarily inspired by the observations of Terra Preta de Indio (Amazonian Dark Earths), anthropogenic soils produced by pre-Columbian indigenous populations that exhibit significantly improved fertility relative to adjacent soils, even after centuries of weathering (Glaser et al., 2001). The chemical characteristics of soil essentially dictate its ability to sustain plant development, manage nutrient cycling, and uphold ecosystem services. The qualities encompass pH, cation exchange capacity (CEC), base saturation, organic matter content, and the accessibility of vital plant nutrients. Traditional agricultural methods, especially in intensive production systems, frequently result in the deterioration of these chemical properties due to acidification, nutrient depletion, and organic matter loss (Lal, 2015). The incorporation of biochar into soil management practices presents potential remedies to these issues due to its distinctive chemical properties and enduring stability in soil ecosystems. Comprehending biochar's influence on soil chemical characteristics is crucial not just for agricultural production but also for wider environmental implications. Biochar application may boost ecosystem services such as climate change mitigation by carbon sequestration, reduction of greenhouse gas emissions from agricultural soils, and better water quality through less nutrient leaching (Woolf et al., 2010). The intricate interactions between biochar and soil, shaped by variables such as biochar production conditions, feedstock attributes, application rates, and intrinsic soil properties, require thorough analysis to enhance its application and forecast results across various agricultural systems.

2. CHEMICAL COMPOSITION AND PROPERTIES OF BIOCHAR

2.1 Structural Characteristics

The chemical composition of biochar is primarily influenced by its feedstock source and production parameters, notably pyrolysis temperature, heating rate, and residence duration. Biochar is predominantly composed of aromatic carbon structures organized in six-membered ring sheets, resembling graphite but exhibiting increased disorder and containing diverse functional groups (Lehmann et al., 2011). This aromatic structure imparts resistance to microbial degradation, allowing biochar to endure in soil for centuries to millennia, thereby functioning as a long-term carbon reservoir and soil enhancer. The extent of carbonization, often represented by the H/C and O/C atomic ratios, offers insight into the aromatic condensation and stability of biochar. Reduced H/C ratios signify increased aromaticity and improved recalcitrance, whereas O/C ratios denote the existence of oxygenated functional groups that affect biochar's reactivity and its interaction with soil constituents (Schimmelpfennig and Glaser, 2012). The production temperature markedly affects these ratios, as elevated temperatures typically provide biochars with diminished H/C and O/C ratios, signifying enhanced stability but possibly decreased immediate reactivity in soil systems.

2.2 Surface Chemistry and Functional Groups

The surface chemistry of biochar is essential for its interaction with soil solution, nutrients, and organic molecules. Biochar surfaces possess several functional groups such as carboxyl (-COOH), hydroxyl (-OH), phenolic, carbonyl (C=O), and quinone groups, which enhance its cation exchange capacity and pH buffering capacity (Mukherjee et al., 2011). The prevalence and allocation of these functional groups are contingent upon production conditions and can be altered through post-production interventions or soil aging. Low-temperature biochar (made below 400°C) generally preserve a greater quantity of oxygen-containing functional groups from the original feedstock, leading to enhanced cation exchange capacity (CEC) and increased rapid reactivity with soil constituents. Conversely, high-temperature biochar (exceeding 600°C) has a reduced number of functional groups while exhibiting increased microporosity and specific surface area, hence augmenting their ability for physical adsorption of nutrients and organic molecules (Weber and Quicker, 2018). The temperature-dependent fluctuation in surface chemistry significantly influences the selection of suitable biochar for particular soil management goals.

2.3 Elemental Composition and Ash Content

In addition to carbon, biochar comprises differing quantities of hydrogen, oxygen, nitrogen, sulfur, and mineral constituents originating from the initial feedstock. The ash concentration, consisting of inorganic elements including calcium, potassium, magnesium, phosphorus, and micronutrients, may vary from under 1% in wood-derived biochar to exceeding 40% in biochar generated from manures or agricultural wastes (Singh et al., 2010). This mineral component immediately enhances soil fertility by supplying nutrients and affects biochar's liming impact on acidic soils. The bioavailability of nutrients in biochar ash significantly varies based on their chemical forms and solubility. Potassium generally occurs in highly soluble forms and becomes readily accessible following biochar application, but phosphorus availability is contingent upon pH and the presence of complexing cations such calcium and iron (Wang et al., 2012). Comprehending the elemental makeup of biochar is crucial for forecasting its short-term and long-term impacts on soil nutrient levels and for formulating application techniques that enhance nutrient use efficiency.

3. MECHANISMS OF BIOCHAR INTERACTION WITH SOIL CHEMICAL PROPERTIES

3.1 Direct Mechanisms

The direct processes by which biochar affects soil chemistry entail the instantaneous addition of chemical elements from biochar to the soil system. This encompasses the liberation of soluble minerals, the introduction of alkalinity, and the provision of exchange sites for nutrient retention. The disintegration of ash constituents offers an instant supply of plant-accessible minerals, especially potassium, calcium, and magnesium, which can markedly influence soil fertility in nutrient-poor areas (Gaskin et al., 2008). The alkaline characteristics of most biochar, due to the presence of carbonates, oxides, and hydroxides of alkali and alkaline earth metals, immediately affect soil pH upon application. The liming effect is especially advantageous in acidic soils, as biochar application can elevate pH, mitigate aluminum toxicity, and enhance nutrient availability (Yuan and Xu, 2011). The extent of pH alteration is contingent upon biochar alkalinity, application rate, and the buffering capability of the recipient soil.

3.2 Indirect Mechanisms

Indirect pathways encompass biochar-induced alterations in soil physical, chemical, and biological processes that ultimately influence chemical characteristics. This encompasses adaptations to soil structure that affect water and nutrient dynamics, differences in microbial community composition and activity that influence nutrient cycling, and variations in redox conditions that effect nutrient transformations (Joseph et al., 2010). Biochar's elevated porosity and surface area enhance soil aggregation and establish microsites with unique chemical conditions. These microsites may contain distinct microbial communities and promote certain biogeochemical processes, such as nitrification or denitrification, which affect total soil fertility (Lehmann et al., 2011). Moreover, biochar can adsorb organic molecules, such as allelopathic agents and phytotoxins, thereby mitigating their adverse effects on plant growth while concurrently affecting nutrient cycling through interactions with soil organic matter.

3.3 Aging and Long-term Transformations

The chemical characteristics of biochar undergo substantial alterations over time through processes commonly referred to as "aging." These processes encompass surface oxidation, interaction with soil organic matter, mineral precipitation, and physical fragmentation. Surface oxidation elevates the prevalence of oxygen-containing functional groups, hence augmenting biochar's cation exchange capacity and its ability to retain nutrients (Cheng et al., 2006).

The creation of organo-mineral complexes between biochar surfaces and soil constituents establishes a dynamic interface for nutrient exchange and transformation activities. Such interactions may result in the formation of a "biochar-mineral complex" that possesses characteristics different from each individual component, potentially improving nutrient retention and availability over prolonged durations (Joseph et al., 2010). Comprehending these aging processes is essential for forecasting the enduring effects of biochar on soil fertility and for formulating management plans that consider temporal variations in biochar functioning.

4. IMPACT ON SOIL PH AND BUFFERING CAPACITY

4.1 pH Modification Effects

The application of biochar to soil generally elevates soil pH, with the extent of change contingent upon original soil pH, biochar characteristics, and application rate. The alteration in pH is mainly due to the alkaline characteristics of biochar ash, which comprises oxides, hydroxides, and carbonates of basic cations (Yuan et al., 2011). In acidic soils, the application of biochar can elevate pH by 0.5 to 2 units, thereby considerably influencing nutrient availability and microbial activity.

Table 1. Effects of different biochar types on soil pH changes

Biochar Feedstock	Production Temperature (°C)	Initial Soil pH	Application Rate (t/ha)	pH Change	Reference
Pine wood	450	4.8	20	+1.2	Novak et al. (2009)
Rice husk	500	5.2	10	+0.8	Masulili et al. (2010)
Poultry litter	350	6.1	15	+1.5	Chan et al. (2008)
Wheat straw	550	5.5	25	+0.9	Van Zwieten et al. (2010)
Bamboo	600	4.5	30	+1.8	Bian et al. (2014)
Sewage sludge	500	5.8	20	+0.6	Hossain et al. (2011)
Corn cob	400	6.3	10	+0.4	Uzoma et al. (2011)
Oak wood	650	5.0	40	+2.1	Major et al. (2010)

The increase of pH subsequent to biochar application significantly affects nutrient availability. In acidic soils, elevating pH diminishes the solubility of potentially harmful aluminum and manganese, while enhancing the availability of phosphorus, molybdenum, and base cations. Nonetheless, an excessive increase in pH levels in neutral or alkaline soils may diminish the availability of micronutrients like iron, zinc, and copper, hence requiring meticulous evaluation of biochar selection and application rates (Ippolito et al., 2012).

4.2 Buffering Capacity Enhancement

In addition to immediate pH alteration, biochar enhances soil buffering capacity via many methods. The existence of functional groups on biochar surfaces, specifically carboxyl and phenolic groups, imparts pH-dependent charge that can either receive or give protons, thus mitigating pH fluctuations (Cheng et al., 2008). The improved buffering capacity is especially crucial in agricultural systems affected by acidifying processes like nitrification and organic matter breakdown. The buffering capability of biochar-amended soils enhances over time as biochar matures and acquires additional functional groups via oxidation. Research indicates that aged biochar possesses a superior buffering ability compared to fresh biochar, due to the development of carboxylic groups on its surfaces via biotic and abiotic oxidation processes (Cheng and Lehmann, 2009). The temporal improvement in buffering capacity indicates that biochar's advantages for pH regulation may amplify with time, ensuring prolonged stability of soil chemical conditions.

4.3 Implications for Acid Soil Management

In areas where soil acidity severely limits agricultural productivity, biochar application provides a sustainable alternative or supplement to conventional liming methods. In contrast to agricultural lime, which necessitates many applications due to its consumption in neutralization processes, biochar offers both rapid and sustained pH enhancement through its stable carbon structure and gradual increase in buffering capacity (Biederman and Harpole, 2013). Biochar offers economic and environmental benefits compared to traditional liming materials, such as decreased shipping expenses when sourced from local feedstocks, further soil enhancements beyond pH adjustment, and the potential for carbon sequestration. The heterogeneity in biochar alkalinity and the requirement for elevated application rates relative to agricultural lime necessitate a site-specific assessment of economic viability and meticulous selection of suitable biochar materials for the amelioration of acidic soils.

5. EFFECTS ON CATION EXCHANGE CAPACITY AND NUTRIENT RETENTION

5.1 Enhancement of Cation Exchange Capacity

Cation exchange capacity is a crucial soil characteristic that dictates the soil's capacity to store and provide vital cationic nutrients. The use of biochar consistently enhances soil cation exchange capacity by both directly providing exchange sites and indirectly influencing soil organic matter dynamics. The cation exchange capacity (CEC) of biochar varies from 10 to over 200 cmol(+)/kg, influenced by feedstock and production circumstances, with low-temperature biochar often demonstrating higher initial CEC owing to an increased retention of oxygen-containing functional groups (Lehmann et al., 2003; Mugabo et al., 2024). Biochar's contribution to soil cation exchange capacity transcends its intrinsic exchange capacity via multiple processes. The elevated specific surface area of biochar offers several locations for cation adsorption via electrostatic attraction and specialized adsorption processes. Secondly, biochar facilitates the development of organo-mineral complexes that generate supplementary exchange sites at the interface of biochar, soil minerals, and organic matter. Third, the gradual oxidation of biochar surfaces produces new functional groups that improve exchange capacity over time (Liang et al., 2006).

5.2 Nutrient Retention Mechanisms

Biochar improves nutrient retention via various physicochemical mechanisms functioning at distinct spatial and temporal scales. The fundamental mechanism for retaining nutrients like ammonium, potassium, calcium, and magnesium is the electrostatic adsorption of cations to negatively charged functional groups. The pH-dependent charge properties of biochar indicate that its retention capacity fluctuates with soil pH, typically augmenting as pH increases within the range common to agricultural soils (Mukherjee et al., 2011).

Figure 1. Conceptual model of nutrient retention mechanisms in biochar-amended soil

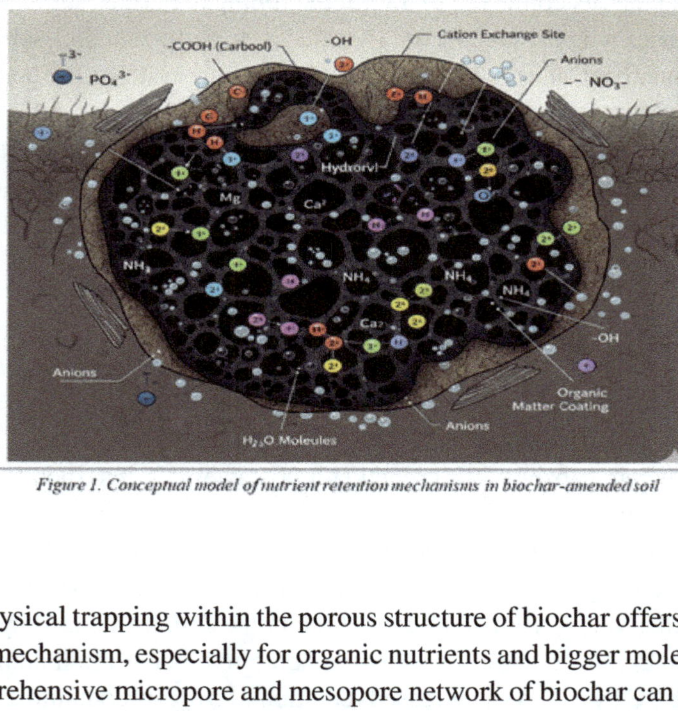

Figure 1. Conceptual model of nutrient retention mechanisms in biochar-amended soil

The physical trapping within the porous structure of biochar offers an additional retention mechanism, especially for organic nutrients and bigger molecular entities. The comprehensive micropore and mesopore network of biochar can physically sequester nutrients, safeguarding them from leaching while ensuring their availability for plant absorption via diffusion mechanisms. This physical retention is crucial for nitrogen molecules, since biochar has demonstrated the potential to diminish nitrate leaching while preserving or enhancing nitrogen availability to plants (Clough et al., 2013).

5.3 Implications for Fertilizer Efficiency

The improved nutrient retention ability of biochar-amended soils has considerable implications for fertilizer efficiency and environmental conservation. Biochar application enhances fertilizer efficiency by minimizing nitrogen leaching losses, perhaps allowing for reduced application rates without compromising crop yields. Research has shown gains of 10-30% in nitrogen utilization efficiency and comparable boosts for other nutrients in soils modified with biochar (Steiner et al., 2008). The interaction between biochar and fertilizers encompasses not just retention but also alterations in nutrient processing pathways. Biochar can affect nitrification rates, potentially decreasing nitrous oxide emissions while preserving nitrogen availability. Biochar may improve phosphorus availability by influencing pH and competing for adsorption sites on soil minerals, albeit responses differ based on soil type and biochar characteristics (DeLuca et al., 2009).

6. INFLUENCE ON SOIL ORGANIC MATTER AND CARBON DYNAMICS

6.1 Carbon Sequestration Potential

The application of biochar gives a substantial opportunity for enduring carbon sequestration in agricultural soils. The obstinate characteristics of biochar carbon, with average residence durations ranging from hundreds to thousands of years, provide significant carbon sequestration in contrast to other organic treatments that degrade more swiftly (Lehmann et al., 2006). The stability of biochar carbon originates from its aromatic structure, which withstands microbial degradation, and its capacity to establish protective connections with soil minerals.

The carbon sequestration capacity of biochar surpasses the mere incorporation of stable carbon by affecting the dynamics of native soil organic matter. The "negative priming effect" noted in certain biochar-amended soils, characterized by a reduction in the breakdown of native organic matter, can augment overall carbon sequestration. This phenomenon may arise from biochar's sorption of enzymes or substrates, alterations in microbial community composition, or modifications in soil physical qualities that safeguard organic matter from degradation (Zimmerman et al., 2011).

6.2 Interactions with Native Soil Organic Matter

Biochar significantly affects the dynamics of indigenous soil organic matter via both physical and chemical processes. Physical protection within biochar pores and aggregates can safeguard organic matter from microbial degradation, while chemical interactions such as sorption and complexation can preserve organic molecules. These interactions establish a "biochar-organic matter continuum," wherein the distinctions between biochar carbon and soil organic carbon progressively diminish over time (Lehmann and Kleber, 2015; Dutta et al., 2024).

Table 2. Effects of biochar on soil organic matter properties

Soil Property	Without Biochar	With Biochar (2% w/w)	Percent Change	Measurement Method
Total Organic C (g/kg)	15.3 ± 1.2	28.7 ± 2.1	+87.6%	Dry combustion
Dissolved Organic C (mg/L)	45.2 ± 3.5	31.8 ± 2.8	-29.6%	Water extraction
Microbial Biomass C (mg/kg)	285 ± 25	342 ± 31	+20.0%	Chloroform fumigation
Particulate Organic Matter (g/kg)	4.8 ± 0.5	6.2 ± 0.6	+29.2%	Size fractionation
Humic Acid C (g/kg)	3.2 ± 0.3	4.1 ± 0.4	+28.1%	Alkaline extraction
C Mineralization Rate (mg CO2-C/kg/day)	12.5 ± 1.1	10.8 ± 0.9	-13.6%	Incubation study
Mean Residence Time (years)	25 ± 3	45 ± 5	+80.0%	Modeling estimate

The alteration of organic matter quality and quantity in biochar-amended soils affects various soil processes, including nutrient cycling, water retention, and aggregate formation. The interaction of biochar and organic matter enhances aggregation, hence improving soil structure, perhaps diminishing organic matter erosion losses and enhancing water infiltration and retention. The physical enhancements augment the chemical advantages of heightened organic matter stability, producing synergistic impacts on overall soil fertility.

6.3 Priming Effects and Microbial Responses

The incorporation of biochar into soil can induce "priming effects," which are alterations in the decomposition rate of indigenous soil organic matter. Numerous investigations indicate negative priming (diminished breakdown), however positive priming has also been noted, especially in low-temperature biochar with labile carbon components. The direction and amplitude of priming effects are contingent upon the properties of biochar, the features of the soil, and the prevailing environmental circumstances (Maestrini et al., 2015).

The microbial responses to biochar application are essential in influencing soil organic matter dynamics. Biochar can modify microbial community composition, promoting specific functional groups while inhibiting others. Alterations in enzyme activity, specifically those related to organic matter decomposition, have been noted in soils modified with biochar. These microbial reactions affect both carbon cycling and nutrient transformations, impacting total soil fertility and ecosystem functionality.

7. IMPACT ON NUTRIENT AVAILABILITY AND CYCLING

7.1 Nitrogen Dynamics

The application of biochar markedly affects nitrogen cycle activities in soil via many ways. The elevated carbon-to-nitrogen ratio of most biochars (about 20-500:1) initially facilitates nitrogen immobilization, potentially diminishing plant-available nitrogen in the short run. This immobilization impact is typically transient and can be mitigated through suitable fertilization techniques or simultaneous treatment with nitrogen-rich amendments (Clough and Condron, 2010). Biochar affects nitrogen transformations across all principal stages in the nitrogen cycle. Ammonification rates may decline due to the sorption of organic nitrogen molecules or the enzymes implicated in mineralization. Nitrification may be either augmented or inhibited based on the pH effects of biochar, as alkaline biochars can potentially promote nitrification in acidic soils, whilst elevated pH may impede the process in neutral to alkaline soils. Denitrification rates frequently diminish in biochar-amended soils owing to enhanced aeration and potential sorption of nitrogen oxides (Cayuela et al., 2014; Singh et al. 2024a). Biochar's ability to retain ammonium via cation exchange and its aptitude to adsorb nitrate through many routes might markedly diminish nitrogen leaching losses. Research indicates reductions in nitrate leaching of 10-40% from biochar-amended soils, with more significant decreases often associated with elevated application rates and coarse-textured soils. This enhanced nitrogen retention increases fertilizer usage efficiency and diminishes environmental consequences linked to nitrogen loss in aquatic ecosystems (Borchard et al., 2019).

7.2 Phosphorus Availability

The effect of biochar on phosphorus availability is intricate and contingent upon various interacting elements, including biochar attributes, soil conditions, and phosphorus sources. Biochar can directly supply phosphorus, especially when derived from phosphorus-rich feedstocks like chicken litter or sewage sludge. The availability of phosphorus fluctuates with pyrolysis temperature, as lower temperatures often retain greater quantities of accessible phosphorus forms (Christel et al., 2016).

Figure 2. Phosphorus transformation pathways in biochar-amended soils

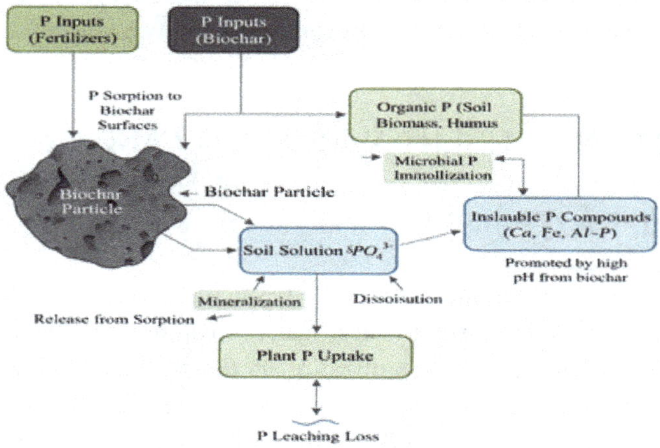

Figure 2. Phosphorus Transformation Pathways in Biochar-Amended Soils

The indirect effects of biochar on phosphorus availability frequently surpass its direct contributions. The elevation in pH resulting from biochar application can improve phosphorus availability in acidic soils by diminishing the precipitation of aluminium and iron phosphates. In calcareous soils, biochar may reduce phosphorus availability by facilitating the production of calcium phosphate. Biochar may compete with phosphorus for sorption sites on soil minerals, thereby elevating solution phosphorus concentrations (Xu et al., 2014).

7.3 Micronutrient Transformations

The application of biochar affects the availability and cycling of vital micronutrients through multiple ways. The ash content of biochar can supply micronutrients such as iron, manganese, zinc, copper, boron, and molybdenum, however concentrations fluctuate significantly based on the feedstock source. Biochars made from wood generally exhibit reduced concentrations of micronutrients in comparison to those produced from crop residues or animal manures (Singh et al., 2010). The application of biochar considerably influences the availability of micronutrients through pH changes. The application of biochar in acidic soils may elevate pH, thereby diminishing the availability of iron, manganese, zinc, and copper due to their reduced solubility at higher pH levels. This effect can be advantageous in situations where micronutrient toxicity is a concern, however it may require vitamin supplementation in poor soils. In contrast, the availability of molybdenum often rises with pH, which may advantage legume crops reliant on molybdenum for nitrogen fixation (Rondon et al., 2007). The extensive surface area and varied functional groups of biochar facilitate the complexation and chelation of micronutrients, affecting their mobility and bioavailability. Organo-metallic complexes formed on charcoal surfaces can preserve micronutrients in plant-accessible forms, while inhibiting leaching losses or precipitation into inaccessible molecules. The chelation effect is particularly significant for iron and zinc, as organic molecules generated from biochar can improve solubility and plant absorption, even at elevated pH levels.

8. BIOCHAR EFFECTS ON CONTAMINATED AND PROBLEM SOILS

8.1 Heavy Metal Immobilization

The application of biochar has emerged as an effective remediation technique for soils contaminated with heavy metals. The mechanisms of heavy metal immobilization by biochar including electrostatic adsorption, ion exchange, complexation with functional groups, precipitation, and physical trapping within pore structures. The efficacy of immobilization differs among metal species, with biochar often exhibiting a strong affinity for lead, cadmium, and copper, a moderate affinity for zinc and nickel, and inconsistent effects on arsenic and chromium (Ahmad et al., 2014; Ghazaryan et al., 2024). The elevation of pH resulting from biochar application is essential for the immobilization of heavy metals, as it facilitates the precipitation of metal hydroxides, carbonates, and phosphates. The development of surface complexes between metals and oxygen-containing functional groups on biochar establishes robust retention mechanisms that diminish metal bioavailability and mobility. Research has shown decreases of 20-90% in extractable heavy metal concentrations after the application of biochar to polluted soils (Beesley et al., 2011).

Table 3. Effectiveness of different biochars for heavy metal immobilization

Heavy Metal	Biochar Type	Application Rate (%)	Reduction in Bioavailability (%)	Mechanism	Reference
Lead (Pb)	Oak wood	5	82	Precipitation, surface complexation	Uchimiya et al. (2010)
Cadmium (Cd)	Rice straw	3	65	Ion exchange, pH effect	Bian et al. (2014)
Copper (Cu)	Bamboo	2	71	Complexation, precipitation	Chen et al. (2011)
Zinc (Zn)	Sewage sludge	4	58	Adsorption, pH increase	Méndez et al. (2012)
Nickel (Ni)	Coconut shell	3	43	Ion exchange, complexation	Shen et al. (2016)
Arsenic (As)	Iron-modified	2	76	Specific adsorption	Vithanage et al. (2017)
Chromium (Cr)	Sugarcane bagasse	5	89	Reduction, precipitation	Mandal et al. (2017)

The enduring stability of heavy metal immobilization in biochar-amended soils is contingent upon the permanence of biochar and alterations in soil conditions. Although initial immobilization may be robust, variables such as pH fluctuations, organic matter degradation, and root exudation can potentially lead to the remobilization of metals over time. Consequently, the monitoring and evaluation of long-term efficacy are crucial for the successful implementation of remediation solutions utilizing biochar.

8.2 Salinity and Sodicity Amelioration

The application of biochar demonstrates potential for improving saline and sodic soil conditions, while the results are contingent upon the qualities of the biochar and the features of the soil. In saline soils, biochar enhances conditions by improving water retention, hence facilitating salt leaching, and by adsorbing excess sodium ions. The porous composition of biochar enhances soil physical characteristics, facilitating drainage and diminishing surface salt buildup (Akhtar et al., 2015). The efficacy of biochar in sodic soils is primarily determined by its ash composition and its capacity to provide calcium and magnesium to displace salt from exchange sites. Biochars derived from feedstocks abundant in divalent cations, such as wood or agricultural residues, can yield advantageous effects akin to those of gypsum application. The enhancement of soil structure and aggregation by biochar promotes the leaching of displaced sodium, aiding in the restoration of sodic soils (Chaganti et al., 2015; singh et al., 2023). The organic chemicals emitted by biochar may also aid in salt reduction by influencing soil biological activity and enhancing plant stress resilience. Augmented microbial activity in biochar-amended saline soils can enhance nutrient cycling and provide chemicals that assist plants in withstanding salt stress. Nevertheless, caution is essential in the selection of suitable biochars, as certain high-ash biochars may introduce salts into the soil, thereby intensifying salinity issues.

8.3 Remediation of Organic Contaminants

Biochar exhibits considerable potential for the immobilization of organic contaminants in soil, such as pesticides, herbicides, polycyclic aromatic hydrocarbons (PAHs), and other persistent organic pollutants. The principal mechanism entails sorption to biochar surfaces via hydrophobic contacts, π-π electron donor-acceptor interactions, hydrogen bonding, and pore-filling processes. The elevated surface area and aromatic configuration of biochar render it very efficient for adsorbing hydrophobic organic molecules (Zhang et al., 2013).

The sorption capacity of biochar for organic pollutants typically enhances with elevated pyrolysis temperatures, attributed to augmented surface area and aromaticity. This improved sorption may have adverse effects, potentially diminishing the bioavailability of pollutants for microbial breakdown, while concurrently restricting their movement and ecological consequences. Research indicates decreases of 50-99% in freely dissolved concentrations of diverse organic pollutants subsequent to biochar application (Gomez-Eyles et al., 2011).

9. INTEGRATED EFFECTS ON SOIL FERTILITY

9.1 Synergistic Interactions

The influence of biochar on soil fertility transcends isolated effects on chemical properties, involving intricate synergistic interactions among the physical, chemical, and biological constituents of soil. These synergies enhance the advantages of biochar application, resulting in a series of beneficial effects on total soil health and productivity. Enhanced soil structure improves water infiltration and retention, subsequently influencing nutrient solubility, transport, and plant uptake (Atkinson et al., 2010). The link between biochar and soil microorganisms illustrates these synergistic interactions. Biochar offers a habitat for advantageous bacteria due to its porous structure, while also delivering nutrients and establishing optimal pH conditions for microbial activity. Augmented microbial variety and activity facilitate nutrient cycling, organic matter decomposition, and the synthesis of plant growth-promoting compounds. These biological advances augment the direct chemical advantages of biochar, fostering a more fruitful and resilient soil ecosystem (Lehmann et al., 2011).

Figure 3. Integrated model of biochar effects on soil fertility components

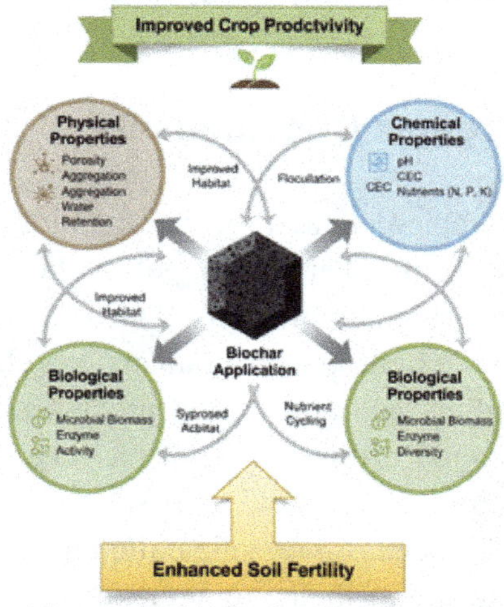

Figure 3. Integrated model of biochar effects on soil fertility components

9.2 Crop Productivity Responses

Meta-analyses of biochar field trials indicate variable yet predominantly positive impacts on crop productivity, with average yield enhancements between 10% and 25% across various cropping systems. The yield response magnitude is influenced by various factors, such as soil type, crop species, biochar characteristics, and management practices. Significant yield enhancements are generally observed in acidic, coarse-textured soils characterized by low inherent fertility, where biochar effectively mitigates several limiting factors concurrently (Jeffery et al., 2017). Crop responses to biochar application indicate that improvements in soil fertility are multifaceted rather than attributable to isolated factors. Improved nutrient retention decreases the need for fertilizers while sustaining yields; enhanced water relations provide resilience against drought stress; and pH adjustment mitigates aluminium toxicity in acidic soils. Biochar-induced alterations in soil biology, such as increased mycorrhizal colonization and the suppression of soil-borne pathogens, enhance crop performance (Elad et al., 2010; Singh et al., 2022a).

9.3 Temporal Evolution of Fertility Effects

The impact of biochar on soil fertility changes over time due to aging processes that alter its physical and chemical characteristics. The initial effects may be primarily influenced by nutrients from ash and alterations in pH, whereas the long-term advantages increasingly stem from greater cation exchange capacity, improved soil structure, and established biological relationships. This temporal evolution requires the assessment of both short-term and long-term effects when analysing biochar as a soil supplement (Major et al., 2010).

Table 4. Temporal changes in biochar effects on key soil fertility parameters

Time Period	pH Effect	CEC Enhancement	Nutrient Availability	Microbial Activity	Soil Structure	Overall Fertility
0-3 months	Strong increase	Moderate	High (ash nutrients)	Variable	Minimal change	Moderate improvement
3-12 months	Stabilizing	Increasing	Moderate	Increasing	Improving	Significant improvement
1-3 years	Stable	High	Balanced	Enhanced	Well-developed	Optimal benefits
3-10 years	Slightly declining	Very high	Sustained	Stable, diverse	Excellent	Sustained high fertility
>10 years	Near original + buffer	Maximum	Long-term cycling	Established ecosystem		

9.4 Economic Considerations of Biochar Application

The economic feasibility of biochar utilization for soil fertility enhancement relies on intricate relationships among manufacturing costs, application charges, yield enhancements, and long-term advantages. Production prices fluctuate significantly depending on feedstock availability, pyrolysis method, and operational scale, ranging from $50 to $500 per metric ton. Economic evaluations of biochar advantages should examine cumulative impacts across numerous growing seasons rather than focusing just on single-year returns (Meyer et al., 2011).

Cost-benefit evaluations increasingly support the application of biochar when several ecosystem services are included, including carbon sequestration, diminished fertilizer needs, enhanced water usage efficiency, and lowered environmental remediation expenses. In areas with plentiful and underutilized agricultural wastes, on-farm biochar production can yield economic benefits by valorizing waste and decreasing disposal expenses. Furthermore, nascent carbon markets and compensation frameworks for ecosystem services may offer supplementary revenue sources that enhance the economic viability of biochar implementation (Roberts et al., 2010; Dutta et al., 2023). The creation of biochar-based fertilizers, in which biochar is fortified with nutrients or amalgamated with traditional fertilizers, signifies a financially advantageous strategy that harnesses the synergies between biochar and nutrient management. These goods can demand elevated pricing while offering improved nutrient utilization efficiency and diminished environmental effects relative to traditional fertilizers alone. Economic modeling indicates that value-added biochar products may attain quicker market penetration and wider adoption compared to bulk biochar uses.

10. ENVIRONMENTAL IMPLICATIONS AND SUSTAINABILITY

10.1 Greenhouse Gas Emissions

The application of biochar affects greenhouse gas emissions from agricultural soils via many mechanisms. Biochar can dramatically diminish nitrous oxide (N_2O) emissions by 10-80%, due to greater soil aeration, improved nitrogen retention, decreased denitrification rates, and potential catalytic conversion of N_2O to N_2 on biochar surfaces. The decrease in N_2O emissions is especially significant due to the substantial global warming potential of this greenhouse gas (Cayuela et al., 2014). The impact on methane (CH_4) emissions is contingent upon soil conditions, with biochar often diminishing emissions from inundated rice paddies but exerting no influence on upland soils. The mechanisms encompass enhanced soil aeration, which diminishes methanogenesis, and the probable augmentation of methanotrophic bacteria that oxidize methane. Carbon dioxide (CO_2) emissions may initially rise due to the decomposition of labile carbon fractions in fresh biochar; however, long-term CO_2 emissions often decline due to negative priming effects on the decomposition of native soil organic matter (Liu et al., 2016).

The overall climate impact of biochar application encompasses not only its direct influence on soil greenhouse gas emissions but also carbon sequestration, diminished fertilizer production and application needs, and the potential replacement of fossil fuels when biochar production is incorporated into bioenergy systems. Life cycle studies demonstrate that biochar systems can attain negative carbon emissions, successfully sequestering CO_2 from the atmosphere while concurrently enhancing soil fertility (Woolf et al., 2010; Singh et al., 2022).

10.2 Water Quality Protection

The application of biochar enhances water quality by minimizing nutrient loss, immobilizing heavy metals, and retaining pesticides. The improved nutrient retention ability of biochar-amended soils markedly decreases nitrogen and phosphorus leaching into surface and groundwater, mitigating primary contributors to eutrophication in aquatic ecosystems. Research has recorded reductions in nutrient leaching losses ranging from 20% to 50%, with more significant decreases observed in coarse-textured soils and during periods of excessive precipitation (Laird et al., 2010). Biochar's ability to trap heavy metals and organic pollutants inhibits their movement to water bodies, safeguarding human health and aquatic ecosystems. This is especially pertinent in agricultural regions with previous pollution or continuous introduction of potentially hazardous compounds by pesticide application or the use of contaminated irrigation water. The enduring stability of pollutant immobilization by biochar ensures ongoing protection of water resources (Ahmad et al., 2014). In addition to contaminant retention, the application of biochar can enhance soil water management, thereby diminishing surface runoff and erosion that carry contaminants to aquatic systems. Improved infiltration and water retention in biochar-amended soils diminish the volume and speed of surface runoff, hence reducing sediment and related pollutant transfer. These hydrological advantages enhance the chemical processes of water quality preservation, establishing several obstacles to pollutant migration.

10.3 Biodiversity and Ecosystem Services

The application of biochar to agricultural soils influences biodiversity at multiple trophic levels, from soil microorganisms to above-ground fauna. Enhanced microbial diversity in biochar-amended soils, attributed to the creation of diverse microhabitats and improved soil conditions, supports more complex and resilient soil food webs. This biological diversity underpins essential ecosystem services including nutrient cycling, pest suppression, and disease resistance (Lehmann et al., 2011).

Soil fauna responses to biochar application are generally positive, with increased abundance and diversity of beneficial organisms such as earthworms, arthropods, and nematodes. The improved soil structure and moisture conditions in biochar-amended soils create favorable habitats for these organisms, which in turn contribute to soil fertility through their activities in organic matter decomposition, nutrient cycling, and soil aggregation. However, responses vary among species and depend on biochar properties, with some studies reporting negative effects at very high application rates (McCormack et al., 2013).

The ecosystem services provided by biochar application extend beyond agricultural productivity to include climate regulation, water purification, and habitat provision. These multiple benefits align with sustainable intensification goals that seek to increase agricultural production while minimizing environmental impacts. The permanence of biochar in soil ensures that these ecosystem services persist over extended periods, providing long-term returns on investment in soil health and environmental quality.

11. APPLICATION STRATEGIES AND BEST MANAGEMENT PRACTICES

11.1 Biochar Selection Criteria

Choosing suitable biochar for certain soil management goals necessitates evaluating several criteria such as feedstock origin, manufacturing conditions, and desired soil characteristics. High-ash biochars derived from nutrient-dense feedstocks at moderate temperatures (400-600°C) offer optimum liming effects for the amelioration of acidic soils while preserving reactive functional groups. Nutrient retention objectives, however, prefer high-temperature biochars (>600°C) characterized by elevated surface area and porosity (Ippolito et al., 2020).

Aligning biochar characteristics with soil limitations optimizes advantages while mitigating possible adverse impacts. In sandy soils characterized by low water and nutrient retention capacity, biochars exhibiting high porosity and cation exchange capacity yield the most significant enhancements. Clay soils derive greater advantages from biochars that enhance aggregation and drainage, particularly those characterized by bigger particle sizes and moderate ash levels. Contaminated soils necessitate biochars with elevated sorption capacity for specific pollutants, hence requiring meticulous assessment of both biochar and contaminant characteristics (Biederman and Harpole, 2013). Quality standards and characterisation techniques for biochar are crucial for ensuring consistent and predictable results. Parameters such as proximal and ultimate analysis, pH, electrical conductivity, cation exchange capacity, surface area, and pollutant levels must be assessed prior to extensive application. Global initiatives like the International Biochar Initiative and the European Biochar Certificate offer standards for biochar characterization and quality assurance, facilitating informed choices in biochar choosing.

11.2 Application Methods and Rates

Optimal biochar application rates fluctuate significantly based on soil parameters, biochar attributes, and management goals, generally spanning from 5 to 50 t/ha for agricultural uses. Reduced rates (5-10 t/ha) may be adequate for enhancing particular soil characteristics like as pH or moisture retention, whereas elevated rates (20-50 t/ha) are frequently required for extensive fertility improvement or pollution cleanup. Economic factors frequently limit application rates, requiring careful targeting of biochar application to optimize return on investment (Major et al., 2010).

Figure 4. Decision framework for biochar application rate determination

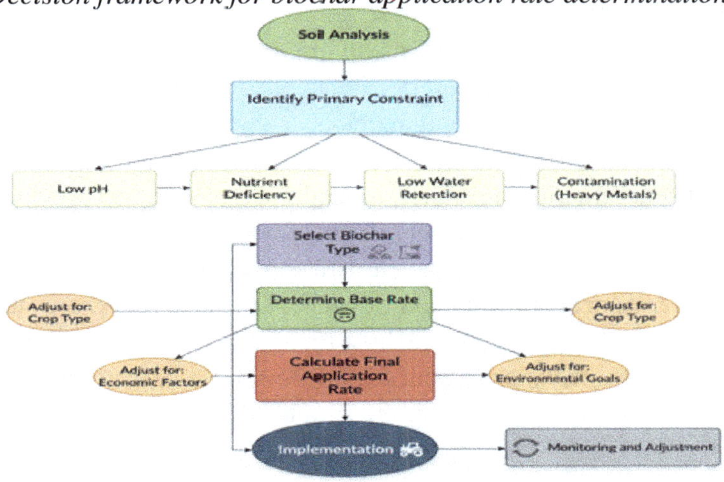

Figure 4. Decision framework for biochar application rate determination

The techniques of application substantially affect the efficacy of biochar additions. Integrating biochar into the root zone via tillage guarantees close interaction among biochar, soil, and roots; however, this may be impractical in no-till systems. The application of substances on the surface, followed by integration via natural processes or minimal disturbance methods, maintains soil structure but may postpone advantages. Banding or spot application optimizes biochar concentration for optimum benefit while minimizing total application needs, especially advantageous for high-value crops or when biochar supply is constrained.

11.3 Integration with Nutrient Management

The effective incorporation of biochar into nutrient management systems necessitates the modification of fertilization tactics to consider biochar's impact on nutrient dynamics. Initial nitrogen immobilization caused by biochars with a high carbon-to-nitrogen ratio may require augmented nitrogen fertilization during the initial growing season, however long-term enhancements in nitrogen retention can diminish fertilizer needs in later years. Co-composting biochar with organic amendments or pre-loading it with nutrients might reduce immobilization effects and increase the fertilizer value of biochar (Schmidt et al., 2017).

The timing of biochar application in relation to fertilizer addition affects nutrient usage efficiency and crop response. Soaking biochar in nutrient solutions or co-applying it with liquid fertilizers might improve nutrient retention and diminish leaching losses. This method is very efficacious for mobile nutrients like nitrate and potassium. Alternatively, applying biochar several weeks prior to fertilization facilitates the development of biochar-soil interactions, potentially enhancing eventual nutrient retention. Precision agriculture technologies provide chances to enhance biochar and nutrient management via variable-rate application informed by geographical variability in soil characteristics. Areas characterized by insufficient organic matter, inadequate structure, or particular nutritional deficiencies may be prioritized for biochar application, whilst regions with sufficient fertility will undergo traditional management practices. This focused strategy optimizes the return on biochar investment while tackling geographical variability in soil limitations.

12. FUTURE RESEARCH DIRECTIONS AND CHALLENGES

Biochar impacts on soil chemical characteristics and fertility are still poorly understood despite major research. The lack of long-term field investigations beyond 10 years limits understanding of biochar permanence and consequences. Research goals include biochar aging mechanisms, biochar-soil interaction predictive models, and biochar's function in emerging pollutants such microplastics and pharmaceutical chemicals (Lehmann et al., 2021).

Biochar-rhizosphere interactions are a key research area. Optimizing biochar application requires understanding how biochar affects root exudation, nutrient uptake kinetics, and plant-microbe interactions in the rhizosphere. Stable isotope tracing, molecular biology methods, and high-resolution imaging are revealing the complexity of biochar-rhizosphere interactions, but much remains unknown. Biochar behaviour under changing climatic conditions must be better understood for climate change adaptation and mitigation. Developing climate-resilient agricultural systems requires biochar performance research under drought, flooding, and temperature extremes. Biochar's ability to attenuate climate change's effects on soil fertility, such as rapid organic matter decomposition and altered nutrient cycling, warrants further study.

Advanced biochar production method allows for customized qualities and lower production costs. Microwave-assisted, hydrothermal, and solar pyrolysis systems may enhance energy efficiency and product quality. Mobile pyrolysis units allow farmers to produce biochar on-farm, lowering transportation costs and enhancing soil fertility (Meyer et al., 2017).

Designer biochars for specialized functions are a new biochar technology frontier. For specific uses, chemical activation, mineral enrichment, or biological inoculation can improve biochar performance. Iron-modified biochars improve phosphorus retention and arsenic immobilization, while biochar-compost composites combine both. Premium prices for engineered biochars may justify processing expenses. Nanotechnology in biochar research and development can improve functioning and create new uses. Nano-scale analysis of biochar surfaces and molecular interactions informs biochar design. Biochar-based nano-composites may also improve plant growth, nutrient release, and pollutant cleanup, although environmental safety must be assessed.

Biochar technology adoption requires concerted efforts from researchers, policymakers, and stakeholders to overcome policy and implementation problems. Many regions lack biochar production and application regulations, leaving producers and users confused. Market development and environmental protection require unambiguous biochar quality, safety, and application requirements (Lehmann and Joseph, 2015).

Economic incentives and support are needed to overcome biochar adoption barriers and acknowledge its many benefits. Carbon credit programs that recognize biochar's sequestration potential, ecosystem service payments, and soil health subsidies could boost adoption. Developing biochar benefit measuring, reporting, and verification standards is difficult, especially for long-term and indirect effects. Biochar application requires knowledge transfer and capacity building, especially in underdeveloped countries with significant soil fertility limits. Farmers, extension services, and local entrepreneurs need biochar production and application training. Local community participation in research and development is needed to address cultural and socioeconomic aspects affecting technological adoption, such as risk perception and traditional farming techniques.

13. CONCLUSIONS

The use of biochar in agricultural soils significantly affects soil chemical properties and fertility through intricate, interconnected mechanisms that function across various spatial and temporal scales. This review provides evidence that biochar effectively mitigates various soil constraints such as acidity, low cation exchange capacity, poor nutrient retention, and contamination, while also offering broader environmental benefits like carbon sequestration and water quality protection. The chemical modifications resulting from biochar application encompass more than merely the addition of nutrients or organic matter. Biochar's distinctive combination of recalcitrant carbon structure, reactive surface chemistry, and extensive porosity establishes new functional domains in soil, thereby enhancing nutrient cycling, supporting beneficial microbial communities, and improving overall soil health. The pH buffering capacity, increased cation exchange capacity, and nutrient retention abilities of biochar-amended soils promote stability and productivity in agricultural systems, enabling sustained fertility under intensive management practices. The variability in biochar effects across studies highlights the necessity for careful selection and application strategies that are aligned with specific soil limitations and management goals. Although general patterns regarding the benefits of biochar are observed, the outcomes are ultimately influenced by site-specific factors such as soil type, climate, cropping system, and the properties of the biochar itself. This complexity requires ongoing research to create predictive frameworks and decision support tools that facilitate optimal biochar use in various agricultural settings. The temporal evolution of biochar effects, encompassing immediate chemical alterations and long-term enhancements in soil quality, underscores the necessity of perceiving biochar as a long-term investment in soil health rather than a transient solution. The durability of biochar in soil, along with enhancements in its functionality over time, offers cumulative advantages that warrant the initial investment. The long-term perspective is especially pertinent when evaluating biochar's function in sustainable intensification and climate change adaptation strategies. The successful integration of biochar into mainstream agricultural practice necessitates ongoing technological innovation, supportive policy frameworks, and efficient knowledge transfer mechanisms. Research priorities must emphasize the comprehension of long-term biochar behaviour, the optimization of production and application technologies, and the formulation of economically viable implementation strategies that acknowledge various ecosystem services. Biochar's potential to enhance global food security, promote environmental sustainability, and mitigate climate change positions it as an essential instrument in the transition to more sustainable agricultural systems. The evidence indicates that biochar application can markedly improve soil chemical properties and fertility when aligned with soil constraints and management goals.

As our comprehension of biochar-soil interactions advances and implementation obstacles are mitigated, biochar technology presents potential solutions to significant challenges confronting global agriculture in the 21st century.

REFERENCES

Ahmad, M., Rajapaksha, A. U., Lim, J. E., Zhang, M., Bolan, N., Mohan, D., Vithanage, M., Lee, S. S., & Ok, Y. S. (2014). Biochar as a sorbent for contaminant management in soil and water: A review. *Chemosphere*, *99*, 19–33. DOI: 10.1016/j.chemosphere.2013.10.071 PMID: 24289982

Akhtar, S. S., Li, G., Andersen, M. N., & Liu, F. (2015). Biochar enhances yield and quality of tomato under reduced irrigation. *Agricultural Water Management*, *138*, 37–44. DOI: 10.1016/j.agwat.2014.02.016

Atkinson, C. J., Fitzgerald, J. D., & Hipps, N. A. (2010). Potential mechanisms for achieving agricultural benefits from biochar application to temperate soils: A review. *Plant and Soil*, *337*(1), 1–18. DOI: 10.1007/s11104-010-0464-5

Beesley, L., Moreno-Jiménez, E., Gomez-Eyles, J. L., Harris, E., Robinson, B., & Sizmur, T. (2011). A review of biochars' potential role in the remediation, revegetation and restoration of contaminated soils. *Environmental Pollution*, *159*(12), 3269–3282. DOI: 10.1016/j.envpol.2011.07.023 PMID: 21855187

Bian, R., Joseph, S., Cui, L., Pan, G., Li, L., Liu, X., Zhang, A., Rutlidge, H., Wong, S., Chia, C., Marjo, C., Gong, B., Munroe, P., & Donne, S. (2014). A three-year experiment confirms continuous immobilization of cadmium and lead in contaminated paddy field with biochar amendment. *Journal of Hazardous Materials*, *272*, 121–128. DOI: 10.1016/j.jhazmat.2014.03.017 PMID: 24685528

Biederman, L. A., & Harpole, W. S. (2013). Biochar and its effects on plant productivity and nutrient cycling: A meta-analysis. *Global Change Biology. Bioenergy*, *5*(2), 202–214. DOI: 10.1111/gcbb.12037

Borchard, N., Schirrmann, M., Cayuela, M. L., Kammann, C., Wrage-Mönnig, N., Estavillo, J. M., Fuertes-Mendizábal, T., Sigua, G., Spokas, K., Ippolito, J. A., & Novak, J. (2019). Biochar, soil and land-use interactions that reduce nitrate leaching and N2O emissions: A meta-analysis. *The Science of the Total Environment*, *651*, 2354–2364. DOI: 10.1016/j.scitotenv.2018.10.060 PMID: 30336425

Cayuela, M. L., van Zwieten, L., Singh, B. P., Jeffery, S., Roig, A., & Sánchez-Monedero, M. A. (2014). Biochar's role in mitigating soil nitrous oxide emissions: A review and meta-analysis. *Agriculture, Ecosystems & Environment*, *191*, 5–16. DOI: 10.1016/j.agee.2013.10.009

Chaganti, V. N., Crohn, D. M., & Šimůnek, J. (2015). Leaching and reclamation of a biochar and compost amended saline-sodic soil with moderate SAR reclaimed water. *Agricultural Water Management*, *158*, 255–265. DOI: 10.1016/j.agwat.2015.05.016

Chan, K. Y., Van Zwieten, L., Meszaros, I., Downie, A., & Joseph, S. (2008). Using poultry litter biochars as soil amendments. *Soil Research (Collingwood, Vic.)*, *46*(5), 437–444. DOI: 10.1071/SR08036

Chen, Z., Chen, B., Zhou, D., & Chen, W. (2011). Bisolute sorption and thermodynamic behavior of organic pollutants to biomass-derived biochars at two pyrolytic temperatures. *Environmental Science & Technology*, *46*(24), 12476–12483. PMID: 23121559

Cheng, C. H., & Lehmann, J. (2009). Ageing of black carbon along a temperature gradient. *Chemosphere*, *75*(8), 1021–1027. DOI: 10.1016/j.chemosphere.2009.01.045 PMID: 19223059

Cheng, C. H., Lehmann, J., & Engelhard, M. H. (2008). Natural oxidation of black carbon in soils: Changes in molecular form and surface charge along a climosequence. *Geochimica et Cosmochimica Acta*, *72*(6), 1598–1610. DOI: 10.1016/j.gca.2008.01.010

Cheng, C. H., Lehmann, J., Thies, J. E., Burton, S. D., & Engelhard, M. H. (2006). Oxidation of black carbon by biotic and abiotic processes. *Organic Geochemistry*, *37*(11), 1477–1488. DOI: 10.1016/j.orggeochem.2006.06.022

Christel, W., Bruun, S., Magid, J., Kwapinski, W., & Jensen, L. S. (2016). Pig slurry acidification, separation technology and thermal conversion affect phosphorus availability in soil amended with the derived solid fractions, chars or ashes. *Plant and Soil*, *401*(1), 93–107. DOI: 10.1007/s11104-015-2519-0

Clough, T. J., & Condron, L. M. (2010). Biochar and the nitrogen cycle: Introduction. *Journal of Environmental Quality*, *39*(4), 1218–1223. DOI: 10.2134/jeq2010.0204 PMID: 20830909

Clough, T. J., Condron, L. M., Kammann, C., & Müller, C. (2013). A review of biochar and soil nitrogen dynamics. *Agronomy (Basel)*, *3*(2), 275–293. DOI: 10.3390/agronomy3020275

DeLuca, T. H., MacKenzie, M. D., & Gundale, M. J. (2009). Biochar effects on soil nutrient transformations. In Lehmann, J., & Joseph, S. (Eds.), *Biochar for environmental management: Science and technology* (pp. 137–182). Earthscan.

Dutta, D., Meena, A. L., Bhanu, C., Ghasal, P. C., Choudhary, J., Kumar, S., Mishra, R. P., Ansari, M. A., Raghavendra, K. J., Prusty, A. K., Jat, P. C., Kashyap, P., Punia, P., Dixit, M., Singh, O., Rai, A. K., Meena, A. K., Rathi, S., & Yadav, P. (2024). Sustainable soil management for climate resilience: Long- term management effects on soil carbon sequestration and nitrogen dynamics in a semi-arid tropical Inceptisol of India. *Journal of Soil Science and Plant Nutrition*, *24*(3), 4407–4426. DOI: 10.1007/s42729-024-01844-4

Dutta, D., & Singh, O. Shivangi (2023). Carbon Footprint of Different Energy-Intensive Systems. In: Rakshit, A., Biswas, A., Sarkar, D., Meena, V.S., Datta, R. (eds) Handbook of Energy Management in Agriculture. Springer, Singapore. DOI: 10.1007/978-981-19-7736-7_5-1

Elad, Y., David, D. R., Harel, Y. M., Borenshtein, M., Kalifa, H. B., Silber, A., & Graber, E. R. (2010). Induction of systemic resistance in plants by biochar, a soil-applied carbon sequestering agent. *Phytopathology*, *100*(9), 913–921. DOI: 10.1094/PHYTO-100-9-0913 PMID: 20701489

Gaskin, J. W., Steiner, C., Harris, K., Das, K. C., & Bibens, B. (2008). Effect of low-temperature pyrolysis conditions on biochar for agricultural use. *Transactions of the ASABE*, *51*(6), 2061–2069. DOI: 10.13031/2013.25409

Ghazaryan, K., Pandey, D., Singh, S., Varagyan, V., Alexiou, A., Petropoulos, D., Kriemadis, A., Rajput, V., Minkina, T., Singh, R., Sousa, J., Kumar, S., El-Ramady, H., Singh, O., & Singh, A. (2024). Enhancing Crop Production: Unveiling the Role of Nanofertilizers in Sustainable Agriculture and Precision Nutrient Management. *Egyptian Journal of Soil Science*, *64*(3), 981–1007. DOI: 10.21608/ejss.2024.282778.1748

Glaser, B., Haumaier, L., Guggenberger, G., & Zech, W. (2001). The 'Terra Preta' phenomenon: A model for sustainable agriculture in the humid tropics. *Naturwissenschaften*, *88*(1), 37–41. DOI: 10.1007/s001140000193 PMID: 11302125

Gomez-Eyles, J. L., Sizmur, T., Collins, C. D., & Hodson, M. E. (2011). Effects of biochar and the earthworm Eisenia fetida on the bioavailability of polycyclic aromatic hydrocarbons and potentially toxic elements. *Environmental Pollution*, *159*(2), 616–622. DOI: 10.1016/j.envpol.2010.09.037 PMID: 21035930

Hossain, M. K., Strezov, V., Chan, K. Y., Ziolkowski, A., & Nelson, P. F. (2011). Influence of pyrolysis temperature on production and nutrient properties of wastewater sludge biochar. *Journal of Environmental Management*, *92*(1), 223–228. DOI: 10.1016/j.jenvman.2010.09.008 PMID: 20870338

Ippolito, J. A., Laird, D. A., & Busscher, W. J. (2012). Environmental benefits of biochar. *Journal of Environmental Quality, 41*(4), 967–972. DOI: 10.2134/jeq2012.0151 PMID: 22751039

Ippolito, J. A., Spokas, K. A., Novak, J. M., Lentz, R. D., & Cantrell, K. B. (2020). Biochar elemental composition and factors influencing nutrient retention. In Lehmann, J., & Joseph, S. (Eds.), *Biochar for environmental management: Science, technology and implementation* (3rd ed., pp. 137–162). Routledge.

Jeffery, S., Abalos, D., Prodana, M., Bastos, A. C., van Groenigen, J. W., Hungate, B. A., & Verheijen, F. (2017). Biochar boosts tropical but not temperate crop yields. *Environmental Research Letters, 12*(5), 053001. DOI: 10.1088/1748-9326/aa67bd

Joseph, S. D., Camps-Arbestain, M., Lin, Y., Munroe, P., Chia, C. H., Hook, J., van Zwieten, L., Kimber, S., Cowie, A., Singh, B. P., Lehmann, J., Foidl, N., Smernik, R. J., & Amonette, J. E. (2010). An investigation into the reactions of biochar in soil. *Soil Research (Collingwood, Vic.), 48*(7), 501–515. DOI: 10.1071/SR10009

Laird, D. A., Fleming, P., Davis, D. D., Horton, R., Wang, B., & Karlen, D. L. (2010). Impact of biochar amendments on the quality of a typical Midwestern agricultural soil. *Geoderma, 158*(3-4), 443–449. DOI: 10.1016/j.geoderma.2010.05.013

Lal, R. (2015). Restoring soil quality to mitigate soil degradation. *Sustainability (Basel), 7*(5), 5875–5895. DOI: 10.3390/su7055875

Lehmann, J., Cowie, A., Masiello, C. A., Kammann, C., Woolf, D., Amonette, J. E., Cayuela, M. L., Camps-Arbestain, M., & Whitman, T. (2021). Biochar in climate change mitigation. *Nature Geoscience, 14*(12), 883–892. DOI: 10.1038/s41561-021-00852-8

Lehmann, J., da Silva, J. P.Jr, Steiner, C., Nehls, T., Zech, W., & Glaser, B. (2003). Nutrient availability and leaching in an archaeological Anthrosol and a Ferralsol of the Central Amazon basin Fertilizer, manure and charcoal amendments. *Plant and Soil, 249*(2), 343–357. DOI: 10.1023/A:1022833116184

Lehmann, J., Gaunt, J., & Rondon, M. (2006). Bio-char sequestration in terrestrial ecosystems–a review. *Mitigation and Adaptation Strategies for Global Change, 11*(2), 403–427. DOI: 10.1007/s11027-005-9006-5

Lehmann, J., & Joseph, S. (2015). *Biochar for environmental management: Science, technology and implementation*. Routledge. DOI: 10.4324/9780203762264

Lehmann, J., & Kleber, M. (2015). The contentious nature of soil organic matter. *Nature, 528*(7580), 60–68. DOI: 10.1038/nature16069 PMID: 26595271

Lehmann, J., Rillig, M. C., Thies, J., Masiello, C. A., Hockaday, W. C., & Crowley, D. (2011). Biochar effects on soil biota–a review. *Soil Biology & Biochemistry, 43*(9), 1812–1836. DOI: 10.1016/j.soilbio.2011.04.022

Liang, B., Lehmann, J., Solomon, D., Kinyangi, J., Grossman, J., O'Neill, B., Skjemstad, J. O., Thies, J., Luizao, F. J., Petersen, J., & Neves, E. G. (2006). Black carbon increases cation exchange capacity in soils. *Soil Science Society of America Journal, 70*(5), 1719–1730. DOI: 10.2136/sssaj2005.0383

Liu, X., Zheng, J., Zhang, D., Cheng, K., Zhou, H., Zhang, A., Li, L., Joseph, S., Smith, P., Crowley, D., Kuzyakov, Y., & Pan, G. (2016). Biochar has no effect on soil respiration across Chinese agricultural soils. *The Science of the Total Environment, 554*, 259–265. DOI: 10.1016/j.scitotenv.2016.02.179 PMID: 26950640

Maestrini, B., Nannipieri, P., & Abiven, S. (2015). A meta-analysis on pyrogenic organic matter induced priming effect. *Global Change Biology. Bioenergy, 7*(4), 577–590. DOI: 10.1111/gcbb.12194

Major, J., Rondon, M., Molina, D., Riha, S. J., & Lehmann, J. (2010). Maize yield and nutrition during 4 years after biochar application to a Colombian savanna oxisol. *Plant and Soil, 333*(1), 117–128. DOI: 10.1007/s11104-010-0327-0

Mandal, S., Thangarajan, R., Bolan, N. S., Sarkar, B., Khan, N., Ok, Y. S., & Naidu, R. (2017). Biochar-induced concomitant decrease in ammonia volatilization and increase in nitrogen use efficiency by wheat. *Chemosphere, 142*, 120–127. DOI: 10.1016/j.chemosphere.2015.04.086 PMID: 25959224

Masulili, A., Utomo, W. H., & Syechfani, M. S. (2010). Rice husk biochar for rice based cropping system in acid soil 1. The characteristics of rice husk biochar and its influence on the properties of acid sulfate soils and rice growth in West Kalimantan, Indonesia. *Journal of Agricultural Science (Toronto), 2*(1), 39–47. DOI: 10.5539/jas.v2n1p39

McCormack, S. A., Ostle, N., Bardgett, R. D., Hopkins, D. W., & Vanbergen, A. J. (2013). Biochar in bioenergy cropping systems: Impacts on soil faunal communities and linked ecosystem processes. *Global Change Biology. Bioenergy, 5*(2), 81–95. DOI: 10.1111/gcbb.12046

Méndez, A., Gómez, A., Paz-Ferreiro, J., & Gascó, G. (2012). Effects of sewage sludge biochar on plant metal availability after application to a Mediterranean soil. *Chemosphere, 89*(11), 1354–1359. DOI: 10.1016/j.chemosphere.2012.05.092 PMID: 22732302

Meyer, S., Bright, R. M., Fischer, D., Schulz, H., & Glaser, B. (2011). Albedo impact on the suitability of biochar systems to mitigate global warming. *Environmental Science & Technology*, *46*(24), 12726–12734. PMID: 23146092

Meyer, S., Genesio, L., Vogel, I., Schmidt, H. P., Soja, G., Someus, E., Shackley, S., Verheijen, F. G., & Glaser, B. (2017). Biochar standardization and legislation harmonization. *Journal of Environmental Engineering and Landscape Management*, *25*(2), 175–191. DOI: 10.3846/16486897.2016.1254640

Mugabo, J. P., Bhople, B. S., Kumar, A. K., Singh, V. K., Mahmoud, E. A., Ullah, F., & Singh, O. (2024). Insights into the tripartite interaction: Effects of Arbuscular mycorrhizae and Rhizobium on root morphology, soil enzymes, and biochemical properties in pea cultivation in alluvial soils of Punjab, India. *Cogent Food & Agriculture*, *10*(1), 2366384. DOI: 10.1080/23311932.2024.2366384

Mukherjee, A., Zimmerman, A. R., & Harris, W. (2011). Surface chemistry variations among a series of laboratory-produced biochars. *Geoderma*, *163*(3-4), 247–255. DOI: 10.1016/j.geoderma.2011.04.021

Novak, J. M., Busscher, W. J., Laird, D. L., Ahmedna, M., Watts, D. W., & Niandou, M. A. (2009). Impact of biochar amendment on fertility of a southeastern coastal plain soil. *Soil Science*, *174*(2), 105–112. DOI: 10.1097/SS.0b013e3181981d9a

Roberts, K. G., Gloy, B. A., Joseph, S., Scott, N. R., & Lehmann, J. (2010). Life cycle assessment of biochar systems: Estimating the energetic, economic, and climate change potential. *Environmental Science & Technology*, *44*(2), 827–833. DOI: 10.1021/es902266r PMID: 20030368

Rondon, M. A., Lehmann, J., Ramírez, J., & Hurtado, M. (2007). Biological nitrogen fixation by common beans (Phaseolus vulgaris L.) increases with bio-char additions. *Biology and Fertility of Soils*, *43*(6), 699–708. DOI: 10.1007/s00374-006-0152-z

Schimmelpfennig, S., & Glaser, B. (2012). One step forward toward characterization: Some important material properties to distinguish biochars. *Journal of Environmental Quality*, *41*(4), 1001–1013. DOI: 10.2134/jeq2011.0146 PMID: 22751042

Schmidt, H. P., Pandit, B. H., Cornelissen, G., & Kammann, C. I. (2017). Biochar-based fertilization with liquid nutrient enrichment: 21 field trials covering 13 crop species in Nepal. *Land Degradation & Development*, *28*(8), 2324–2342. DOI: 10.1002/ldr.2761

Shen, Z., Som, A. M., Wang, F., Jin, F., McMillan, O., & Al-Tabbaa, A. (2016). Long-term impact of biochar on the immobilisation of nickel (II) and zinc (II) and the revegetation of a contaminated site. *The Science of the Total Environment, 542*, 771–776. DOI: 10.1016/j.scitotenv.2015.10.057 PMID: 26551277

Singh, B., Singh, B. P., & Cowie, A. L. (2010). Characterisation and evaluation of biochars for their application as a soil amendment. *Soil Research (Collingwood, Vic.), 48*(7), 516–525. DOI: 10.1071/SR10058

Singh, O., Shahi, U.P., Dhyani, B.P., Kumar, S., Vivek, S., & Sengar, R.S., Shivangi, and Singh, A. (2023). Effect of zinc oxide nanoparticles application on growth and yield of basmati rice (Oryza sativa) in alkaline soil. *Indian Journal of Agronomy, 68*(3), 1–6.

Singh, O., Singh, S., Singh, V. K., & Singh, A. (2023). Biochar: An organic amendment for sustainable soil health. In Baskar, C. (Eds.), *Encyclopedia of Green Materials* (pp. 265–271). Springer Nature Singapore Pte Ltd., DOI: 10.1007/978-981-16-4921-9_265-1

Singh, P., Singh, A., Singh, V. K., & Singh, O. (2022). Green Manuring for Sustainable Agriculture. In Baskar, C. (Eds.), *Encyclopedia of Green Materials*. Springer Nature Singapore Pte Ltd., DOI: 10.1007/978-981-16-4921-9_261-1

Singh, S., Singh, O., Shahi, U. P., Singh, P. K., Singh, A., Rajput, V. D., Minkina, T., El-Ramady, H., & Ghazaryan, K. (2024a). Carbon sequestration through organic amendments, clay mineralogy and agronomic practices: A review. *Egyptian Journal of Soil Science, 64*(2), 581–598. DOI: 10.21608/ejss.2024.260719.1707

Singh, S., Singh, V., Singh, A., & Singh, O. (2022). Zero-Budget Natural Farming: Way to Sustainable Future. In Baskar, C. (Eds.), *Encyclopedia of Green Materials*. Springer Nature Singapore Pte Ltd., DOI: 10.1007/978-981-16-4921-9_264-1

Singh, V. K., Gill, A. A. S., Singh, A., Singh, O., & Singh, T. (2022a). Natural Potassium Fertilizers for Sustainable Agriculture. In Baskar, C. (Eds.), *Encyclopedia of Green Materials*. Springer Nature Singapore Pte Ltd., DOI: 10.1007/978-981-16-4921-9_262-1

Steiner, C., Glaser, B., Teixeira, W. G., Lehmann, J., Blum, W. E., & Zech, W. (2008). Nitrogen retention and plant uptake on a highly weathered central Amazonian Ferralsol amended with compost and charcoal. *Journal of Plant Nutrition and Soil Science, 171*(6), 893–899. DOI: 10.1002/jpln.200625199

Uchimiya, M., Lima, I. M., Klasson, K. T., & Wartelle, L. H. (2010). Contaminant immobilization and nutrient release by biochar soil amendment: Roles of natural organic matter. *Chemosphere, 80*(8), 935–940. DOI: 10.1016/j.chemosphere.2010.05.020 PMID: 20542314

Uzoma, K. C., Inoue, M., Andry, H., Fujimaki, H., Zahoor, A., & Nishihara, E. (2011). Effect of cow manure biochar on maize productivity under sandy soil condition. *Soil Use and Management, 27*(2), 205–212. DOI: 10.1111/j.1475-2743.2011.00340.x

Van Zwieten, L., Kimber, S., Morris, S., Chan, K. Y., Downie, A., Rust, J., Joseph, S., & Cowie, A. (2010). Effects of biochar from slow pyrolysis of papermill waste on agronomic performance and soil fertility. *Plant and Soil, 327*(1), 235–246. DOI: 10.1007/s11104-009-0050-x

Vithanage, M., Herath, I., Joseph, S., Bundschuh, J., Bolan, N., Ok, Y. S., Kirkham, M. B., & Rinklebe, J. (2017). Interaction of arsenic with biochar in soil and water: A critical review. *Carbon, 113*, 219–230. DOI: 10.1016/j.carbon.2016.11.032

Wang, T., Camps-Arbestain, M., Hedley, M., & Bishop, P. (2012). Predicting phosphorus bioavailability from high-ash biochars. *Plant and Soil, 357*(1), 173–187.

Weber, K., & Quicker, P. (2018). Properties of biochar. *Fuel, 217*, 240–261. DOI: 10.1016/j.fuel.2017.12.054

Woolf, D., Amonette, J. E., Street-Perrott, F. A., Lehmann, J., & Joseph, S. (2010). Sustainable biochar to mitigate global climate change. *Nature Communications, 1*(1), 1–9. DOI: 10.1038/ncomms1053 PMID: 20975722

Xu, G., Sun, J. N., Shao, H. B., & Chang, S. X. (2014). Biochar had effects on phosphorus sorption and desorption in three soils with differing acidity. *Ecological Engineering, 62*, 54–60. DOI: 10.1016/j.ecoleng.2013.10.027

Yuan, J. H., & Xu, R. K. (2011). The amelioration effects of low temperature biochar generated from nine crop residues on an acidic Ultisol. *Soil Use and Management, 27*(1), 110–115. DOI: 10.1111/j.1475-2743.2010.00317.x

Yuan, J. H., Xu, R. K., & Zhang, H. (2011). The forms of alkalis in the biochar produced from crop residues at different temperatures. *Bioresource Technology, 102*(3), 3488–3497. DOI: 10.1016/j.biortech.2010.11.018 PMID: 21112777

Zhang, X., Wang, H., He, L., Lu, K., Sarmah, A., Li, J., Bolan, N. S., Pei, J., & Huang, H. (2013). Using biochar for remediation of soils contaminated with heavy metals and organic pollutants. *Environmental Science and Pollution Research International, 20*(12), 8472–8483. DOI: 10.1007/s11356-013-1659-0 PMID: 23589248

Zimmerman, A. R., Gao, B., & Ahn, M. Y. (2011). Positive and negative carbon mineralization priming effects among a variety of biochar-amended soils. *Soil Biology & Biochemistry*, *43*(6), 1169–1179. DOI: 10.1016/j.soilbio.2011.02.005

Chapter 5

Exploring Microbial Responses to Nanomaterials in Soil Environments:
Balancing Act

Balwant Singh
http://orcid.org/0000-0002-6224-7194
Dr. Ram Manohar Lohia Avadh University, Ayodhya, India

Asha K. Lakra
http://orcid.org/0009-0007-6505-667X
North-Eastern Hill University, Shillong, India

Mukul M. Barwant
http://orcid.org/0000-0002-5639-4251
Sanjivani Arts Commerce and Science College, Kopargoan, India

Shalagha A. B. Sharma
http://orcid.org/0000-0001-8223-1774
School of Biological Engineering, Shobhit University, Meerut, India

Femina Sobin
http://orcid.org/0000-0003-4563-8449
St. Aloysius College, Jabalpur, India

Alka Rajput
School of Bioscience, RIMT University, Mandi Gobindgarh, India

ABSTRACT

As nanotechnology advances, the deployment of nanomaterials in various applications increases, inevitably leading to their release into the environment, particularly soil. By elucidating the mechanisms underlying microbial responses to nanomaterials, researchers can develop strategies to mitigate potential adverse effects and harness the beneficial roles of nanomaterials in soil environments. Moreover, the chapter explores the concept of a "balancing act" in microbial-nanomaterial interactions, highlighting the delicate equilibrium between nanomaterial-induced stress and

DOI: 10.4018/979-8-3693-7006-3.ch005

microbial adaptation. It discusses how microbial communities may undergo shifts in composition and function in response to nanomaterial exposure, potentially affecting nutrient cycling, soil fertility, and ecosystem resilience. By advancing our understanding of these interactions, we can develop sustainable practices for the safe deployment of nanomaterials in agriculture, remediation, and other soil-related applications, ensuring the preservation of soil ecosystem health and function.

INTRODUCTION

The field of nanotechnology, which studies sizes between 1 and 100 nm, has made it possible to create manufactured nano-objects (MNOs) with unique and useful characteristics. Due to their distinctive qualities—such as their size, surface and charge area, target specificity, sensitivity, degradability, ecological sustainability, and potential to lower environmental pollution—MNOs have drawn a lot of interest. These characteristics set them apart from both individual particles and bulk material. Manufactured nano-objects (MNOs) with a connection to agriculture have the potential to transform agricultural productivity and advance sustainable development objectives. MNOs can increase efficiency while reducing collateral harm to non-target creatures and the environment. These MNOs have a variety of physico-chemical properties and the capacity to encapsulate and distribute active components in a controlled, targeted, and stimuli-responsive manner. It is known that applying MNOs as nano-pesticides and nano-fertilizers (Ahmed, et al,.2023).

Because of these special properties, the field of artificial nanoparticles (NPs) research has experienced exponential growth in several domains. The many uses of nanotechnology in the agricultural, environmental, and related fields—including industry and medicine—have drawn attention to the technology. Alongside this, NPs are supplemented into soils for remedial purposes, which ultimately raises problems regarding microbial populations. They eventually impede the physiological and metabolic processes of living things because of their resistant character, which causes them to accumulate in the environment. Additionally, because of their many advantages in terms of reducing industrial and heavy metal pollution, enhancing agricultural practices, and advancing other medical uses, their continuous buildup in soils has been noted as a result of misinformation (Khanna et al., 2021).

Carbon nanomaterials (CNMs) have a remarkable capacity to absorb hydrophobic organic pollutants (HOCs) due to their extremely hydrophobic nature and enormous reactive surface areas. Because of these characteristics, CNMs may have a positive or negative effect on the persistence, mobility, and bioavailability of pollutants in soils. Positive effects could result from sorption and sequestration, which would lower the bioavailability of the contaminants, or negative effects from increased contaminant dissemination (Riding, et al., 2015).

Applying nano-agro-chemicals as nano-pesticides or nano-fertilizers has the potential to completely transform crop protection and production in agriculture because of their positive impacts on plant health and vigor. They can overcome biological barriers and find new, effective paths to the target location thanks to their special physico-chemical characteristics. Agrochemicals' structure and function are also improved by them, leading to increased solubility, protection against hydrolysis and photodecomposition, and more targeted and regulated release (An et al,.2022). Like other industrial goods, nanomaterials can leak into the environment accidentally or on purpose. It has been shown that NMs may affect microbial activity, abundances, and diversity utilizing various techniques, indicators, and natural soils. For microbial-dependent crop development, effects on beneficial soil processes including the microbial N2 cycle, symbiotic relationships (like AMF), Fe metabolism (siderophores), and the synthesis of antifungal chemicals are especially concerning. Nonetheless, the features of the soil and the NMs taken into consideration have a significant impact on the harmful effects on the microbial community (Karimi & Mohseni,2017).

Nanotechnology is currently a cutting-edge, creative technology being used in environmental remediation. Heavy metals, undesired microbes, and other contaminants can be eliminated from soil and water by using nanomaterials (NMs). The potential of NMs to enhance agricultural soil quality and, consequently, the adoption of more sustainable agricultural methods, is presently being investigated. Due to their huge surface area and compact size, they have numerous unique qualities (e.g., improved cation exchange capacity, long-lasting nutrient release, and nutrient delivery) that make them potentially useful for solving soil restoration issues. They are also highly reactive. When it comes to managing the quality of the contaminated water, NMs stop the creation of biofilms and eliminate unwanted microbes (Raja, & Husen,2020).

Because engineering nanoparticles (ENP) are used in industry and human goods, it is typical to find wastes and by-products from industrial chemical processes containing these particles. Incidental nanoparticles can also be found in the environment. The application of nanotechnologies to the remediation of contaminated soils is currently a developing field that has the potential to significantly enhance the efficacy of conventional remediation techniques. However, the release of nanotechnologies into ecosystems has also given rise to environmental concerns about the health of humans and the environment (Medina-Pérez et al,. 2019). The need for food and the growing global population have led to the development of new agricultural techniques that increase food production by using fertilizers and insecticides that are more effective. These technologies carry the risk of causing the uncontrolled release of undesirable materials into the environment, which could contaminate groundwater and soil. Nowadays, the use of nanotechnology holds great promise for raising agricultural yield and cleaning up contaminated areas (Mura et al,.2013).

NANOMATERIALS IN SOIL ENVIRONMENTS

In this modern era, nanotechnology has been evolving immensely in various arenas of agriculture, (Chhipa, 2017) industrial inventions, etc. and thus, their disposal into environment is also growing (Gonçalves et al., 2019) Some of the growing applications of nanoparticles in the field of agriculture includes as nano-pesticides nano-fertilizers, wireless nano-sensors and nanonetworks being used to monitor water quality, incidences of crop disease, efficiency of nutrient use, growth rates and pesticides; nano encapsulated agrochemicals and nanotubes can possibly boost the nutritive value and yield of a crop, prompt detection of pollutants and plant diseases; development of an resistant varieties of crops (pest insect and disease resistant) via suitable breeding programs that incorporate nano-based materials like nanoparticle facilitated genes; development of genetically modified (GM) crops by means of spontaneous or induced mutation, transgenic breeding, etc. (Mishra et al., 2017; Singh et al., 2020; Yashveer et al., 2014; Kumar, 2020; Prasad et al., 2017). Such application of nanomaterials lessens chemical hazards, crop yields, loss of nutrients, and outbreaks of pest, despite also helping to stabilize fluctuating climatic conditions and food security levels (Omanovic-miklicanin and Science, 2017; Elemike et al., 2019).

According to Rajput et al. (2020), the transport and fate of NPs in soil have recognized very less attention as compared to other sources. He also states that, the soil delivers nutrition to crops, and act as a accumulator sink for NPs. NPs can enter soil via sedimentation in the form of aerosols and dust, precipitation, absorption of gaseous compounds directly by soil and fallen leaves or as a result of human activity (Gladkova and Terekhova, 2013).

Besides bounteous benefits, nanotechnology also has several limitations (Prasad et al., 2017; Bindraban et al., 2015; Ghouri et al., 2020) like -

- Adverse effect on plants which are not targeted and associated organisms
- Food products containing nanomaterials may lead to numerous human health concerns
- Phytotoxicity
- Cytotoxicity
- Genotoxicity

Hence, substantial efforts should be made to investigate the interactions and behaviour of nanoparticles in the environment to safeguard and make available a safe future (Kumar et al., 2018; Mishra et al., 2017)

TYPES OF NANOMATERIALS USED

Herein are discussed few examples of nanomaterials used in different forms, their applications and results.

Table 1. Few examples of Nanomaterials: Application in soil and obtained results

S. No.	Nanomaterials used	Application in Soil	Result	Reference
1.	Starch nanocrystals	Pesticide delivery	*Diaphorina citri* - high insecticidal activity, safe to bees	Liu et. al., 2024
2.	Zein-FA and CPT@Zein-FA	Natural product (Camptothecin) pesticides	Lepidopteran pests (Spodoptera frugiperda, Ostrinia furnacalis and Mythimna separate) - Improved bioactivity, plays role in sustainable development	Huang et. al., 2024
3.	Alginate Nanocomposite	developing controlled release preparations for herbicide application	Extended release of diuron and atrazine	Paul et. al., 2024
4.	Chitosan Nanoparticles		Dichlorprop release, while interacting with chiral receptors the enantioselective compound's behaviour may change	
5.	Poly(ε-caprolactone) Nanogranules		For encapsulation of Atrazine, encapsulation efficiency 65%, stable when chitosan coated, potential efficiency for weed control	
6.	Nanocopper	Antibacterial properties against Xanthomonas axonopodis pv. punicae, the incitant of pomegranate bacterial blight	Cell wall degradation of bacteria (*X. axonopodis*) colonization of plant tissues failed, failure to produce water soaking ability, potential environment friendly for management of bacterial blight of pomegranate	Mondal and Mani, (2012)

continued on following page

Table 1. Continued

S. No.	Nanomaterials used	Application in Soil	Result	Reference
7.	Silver nanoparticles	Antifungal properties	Against soil-borne phytopathogens viz., *Sclerotium rolfsii, Rhizoctonia solani, Fusarium oxysporum* and *Sclerotinia sclerotiorum*	Kaman and Dutta (2019)
8.	Zinc oxide nanoparticles	Antifungal properties	mediating *Trichoderma* spp., against soil-borne pathogens like., *Rhizoctonia solani, Macrophomina phseolina* and *Fusarium fujikuroi*	Zaki et. al. (2022)
9.	Gold nanoparticles	Antiviral properties	Against soil-borne Barley yellow dwarf virus (BaYDV)	Alkubaisi and Aref (2017)
10.	Silicon dioxide nanoparticles	Induction of plant defense mechanism	activates antioxidant system and innate defense responses in wheat seedlings against the pathogen *R. solani*	Fu et al., 2014
11.	Lignin nanoparticle	Increased seed germination	positive effect on seed germination of primed maize seeds	Del Buono et. al., (2021)
12.	Selenium nanoparticles	Enhancers of Phytohormones and plant signaling molecules	Stress signaling molecules abscisic acid and indole-3-acetic acid accumulation, strawberry plants - increased secretion of organic acid, nutritional values and fruit quality improvement	Zahedi et al. (2019)
13.	Nanosized zerovalent iron	Pesticide degradation	Pesticidees such as molinate, atrazine, and chlorpyrifos are susceptible to degradation	Hee Joo and Cheng, 2006
14.	Titanium oxide	Pesticide degradation	Degradation of Imidacloprid	Guan et al., 2008
15.	Nano-coating of sulfur	Fertilizer delivery	NPK controlled delivery	Wilson et. al., 2008
16.	Iron oxide and Zirconium oxide nanoparticles	Pesticide sensor	Senses Dimethoate	Gan et al., 2010
17.	Gold nanoparticles	Genetic material delivery - DNA		Vijayakumar et al., 2010

MECHANISMS OF INTERACTION WITH SOIL COMPONENTS

Entry of NPs to soil and or aquatic environment may be through landfill sites, industrial spills, in the course of sewage sludge fertilizer application or sewage from industrial and residential areas, overuse of fertilizers and pesticides and NPs as a means to clean the soil i.e. soil remediation (Bashir et. al., 2020). Nanomaterial's application as solitary carriers of agrochemicals possibly will ease its unregulated interaction with components of soil causing severe variations in soil microbiota activity and in some circumstances could also have negative impact on soil fertility, consequently crop productivity (Rajput et. al., 2020).

Keller et al. (2010) advocates that natural organic matter of soil gets adsorbed onto surfaces NP, thus reducing their aggregation significantly and finally stabilization under usual conditions of the environment. Beneficial microbial consortia (archaea, bacteria and fungi) deliver indispensable ecosystem services like crop production, nutrient cycling, and climate regulation (Dutta and Dutta, 2016). Researches have shown that various NPs, precisely those prepared with copper, silver and zinc, possess antimicrobial properties (Vimbela et. al., 2017; Wang et. al., 2017). Such NPs on entering soil system, could be a threat to beneficial microbial consortia. Few of the findings are enlisted as follows –

- Fajardo et al. (2014) reports morphological alterations in cells and bactericidal effects on the soil bacterias namely *Pseudomonas stutzeri* and *Bacillus cereus* by Ag NPs. These effects are considerably influenced by including concentration, functionalization, exposure time, and soil texture (Grün et al., 2019).
- Shen et. al., (2015); Frenk et. al., (2013) and Peyrot et. al., (2014) reports that Ag NPs also reduce enzymatic activities, whereas bacterial biomass and growth were reduced by Zn- and Cu- based NPs.

  Release of toxic ions from NMs is influenced by organic material and pH of soil therefore, their toxic effects on microbes (Reidy et al., 2013; Klaine et al., 2008). The toxic effect of ZnO NPs against *Folsomia candida* was augmented in acidic soil (Waalewijn-Kool et al., 2014).

As compared to air and water, soil is the most important receptor of NPs. Soil is becoming a potential sink as a result of surplus applications of diverse NMs being introduced in agriculture (Gogos et al., 2016). In spite of increasing alarm of NPs in the environment and consequent effects on ecosystem, very less data is available regarding the amount of NPs entering the agricultural soil through either direct or indirect inputs. Researches on the transport, bioavailability, fate and possible environmental risks of NPs is still in its embryonic stage.

ENVIRONMENTAL FATE AND BEHAVIOUR OF NANOMATERIALS

Tarafdar and Adhikari, (2015) briefs that largely, in soil nanoparticles exists in two categories namely -

1. Nano-minerals: The minerals that only exist in this size range, like Mn and Fe oxides and some clays.
2. Mineral NPs: The minerals that can also exist in larger sizes, like almost all the known minerals.

Natural formation of nanoparticles in water bodies, volcanic dust, sediments and soils is being ruled by the diverse soil, biological and geological processes. Three classes of natural nanoparticles are present in the soil system namely -

1. Nanosheets: They occur as a fine coat on surfaces of primary minerals of the soil matrix, promoting accelerated and intensive splitting of existing minerals and subsequent precipitation of neophases.
2. Nanorods: They are formed due to forceful dissolution of prime soil minerals subsequently precipitate and develop secondary phases.
3. Nanoparticles: May have been formed in soils by means of biotic processes

Lately momentous amounts of various NPs have been produced for use in different field like industries, cosmetics, medicines, agriculture and in research and development. These nanoparticles will indeed end up being deposited or castoff into the soil. The major, point or diffused sources of NP in soil system is presently through the NPs being released from consumer products, dumped into the wastewater treatment plant and sewage sludge. NP will pass into the environment at some point in their life cycle by means of release during:

1. Production of nano-enabled products and raw material for it
2. Using of the products containing NP
3. After disposal of products containing NP (waste handling) (Gottschalk et. al., (2009); Gottschalk et. al., (2013); Tolaymat et. al., (2017))

Quite a few crucial processes are expected to affect the bioavailability and fate of nanoparticles in the soil environment. Processes of transformation of nanomaterials can be categorized into –

1. Physical processes –including agglomeration, sedimentation, aggregation and deposition in the porous media.
2. Chemical processes – includes redox reactions (sulfidation and oxidation), dissolution followed by speciation changes, corona formation and photochemical reactions.
3. Biological processes - It includes probably, microbial biomodification and biodegradation (Lead et. al. 2018; Hartmann et. al. 2014; Lowry et. al. 2012)

Some nanomaterials dissolutes so that fate and bioavailability become more aligned with the soluble components Like, ZnO in bulk dissolves quickly in soils (Smolders and Degryse, 2002; McBeath and McLaughlin, 2014), making ZnO NMs short-lived in soils unless they are covered with agents limiting dissolution. They may also be transformed into a blend of species like $Zn_3 (PO_4)_2$, ZnS, Zn-cysteine, Zn_2^+ and Zn-substituted ferrihydrate which gets adsorbed onto surface of minerals (Judy et al., 2015; Lombi et. al., 2012; Wang et al., 2013; Ma et. al., 2013; Brunetti et al., 2015;).

Metal oxides dissolutes much quicker in soils than in water as soil surfaces make available robust pH buffering andacts as a sink for cations or anions getting released from the NMs. Usually, Carbon-based NMs are water insoluble but possibly will degrade via microbial decomposition or photolysis (Chouhan et. al. 2016; Navarro et. al. 2016).

Homo-aggregation of NMs would be slow in most soils because for most NMs, the pH and ionic strength of most soil solutions would lie below the critical coagulation concentration (Cornelis et. al. 2014). Hetero-aggregation is expected to be very significant in soils, as often, pore waters of soil have higher natural colloids' concentrations of in suspension (Klitzke et. al. 2015; Hoppe et. al. 2015). Aggregation of NMs has considerable consequences for controlling its transport through soils as it will enhance straining. Conversely, NOM present in porewaters of soil has frequently shown to prevent homo- and heteroaggregation and stabilization of NMs (Praetorius et. al. 2014). Thus suggesting that accumulation of NMs which are less soluble, will probably take place in topsoils and have less in-depth movement in most of the soils. Less studies have been conducted about NMs movement in soil conditions. This perhaps will be a significant gap for more detailed valuation of risk of NM transport in soils (Kasel et. al. 2013).

The uptake and accumulation of nanoparticles by living organisms may have cumulative toxic effects; however, organisms may counteract these effects either by storage or excretion of the nanoparticles in a benign form (Shi et. al. 2013).

Table 2. Some examples of Nanomaterials, their applications and bioaccumulation

Nanomaterials	Application	Bioaccumulation	References
Gold	Bio-medical applications – diagnosis, drug delivery and treatment	Accumulated by tissues – brain, spleen, liver and kidney	Lasagna-Reeves et. al. 2010; Xing et. al. 2016; Kahlon et. al. 2018
Silver	Industrial applications	Infiltrate fish organelles – mitochondria, nucleus and blood vessels; accumulates in liver and gills	Yeo and Pak 2008; Thio et. al. 2012
Titanium dioxide	Household and industrial applications	Humans - nanoparticles pool up with plasma proteins, platelets and coagulation variables; penetrate erythrocytes; In soil organisms - direct uptake via skin is a possible course of exposure	Shi et. al. 2013
Cerium oxide	Cell electrolytes, Catalysts,, semiconductors, coatings, antioxidants, and polishing chemicals	The biota tested - absorbed then accumulated and diluted in the food web built; aquatic hydrophytes - morphological abnormalities; induce DNA damage, cell death and oxidative stress	Pulido-Reyes et. al. 2015; Khan et. al. 2011; Arnold et al. 2013; Zhao et. al. 2017; Zhang et. al. 2016
Carbon	Biomedical applications – diagnosis, bioimaging, gene therapy, drug delivery, photothermal therapy, etc.	Mice – (intravenous injection) accumulated in the spleen and liver, minor contamination in lungs	Son et. al. 2016; Xie et. al. 2017; Luo et. al. 2014
Zinc oxide	Cosmetics, Sunscreens, paper, paint, plastics, building materials and ceramics	Test fish - exhibited higher oxidative effects and bioaccumulation, more severe histopathological changes than bulk ZnO (subacute exposure of 30 days)	Hao et. al. 2013; Osmond and McCall, 2010

Unexpected biological reactions with nanomaterials produce by-products that saturate lysosomal compartments and perturb degradative and autophagic path-ways that are essential for cells to degrade proteins (Sternet al. 2012).

MICROBIAL COMMUNITY DYNAMICS IN NANOMATERIAL-TREATED SOILS

The unrestricted accumulation of metal-based nanoparticles in the soil system has adversely affected the soil environment as well as microbiota. The impact on soil microbial population is also linked with alteration of soil properties and complexities (Zhang et al. 2020) which eventually affect the growth and functions of microorganisms particularly those species which are beneficial for soil and plants.

Impact of Nanomaterials on Soil Microbiota

Soil microorganism renders a crucial role in sustaining soil and nurturing plant heath (Lambers et al. 2009; Jacoby et al. 2017). Use of metal-based nanomaterials as an agricultural product has the affinity to manipulate the microbial population. For instance, Fe and TiO_2 based nanomaterials can elevate the population of the targeted species while it might proof toxic to other group of microbes and other biotic entities (Lecoanet and Wiesner 2004; Yavuz et al. 2006; Mueller and Nowack 2010). While some other nanomaterials like nZVI are only known to have negative impact on microorganisms (Cullen et al. 2011). Nanopartciles (NPs) which have documented to possess severe and mild impacts on soil microbes includes Ag-NPs, CuO-NPs, FeO-NPs, TiO_2–NPs and ZnO-NPs (He et al. 2016; Zhang et al. 2018; Asadishad et al. 2018). These nanomaterials invade the microbial cells via endocytosis or perhaps by penetration (Kumar et al. 2018) and its impact is greatly influenced by its size, surface charges, capping agents, presence of divalent cations/ anions, functions, concentrations, exposure time and also soil properties (Sondi and Salopek-Sondi 2004; Acharya et al. 2018; Grün et al. 2019). The attribution of its concentration is more pronounced (Peyrot et al. 2014). The features, bioavailability and speciation of the nanomaterials are also influenced by the hormones, enzymes, organic acids and extracellular polymeric substances (Ameen et al. 2021).

The toxicity on soil microorganisms by the metallic ions released from the nanomaterials can also be manipulated by the soil pH and organic matter content (Klaine et al. 2008; Reidy et al. 2013). For example, in a study conducted by Schlich and Hund-Rike (2015) it was found that the sorption of Ag was hindered by low pH and organic matter content of the soil which increased its toxicity however high pH and organic matter of the soil eased the sorption of Ag thereby increasing its bioavailability and reducing its toxic impact. Similar observation was also reported by Mcgee et al. (2020), AgNP significantly reduced the microbial population in acidic soil whereas the population in pH neutral soil remained unaffected.

The impact of nanomaterials differs between the different microbial communities. For example, TiO_2 NPs was found to adversely affect the soil microbial prokaryotes as compared to the fungal community (Moll et al. 2017). Exposure to nanomaterials causes adverse morphological and physiological changes on the soil microorganisms which not only affect the composition and diversity of soil microbiota but also cause significant disturbances on the nutrient cycles and natural processes. Ag-NPs have been reported to alter the morphology of *Bacillus cereus* and affect the ability of *Pseudomonas* sp. to synthesize biofilms which is highly needed to establish root colonization and even decreased the viability of cells in preformed biofilms (Rudrappa et al. 2008; Hartmann et al. 2013; Fajardo et al. 2014). The trans/cis ratios of unsaturated fatty acids in *P. putida* was found to increase linearly when treated with citrate capped AgNPs which led to the reduction in membrane fluidity (Hachicho et al. 2014). Application of CuO NPs on flooded paddy soils has been found to be associated with decrease in microbial composition and diversity. Its excess concentration in the soil alters the microbial nutrients availability (Xu et al. 2015). In a study conducted by Cao et al. (2018) it was found that even after 94 days of application of PVP-AgNPs (100 mgL^{-1}) the concentration of silver still remained 95.7% in the soil which adversely affected the phosphorus and nitrogen cycling efficiency in the wetland and also altered the structure of microbial community. Exposure to CuO and ZnO NPs was reported to cause decrease in microbial biomass, stimulation of microbial metabolic quotient, and also inhibits functions of enzymes such as urease, phosphatase and dehydrogenase (Du et al. 2011; Xin et al. 2020). The Cu ions discharged from NPs were reported to cause cavities, membrane degradation, blebs and cellular collapse on soil bacteria (Concha-Guerrero et al. 2014). The decrease in soil microbial composition and diversity when treated with NPs such as CuO and Ag have also been found to be linked with decrease in microbial decomposition (Pradhan et al., 2011). ZnO-NPs stabilized with acetate inhibited the growth of *Pseudomonas* species (Soni et al. 2017) and was also found to inhibit biofilm formation in P. aeruginosa and P. putida (Ouyang et al. 2017).

The impact of nanomaterials in soil fungi is known to be depended on its type and size, which facilitates its penetration to fungal hyphae and distort its morphology. Smaller size of PdS-NPs showed higher toxicity on *Saccharomyces cerevisiae* by reducing the viability and growth of cells leading to subsequent decrease in cell wall stress. To counteract the stress induced by PdS-NPs, chitin concentration hikes in the *S. cerevisiae* cell wall which triggered over expression of integrity signalling genes (FKS2 and PRM5) in the cell wall enhancing intracellular accumulation of ROS which caused mitochondrial dysfunction and ultimate cell apoptosis (Sun et al. 2014). Thus it can be summarized that nanomaterials exposure to certain fungal species can lead to antifungal effects.

Studies have showed positive as well negative impact of nanomaterials in arbuscular mycorrhizal fungi. Some nanomaterials like Ag and FeO NPs showed dose dependent beneficial impact on AMF colonization and boosted growth promoting traits. In contrary, the bulk application of nanomaterials adversely affected the AMF colonization in plant host (Feng et al. 2013). Li et al. (2015) reported that individual treatment of ZnO NPs and $ZnSO_4$ had restricted AMF colonization in maize plants whereas when these treatments were used in combination it enhanced the colonization and the growth.

Recent studies on interaction between nanomaterials and lignin degrading fungi have shown that particulates release from nanomaterials penetrates the hyphase of fungi and accumulates in within the fungi. This causes alteration in the fungi metabolic processes by reducing the production of ligninolytic enzymes thereby retarding decomposition (Ma et al. 2020).

Nanomaterials have also been found to affect the symbiotic relationship between the host plants and concerned microorganisms. For instance, the symbiotic relationship between *Vicia faba* and *R. leguminosarum* and *Glomus aggregatum* was hindered by the application of AgNPs. It halt the nitrogenase activity and processes such as nodule formation and mycorrhizal colonization. The glomalin and cytoplasmic contents were also found to be adversely affected (Abd-Alla et al. 2016). ZnO-NPs and TiO_2-NPs were reported to cause imbalance in the Rhizobium-Legume symbiosis. When exposed to ZnO-NP and TiO_2-NPs, *Rhizobium leguminosarum* showed morphological changes and damaged surface. These changes were accompanied with delayed nitrogen fixation disturbed nodulation (Fan et al., 2014). Similar effect was observed for *S. meliloti* (Bandyopadhyay et al. 2015). The ZnO-NPs were also traced within the periplasmic spaces of the bacterial cells which resulted in the structural changes of polysaccharides and proteins of the cell walls. The toxicity of Ag and ZnO NPs may be attributed by loss of respiration by bacterial cells (Choi et al. 2008), disintegration of cells and damage by oxidation (Zhang et al. 2018), retardation of production of bioactive molecules (Seneviratne et al. 2016) crucial for cells' survival.

Response of Soil Microbes to Nanomaterial Exposure

Frequent exposure to nanomaterials can lead to development of some resistant strains in few microbes against selected nanomaterials. For instance, some plant growth enhancing bacteria such as *Azotobacter chroococcum*, *Bacillus thuringiensis*, *Pseudomonas mosselii* and *S. meliloti* showed resistance to higher dosage of CuO, TiO_2 and Al_2O_3 NPs while they showed sensitivity towards NPs of Ag and ZnO (Ahmed et al. 2020). In response to AgNP some free-living bacteria produces reactive oxygen species in the biofilm which accounts for the activation of scavenging system. Expressions of proteins linked with quorum and redox sensing, and stress response have also been revealed in proteomic studies of bacteria subjected to AgNP. Stress induced by AgNP may also enhance the production of EPS and inorganic phosphate solubilisation (Gambino et al. 2015). EPS have the ability to mitigate the toxic effect of nanomaterials. In a study, mutant strain (overproducing EPS) of Sinorhizobium meliloti showed high resistivity against the toxicity of Ag-

NPs stabilized by Na-citrate (Joshi et al. 2008). Increase production of EPS is also associated with biofilm formation (Ouyang et al. 2017).

Increase in tolerance or resistance can also account due to evolution of defensive mechanisms, changes in envelope structures, up-regulation of beneficial physiological processes, efflux of metal ions produced by nanomaterials as well as point mutation (Sedlak et al. 2012; Tripathi et al. 2017; Niño-Martínez et al. 2019). Andries et al. (2016) reported that *Phanerochaete chryosporium*, a lignin degrading fungus, defends itself against various chemicals released by AuNPs stressors (light sensitive) by enhancing activities of enzymes such as catalase, lipid peroxidase, superoxide dismutase and malondialdehyde enzymes.

Evolutions of pathways in response to stress have been reported by Hobman and Crossman (2015) in some bacteria. This involves elimination of metal ions from the cells by means of efflux pumps (e.g. copper) or expression of genes which renders metal resistance (e.g. *sil* genes on pMG101). Meier et al. (2020) reported in their study that *Rhodanobacter* and *Bradyrhizobium* were resistant to AgNPs. Several genes (resistant to metals) were found to be involved.

EFFECTS ON SOIL FUNCTIONS AND ECOSYSTEM SERVICES

Nanomaterials (NMs) into the soil environment has raised concerns regarding their potential impact on soil functions and ecosystem services. Nanomaterials, due to their unique physicochemical properties, can interact with soil microbes in ways that may alter the fundamental processes that underpin soil health and ecosystem stability (Singh et al., 2020). Understanding these effects is crucial for balancing the benefits of nanotechnology with the need to maintain soil functionality and ecosystem services (Mueller and Nowack, 2010).

Soil Nutrient Cycling

Soil nutrient cycling is a critical function performed by microbial communities. Microbes facilitate the decomposition of organic matter, release nutrients, and contribute to the formation of humus (Mishra et al., 2017). The presence of nanomaterials can influence these processes in several ways. Some studies have shown that certain nanomaterials, such as nano-sized fertilizers, can enhance microbial activity and nutrient availability (Kumar et al., 2018). These materials can increase the efficiency of nutrient uptake by plants and stimulate microbial growth by providing essential nutrients. Conversely, other nanomaterials, particularly those containing heavy metals or exhibiting high reactivity, can be toxic to soil microbes (Ahmed et al., 2023). They can inhibit microbial growth, reduce enzymatic activity, and disrupt the microbial communities responsible for nutrient cycling. For instance, silver nanoparticles are known for their antimicrobial properties, which can significantly impact microbial populations and functions (Ameen et al., 2021).

Soil Structure and Stability

Soil structure and stability are essential for water infiltration, root penetration, and preventing soil erosion. Microbial activities, such as the production of extracellular polymeric substances (EPS), play a key role in maintaining soil structure (Kumar et al., 2018). Certain nanomaterials can enhance soil aggregation and stability by promoting microbial EPS production. This can improve soil structure, leading to better water retention and reduced erosion (Mishra et al., 2017). Alternatively, nanomaterials that are toxic to microbes can hinder EPS production and disrupt soil aggregation. This can lead to a decline in soil stability, increased erosion, and poorer soil health (Ahmed et al., 2023).

Soil Organic Matter Decomposition

The decomposition of soil organic matter (SOM) is another critical function facilitated by microbial communities. In some cases, nanomaterials may stimulate the microbial breakdown of organic matter by providing catalytic surfaces or by enhancing microbial enzyme activity (Kumar et al., 2018). This can accelerate the decomposition process and increase the availability of nutrients. Toxic nanomaterials can impair microbial activity and enzyme function, slowing down the decomposition of organic matter (Ameen et al., 2021). This can lead to the accumulation of undecomposed organic matter, altering soil carbon dynamics and potentially leading to reduced soil fertility over time (Mueller and Nowack, 2010).

Impact on Ecosystem Services

The alterations in soil functions due to nanomaterial exposure can have broad implications for ecosystem services. Ecosystem services provided by soils include supporting plant growth, regulating water cycles, and sequestering carbon (Kumar et al., 2018). Positive interactions between nanomaterials and soil microbes can enhance nutrient availability and promote plant growth. However, negative interactions can reduce nutrient cycling and soil fertility, adversely affecting plant health (Mueller and Nowack, 2010). Improved soil structure and stability through positive microbial interactions can enhance water infiltration and retention (Mishra et al., 2017). Conversely, disruption of microbial communities can lead to poor soil structure, reduced water infiltration, and increased runoff. Soil microbes play a crucial role in carbon sequestration by decomposing organic matter and stabilizing carbon in soil aggregates (Kumar et al., 2018). Nanomaterials that enhance microbial activity can promote carbon sequestration, while those that inhibit microbial functions can disrupt carbon storage processes, leading to increased carbon release and contributing to climate change (Ameen et al., 2021).

BALANCING RISKS AND BENEFITS

Ecotoxicological Concerns of Nanomaterials

Ecotoxicology concerns the study of the noxious effects of substances on ecosystems and their inhabitants. When it comes to nanomaterials, these concerns become especially significant due to their unequalled properties and potential interactions with biological systems. Nanomaterials often have high surface area-to-volume ratios, which can heighten their responsiveness and interactions with organisms (Kumar et al., 2022). Nanomaterials may heap in the environment, affecting their bioavailability to organisms (Lead et al., 2018). Nanomaterials can directly impact organisms through physical interactions e.g., cell membrane penetration or chemical toxicity e.g., the release of toxic ions (Sukhanova et al., 2018). Even at low concentrations, nanomaterials can bring on sub-lethal effects such as oxidative stress, genotoxicity and disruption of cellular processes (Canesi et al., 2019). Nanomaterials may enter food chains through various pathways e.g., plant uptake and ingestion by filter feeders, potentially leading to bio-magnification at higher trophic levels (Wong et al., 2013). Their persistence and accumulation in organisms can magnify their toxic effects over time (Abbas et al., 2024). Nanomaterials may move through environmental compartments (air, water, soil) differently than bulk materials, affecting their distribution and exposure pathways (Keller et al., 2024).

Current ecotoxicological testing methods and regulatory frameworks may not adequately address nanomaterial-specific risks (Nederstigt et al., 2024). Understanding interactions between nanomaterials and biological molecules (proteins, DNA) is crucial for assessing their toxicity (Medici et al., 2021). The rapid development of novel nanomaterials requires ongoing assessment to anticipate and mitigate potential environmental impacts (Isibor, 2024).

Mitigation Strategies for Minimizing Adverse Effects

Mitigation strategies for minimizing adverse effects of nanomaterials on the environment and organisms are important for ensuring sustainable development and safe deployment of nanotechnology. These strategies aim to proactively address potential risks and reduce environmental and human health impacts. Tailoring nanomaterial properties (e.g., size, shape, surface chemistry) to reduce toxicity while maintaining functionality (Sajid et al., 2022). Encapsulating nanomaterials within inert shells to minimize direct contact with biological systems (Stiegler et al., 2020). Developing biodegradable nanomaterials that break down into non-toxic components in the environment (Aswathi et al., 2023). Implementing engineering controls to prevent nanomaterial release during production, use and disposal (Mazari et al., 2021). Using filters and barriers designed to capture nanomaterials in wastewater and air filtration systems (Khan et al., 2021).

Developing specific guidelines and regulatory frameworks for nanomaterial safety assessments, labeling and permissible limits is important for minimizing adverse effects (Subhan et al., 2022). Harmonizing standards and sharing best practices internationally to ensure consistent and effective regulation (Subhan et al., 2024). Educating stakeholders (industry, researchers, policymakers and public) about nanomaterial risks, safe handling practices, and responsible use (Kirkegaard et al., 2020). Communicating risks, benefits and uncertainties of nanomaterials to the public in a clear and accessible manner (Drobne et al., 2023). Proactively assessing the environmental implications of new and emerging nanomaterials before widespread deployment (Garvey, 2016). Effective mitigation requires collaboration across disciplines e.g., chemistry, biology and engineering to address multifaceted challenges (Muneer, 2023).

Opportunities for Sustainable Nanotechnology in Soil Science

Sustainable nanotechnology in soil science" refers to applying nanotechnology principles and materials to enhance soil health, productivity and sustainability while minimizing adverse environmental impacts. This approach leverages nanomaterials unique properties to address challenges in agriculture, environmental remediation and soil management (Kumar et al., 2018).

Nanomaterials such as nanoparticles and Nanocomposites can be engineered to efficiently adsorb, degrade or immobilize contaminants e.g., heavy metals and organic pollutants in the soil (Mukhopadhyay et al., 2022). Their high surface area-to-volume ratio and reactivity can enhance remediation efficiency compared to traditional methods, reducing the need for large-scale soil excavation and disposal (Prakash et al., 2023). Nanoencapsulation of nutrients can provide controlled release, optimizing plant nutrient uptake and reducing fertilizer leaching into groundwater (Singh et al., 2024). Nanomaterials can positively influence soil microbial communities, promoting beneficial microbial activities e.g. nitrogen fixation and organic matter decomposition (Kalwani et al., 2022). Nanostructures can improve soil aggregation and porosity, enhancing water infiltration, root penetration and overall soil aeration

(Attia & Elsheery, 2020). Nanosensors can monitor soil parameters (e.g., moisture content, nutrient levels, pH) with high sensitivity and spatial resolution, enabling precision agriculture practices (Beegum & Das, 2022). Nanomaterials can be used in smart delivery systems for targeted application of agrochemicals (e.g., pesticides and herbicides), minimizing environmental impact and optimizing effectiveness (de Oliveira et al., 2020; An et al., 2022).

Nanomaterials e.g., carbon nanotubes and graphene can enhance soil carbon sequestration by stabilizing organic matter and improving soil carbon storage capacity (Ferreira et al., 2022). Nanotechnology-based approaches can mitigate greenhouse gas emissions from soils through improved nutrient management and reduced soil disturbance (Yadav et al., 2023). Understanding the environmental behavior, fate and potential toxicity of nanomaterials in soils is critical to ensuring their safe and sustainable deployment (Tran et al., 2023). Collaboration between nanotechnologists, soil scientists, agronomists, ecologists and policymakers is essential for advancing sustainable nanotechnology in soil science (Dhanapal et al., 2024). Educating farmers, industry stakeholders and the public about the potential benefits and risks of nanotechnology in soil science to foster informed decision-making and responsible use (Rajput et al., 2024).

APPLICATIONS OF MICROBIAL RESPONSES TO NANOMATERIALS IN SOIL ENVIRONMENTS

Microbial responses to nanomaterials in soil environments present a complex and dynamic interplay, critical for understanding the ecological impact and potential applications of nanotechnology in agriculture and environmental management (Mishra et al., 2017). Nanomaterials, due to their unique physicochemical properties, can influence soil microbial communities by altering microbial activity, diversity, and functions. These interactions can be beneficial or detrimental depending on the type, concentration, and characteristics of the nanomaterials (Singh et al., 2020).

For instance, certain nanoparticles can enhance microbial processes such as nutrient cycling, pollutant degradation, and plant growth promotion. Silver nanoparticles, for example, have been shown to possess antimicrobial properties, potentially reducing harmful microbial populations and promoting plant health (Singh et al., 2020). Conversely, excessive accumulation of nanomaterials can disrupt microbial communities, leading to toxicity and impaired soil functions (Mishra et al., 2017).

Understanding these microbial responses is essential for the safe and sustainable application of nanotechnology in soil environments (Mishra et al., 2017). Research in this area focuses on identifying the mechanisms of interaction, assessing the ecological risks, and developing guidelines for the appropriate use of nanomaterials (Singh et al., 2020). Balancing the beneficial and adverse effects requires a comprehensive evaluation of long-term impacts, regulatory frameworks, and the development of nanomaterials that are environmentally benign and compatible with soil microbial ecosystems (Ahmed et al., 2023).

CONCLUSION

The exploration of microbial responses to nanomaterials in the soil environment reveals a complex and dynamic interaction that is critical for understanding the broader implications of nanotechnology in soil science. This chapter highlights the dual role of nanomaterials as both potential threats and beneficial agents in soil ecosystems. While certain nanomaterials can adversely affect microbial communities, leading to disruptions in soil health and fertility, others can enhance microbial activity, promoting soil remediation and agricultural productivity.

Balancing these effects requires a nuanced approach, integrating comprehensive risk assessments with innovative applications. Future research should focus on elucidating the mechanisms underlying microbial responses to different types and concentrations of nanomaterials, alongside long-term studies to assess their cumulative impacts. By achieving this balance, we can harness the benefits of nanotechnology while safeguarding the vital microbial diversity that underpins soil health and sustainability.

REFERENCES

Abbas, Q., Liu, G., Ali, M. U., Han, J., El-Naggar, A., & Yousaf, B. (2024). Nanomaterials in environment: release, compartmental transformation, and nanotoxicity. In *Handbook of Nanomaterials* (Vol. 2, pp. 745–789). Elsevier. DOI: 10.1016/B978-0-323-95513-3.00010-1

Abd-Alla, M. H., Nafady, N. A., & Khalaf, D. M. (2016). Assessment of silver nanoparticles contamination on faba bean-Rhizobium leguminosarum bv. viciae-Glomus aggregatum symbiosis: Implications for induction of autophagy process in root nodule. *Agriculture, Ecosystems & Environment*, *218*, 163–177. DOI: 10.1016/j.agee.2015.11.022

Acharya, D., Singha, K. M., Pandey, P., Mohanta, B., Rajkumari, J., & Singha, L. P. (2018). Shape dependent physical mutilation and lethal effects of silver nanoparticles on bacteria. *Scientific Reports*, *8*(1), 201. DOI: 10.1038/s41598-017-18590-6 PMID: 29317760

Ahmed, A., He, P., He, P., Wu, Y., He, Y., & Munir, S. (2023). Environmental effect of agriculture-related manufactured nano-objects on soil microbial communities. *Environment International*, *173*, 107819. DOI: 10.1016/j.envint.2023.107819 PMID: 36842382

Ahmed, B., Ameen, F., Rizvi, A., Ali, K., Sonbol, H., Zaidi, A., Khan, M. S., & Musarrat, J. (2020). Destruction of cell topography, morphology, membrane, inhibition of respiration, biofilm formation, and bioactive molecule production by nanoparticles of Ag, ZnO, CuO, TiO2, and Al2O3 toward beneficial soil bacteria. *ACS Omega*, *5*(14), 7861–7876. DOI: 10.1021/acsomega.9b04084 PMID: 32309695

Alkubaisi, N. A., & Aref, N. M. A. (2017). Dispersed gold nanoparticles potentially ruin gold barley yellow dwarf virus and eliminate virus infectivity hazards. *Applied Nanoscience*, *7*(1-2), 31–40. DOI: 10.1007/s13204-016-0540-0

Ameen, F., Alsamhary, K., Alabdullatif, J. A., & ALNadhari, S. (2021). A review on metal-based nanoparticles and their toxicity to beneficial soil bacteria and fungi. *Ecotoxicology and Environmental Safety*, *213*, 112027. DOI: 10.1016/j.ecoenv.2021.112027 PMID: 33578100

An, C., Sun, C., Li, N., Huang, B., Jiang, J., Shen, Y., Wang, C., Zhao, X., Cui, B., Wang, C., Li, X., Zhan, S., Gao, F., Zeng, Z., Cui, H., & Wang, Y. (2022). Nanomaterials and nanotechnology for the delivery of agrochemicals: Strategies towards sustainable agriculture. *Journal of Nanobiotechnology*, *20*(1), 11. DOI: 10.1186/s12951-021-01214-7 PMID: 34983545

Andries, M., Pricop, D., Oprica, L., Creanga, D. E., & Iacomi, F. (2016). The effect of visible light on gold nanoparticles and some bioeffects on environmental fungi. *International Journal of Pharmaceutics*, *505*(1-2), 255–261. DOI: 10.1016/j.ijpharm.2016.04.004 PMID: 27063667

Arnold, M. C., Badireddy, A. R., Wiesner, M. R., Di Giulio, R. T., & Meyer, J. N. (2013). Cerium oxide nanoparticles are more toxic than equimo-lar bulk cerium oxide in Caenorhabditis elegans. *Archives of Environmental Contamination and Toxicology*, *65*(2), 224–233. DOI: 10.1007/s00244-013-9905-5 PMID: 23619766

Asadishad, B., Chahal, S., Akbari, A., Cianciarelli, V., Azodi, M., Ghoshal, S., & Tufenkji, N. (2018). Amendment of agricultural soil with metal nanoparticles: Effects on soil enzyme activity and microbial community composition. *Environmental Science & Technology*, *52*(4), 1908–1918. DOI: 10.1021/acs.est.7b05389 PMID: 29356510

Aswathi, V. P., Meera, S., Maria, C. A., & Nidhin, M. (2023). Green synthesis of nanoparticles from biodegradable waste extracts and their applications: A critical review. *Nanotechnology for Environmental Engineering*, *8*(2), 377–397. DOI: 10.1007/s41204-022-00276-8

. Attia, T. S., & Elsheery, N. I. (2020). Nanomaterials: scope, applications, and challenges in agriculture and soil reclamation. Sustainable Agriculture Reviews 41: Nanotechnology for Plant Growth and Development, 1-39.

Bandyopadhyay, S., Plascencia-Villa, G., Mukherjee, A., Rico, C. M., José-Yacamán, M., Peralta-Videa, J. R., & Gardea-Torresdey, J. L. (2015). Comparative phytotoxicity of ZnO NPs, bulk ZnO, and ionic zinc onto the alfalfa plants symbiotically associated with Sinorhizobium meliloti in soil. *The Science of the Total Environment*, *515*, 60–69. DOI: 10.1016/j.scitotenv.2015.02.014 PMID: 25698520

Bashir, I., Lone, F. A., Bhat, R. A., Mir, S. A., Dar, Z. A., & Dar, S. A. (2020). Concerns and threats of contamination on aquatic ecosystems. In Hakeem, K., Bhat, R., & Qadri, H. (Eds.), *Bioremediation and Biotechnology* (Vol. 27, pp. 1–26). Springer. DOI: 10.1007/978-3-030-35691-0_1

Beegum, S., & Das, S. (2022). Nanosensors in agriculture. In *Agricultural Nanobiotechnology* (pp. 465–478). Woodhead Publishing. DOI: 10.1016/B978-0-323-91908-1.00012-2

Bindraban, P. S., Dimkpa, C., Nagarajan, L., Roy, A., & Rab-binge, R. (2015). Revisiting fertilisers and fertilisation strategies for improved nutrient uptake byplants. *Biology and Fertility of Soils*, *51*(8), 897–911. DOI: 10.1007/s00374-015-1039-7

Brunetti, G., Donner, E., Laera, G., Sekine, R., Scheckel, K. G., Khaksar, M., Vasilev, K., De Mastro, G., & Lombi, E. (2015). Fate of zinc and silver engineered nanoparticles in sewerage networks. *Water Research*, *77*, 72–84. DOI: 10.1016/j.watres.2015.03.003 PMID: 25841090

Canesi, L., Auguste, M., & Bebianno, M. J. (2019). Sublethal effects of nanoparticles on aquatic invertebrates, from molecular to organism level. In *Ecotoxicology of Nanoparticles in Aquatic Systems* (pp. 38–61). CRC press. DOI: 10.1201/9781315158761-2

Cao, C., Huang, J., Yan, C., Liu, J., Hu, Q., & Guan, W. (2018). Shifts of system performance and microbial community structure in a constructed wetland after exposing silver nanoparticles. *Chemosphere*, *199*, 661–669. DOI: 10.1016/j.chemosphere.2018.02.031 PMID: 29471236

Chhipa, H. (2017). Nanofertilizers and Nanopesticides for Agriculture. *Environmental Chemistry Letters*, *15*(1), 15–22. DOI: 10.1007/s10311-016-0600-4

Choi, O., Deng, K. K., Kim, N. J., Ross, L.Jr, Surampalli, R. Y., & Hu, Z. (2008). The inhibitory effects of silver nanoparticles, silver ions, and silver chloride colloids on microbial growth. *Water Research*, *42*(12), 3066–3074. DOI: 10.1016/j.watres.2008.02.021 PMID: 18359055

Chouhan, R. S., Qureshi, A., Yagci, B., Gulgun, M. A., Ozguz, V., & Niazi, J. H. (2016). Biotransformation of multi-walled carbon nanotubes mediated by nanomaterial resistant soil bacteria. *Chemical Engineering Journal*, *298*, 1–9. DOI: 10.1016/j.cej.2016.04.019

Concha-Guerrero, S. I., Brito, E. M. S., Piñón-Castillo, H. A., Tarango-Rivero, S. H., Caretta, C. A., Luna-Velasco, A., Duran, R., & Orrantia-Borunda, E. (2014). Effect of CuO nanoparticles over isolated bacterial strains from agricultural soil. *Journal of Nanomaterials*, *2014*(1), 148743. DOI: 10.1155/2014/148743

Cornelis, G., Hund-Rinke, K., Kuhlbusch, T., Van den Brink, N., & Nickel, C. (2014). Fate and bioavailability of engineered nanoparticles in soils: A review. *Critical Reviews in Environmental Science and Technology*, *44*(24), 2720–2764. DOI: 10.1080/10643389.2013.829767

Cullen, L. G., Tilston, E. L., Mitchell, G. R., Collins, C. D., & Shaw, L. J. (2011). Assessing the impact of nano-and micro-scale zerovalent iron particles on soil microbial activities: Particle reactivity interferes with assay conditions and interpretation of genuine microbial effects. *Chemosphere*, *82*(11), 1675–1682. DOI: 10.1016/j.chemosphere.2010.11.009 PMID: 21122887 de Oliveira, C., Mulinari, J., Reichert, F., & Júnior, A. (2020, September). Nano-delivery systems of pesticides active agents for agriculture applications-An overview. In Ciência, Tecnologia e Inovação: Do Campo à Mesa, Proceeding of International Agribusiness Congress, Recife, Brazil (pp. 25-27). DOI: 10.31692/ICIAGRO.2020.0051

Del Buono, D., Luzi, F., & Puglia, D. (2021). Lignin nanoparticles: A promising tool to improve maize physiological, biochemical, and chemical traits. *Nanomaterials (Basel, Switzerland)*, *11*(4), 846. DOI: 10.3390/nano11040846 PMID: 33810279

Dhanapal, A. R., Thiruvengadam, M., Vairavanathan, J., Venkidasamy, B., Easwaran, M., & Ghorbanpour, M. (2024). Nanotechnology approaches for the remediation of agricultural polluted soils. *ACS Omega*, *9*(12), 13522–13533. DOI: 10.1021/acsomega.3c09776 PMID: 38559935

Drobne, D., Ciornii, D., Hodoroaba, V. D., Bohmer, N., Novak, S., Kranjc, E., Kononenko, V., & Reuther, R. (2023). Knowledge, Information, and Data Readiness Levels (KaRLs) for Risk Assessment, Communication, and Governance of Nano-, New, and Other Advanced Materials. *Global Challenges (Hoboken, NJ)*, *7*(7), 2200211. DOI: 10.1002/gch2.202200211 PMID: 37483420

Du, W., Sun, Y., Ji, R., Zhu, J., Wu, J., & Guo, H. (2011). TiO 2 and ZnO nanoparticles negatively affect wheat growth and soil enzyme activities in agricultural soil. *Journal of Environmental Monitoring*, *13*(4), 822–828. DOI: 10.1039/c0em00611d PMID: 21267473

Dutta, H., & Dutta, A. (2016). The microbial aspect of climate change. *Energy, Ecology & Environment*, *1*(4), 209–232. DOI: 10.1007/s40974-016-0034-7

Elemike, E., Uzoh, I., Onwudiwe, D., & Babalola, O. (2019). The role of nanotechnology in the fortifica-tion of plant nutrients and improvement of crop production. *Applied Sciences (Basel, Switzerland)*, *9*(3), 499. DOI: 10.3390/app9030499

Fajardo, C., Saccà, M. L., Costa, G., Nande, M., & Martin, M. (2014). Impact of Ag and Al2O3 nanoparticles on soil organisms: In vitro and soil experiments. *The Science of the Total Environment*, *473*, 254–261. DOI: 10.1016/j.scitotenv.2013.12.043 PMID: 24374587

Fan, R., Huang, Y. C., Grusak, M. A., Huang, C. P., & Sherrier, D. J. (2014). Effects of nano-TiO2 on the agronomically-relevant Rhizobium–legume symbiosis. *The Science of the Total Environment*, *466*, 503–512. DOI: 10.1016/j.scitotenv.2013.07.032 PMID: 23933452

Feng, Y., Cui, X., He, S., Dong, G., Chen, M., Wang, J., & Lin, X. (2013). The role of metal nanoparticles in influencing arbuscular mycorrhizal fungi effects on plant growth. *Environmental Science & Technology*, *47*(16), 9496–9504. DOI: 10.1021/es402109n PMID: 23869579

Ferreira, G. M. D., Ferreira, G. M. D., Franca, J. R., & Soares, J. R. (2022). Carbonaceous Materials for Nanoremediation of Polluted and Nutrient-Depleted Soils. In Nanotechnology for Environmental Pollution Decontamination (pp. 345-408). Apple Academic Press. DOI: 10.1201/9781003279563-15

Frenk, S., Ben-Moshe, T., Dror, I., Berkowitz, B., & Minz, D. (2013). Effect of metal oxide nanoparticles on microbial community structure and function in two different soil types. *PLoS One*, *8*(12), e84441. DOI: 10.1371/journal.pone.0084441 PMID: 24349575

Fu, P. P., Xia, Q., Hwang, H. M., Ray, P. C., & Yu, H. (2014). Mechanisms of Nanotoxicity: Generation of reactive oxygen species. *Yao Wu Shi Pin Fen Xi*, *22*(1), 64–75. DOI: 10.1016/j.jfda.2014.01.005 PMID: 24673904

Gambino, M., Marzano, V., Villa, F., Vitali, A., Vannini, C., Landini, P., & Cappitelli, F. (2015). Effects of sublethal doses of silver nanoparticles on Bacillus subtilis planktonic and sessile cells. *Journal of Applied Microbiology*, *118*(5), 1103–1115. DOI: 10.1111/jam.12779 PMID: 25702880

Gan, N., Yang, X., Xie, D., Wu, Y., & Wen, W. A. (2010). Disposable organophosphorus pesticides enzyme biosensor based on magnetic composite nanoparticles modified screen printed carbon electrode. *Sensors (Basel)*, *10*(1), 625–638. DOI: 10.3390/s100100625 PMID: 22315558

Garvey, T. (2016). *Identifying the Potential environmental impacts of engineered nanomaterials*. Rochester Institute of Technology.

Ghouri, M. Z., Khan, Z., Khan, S. H., Ismail, M., Aftab, S. O., Sultan, Q., & Ahmad, A. (2020). Nanotechnology: Transformation of agriculture and food security. *Bioscience*, *3*, 19.

Gladkova, M. M., & Terekhova, V. A. (2013). Engineered nanomaterials in soil: Sources of entry and migration pathways. *Moscow University Soil Science Bulletin*, *68*(3), 129–134. DOI: 10.3103/S0147687413030046

Gogos, A., Moll, J., Klingenfuss, F., van der Heijden, M., Irin, F., Green, M. J., Zenobi, R., & Bucheli, T. D. (2016). Vertical transport and plant uptake of nanoparticles in a soil mesocosm experiment. *Journal of Nanobiotechnology*, *14*(1), 40. DOI: 10.1186/s12951-016-0191-z PMID: 27278090

Gonçalves, S. P. C., Delite, F. D., Côa, F., Neto, L. L. R., da Silva, G. H., Bortolozzo, L. D., Ferreira, A. G., de Medeiros, A. M. Z., Strauss, M., & Martinez, D. S. T. (2019). Chapter 7 - Biotransformation of Nanomaterials in the Soil Environment: Nanoecotoxicology and Nanosafety Implications, Editor(s): do Nascimento RF, Ferreira OP, De Paula AJ, Neto VDOS, In Advanced Nanomaterials, Nanomaterials Applications for Environmental Matrices, Elsevier, pp 265-304, DOI: 10.1016/B978-0-12-814829-7.00007-C

Grün, A. L., Manz, W., Kohl, Y. L., Meier, F., Straskraba, S., Jost, C., Drexel, L., & Emmerling, C. (2019). Impact of silver nanoparticles (AgNP) on soil microbial community depending on functionalization, concentration, exposure time, and soil texture. *Environmental Sciences Europe*, *31*(1), 1–22. DOI: 10.1186/s12302-019-0196-y

Guan, H., Chi, D., Yu, J., & Li, X. (2008). A novel photodegradable insecticide: Preparation, characterization and properties evaluation of nano-Imidacloprid. *Pesticide Biochemistry and Physiology*, *92*(2), 83–91. DOI: 10.1016/j.pestbp.2008.06.008

Hachicho, N., Hoffmann, P., Ahlert, K., & Heipieper, H. J. (2014). Effect of silver nanoparticles and silver ions on growth and adaptive response mechanisms of Pseudomonas putida mt-2. *FEMS Microbiology Letters*, *355*(1), 71–77. DOI: 10.1111/1574-6968.12460 PMID: 24801753

Hao, L., Chen, L., Hao, J., & Zhong, N. (2013). Bioaccumulation and sub-acutetoxicity of zinc oxide nanoparticles in juvenile carp (Cyprinuscarpio): A comparative study with its bulk counterparts. *Ecotoxicology and Environmental Safety*, *91*, 52–60. DOI: 10.1016/j.ecoenv.2013.01.007 PMID: 23375439

Hartmann, N. I. B., Skjolding, L. M., Hansen, S. F., Baun, A., Kjølholt, J., & Gottschalk, F. (2014). *Environmental fate and behaviour of nanomaterials: New knowledge on important transfomation processes*. Danish Environmental Protection Agency.

Hartmann, T., Mühling, M., Wolf, A., Mariana, F., Maskow, T., Mertens, F., Neu, T. R., & Lerchner, J. (2013). A chip-calorimetric approach to the analysis of Ag nanoparticle caused inhibition and inactivation of beads-grown bacterial biofilms. *Journal of Microbiological Methods*, *95*(2), 129–137. DOI: 10.1016/j.mimet.2013.08.003 PMID: 23968644

He, S., Feng, Y., Ni, J., Sun, Y., Xue, L., Feng, Y., Yu, Y., Lin, X., & Yang, L. (2016). Different responses of soil microbial metabolic activity to silver and iron oxide nanoparticles. *Chemosphere*, *147*, 195–202. DOI: 10.1016/j.chemosphere.2015.12.055 PMID: 26766356

Hee Joo, S., & Cheng, I. F. (2006). *Nanotechnology for environmental remediation* (Hee Joo, S., & Cheng, I. F., Eds.). Springer. DOI: 10.1007/b137366

Hobman, J. L., & Crossman, L. C. (2015). Bacterial antimicrobial metal ion resistance. *Journal of Medical Microbiology*, *64*(5), 471–497. DOI: 10.1099/jmm.0.023036-0 PMID: 25418738

Hoppe, M., Mikutta, R., Utermann, J., Duijnisveld, W., Kaufhold, S., Stange, C. F., & Guggenberger, G. (2015). Remobilization of sterically stabilized silver nanoparticles from farmland soils determined by column leaching. *European Journal of Soil Science*, *66*(5), 898–909. DOI: 10.1111/ejss.12270

Huang, D., Liu, H., Yin, X., Yuan, F., Song, B., Teng, Y., Ding, Y., Chen, Q., & Liu, T. (2024). A nanoformulation that enhance the efficiency of camptothecin against lepidopteran pests by controlled release and 'poison bait' strategies. *Chemical Engineering Journal*, *486*, 150358. DOI: 10.1016/j.cej.2024.150358

Isibor, P. O. (2024). Regulations and policy considerations for nanoparticle safety. In *Environmental Nanotoxicology: Combatting the Minute Contaminants* (pp. 295–316). Springer Nature Switzerland. DOI: 10.1007/978-3-031-54154-4_14

Jacoby, R., Peukert, M., Succurro, A., Koprivova, A., & Kopriva, S. (2017). The role of soil microorganisms in plant mineral nutrition—Current knowledge and future directions. *Frontiers in Plant Science*, *8*, 1617. DOI: 10.3389/fpls.2017.01617 PMID: 28974956

. Joshi, M., Bhattacharyya, A., & Ali, S. W. (2008). Characterization techniques for nanotechnology applications in textiles.

Judy, J. D., Kirby, J. K., Creamer, C., McLaughlin, M. J., Fiebiger, C., Wright, C., Cavagnaro, T., & Bertsch, P. M. (2015). Effects of silver sulfide nanomaterials on mycorrhizal colonization of tomato plants and soil microbial communities in biosolid-amended soil. *Environmental Pollution*, *206*, 256–263. DOI: 10.1016/j.envpol.2015.07.002 PMID: 26196315

Kahlon, S. K., Sharma, G., Julka, J. M., Kumar, A., Sharma, S., & Stadler, F. J. (2018). Impact of heavy metals and nanoparticles on aquatic biota. *Environmental Chemistry Letters*, *16*(3), 919–946. DOI: 10.1007/s10311-018-0737-4

Kalwani, M., Chakdar, H., Srivastava, A., Pabbi, S., & Shukla, P. (2022). Effects of nanofertilizers on soil and plant-associated microbial communities: Emerging trends and perspectives. *Chemosphere*, *287*, 132107. DOI: 10.1016/j.chemosphere.2021.132107 PMID: 34492409

Kaman, P. K., & Dutta, P. (2019). Synthesis, characterization and antifungal activity of biosynthesized silver nanoparticle. *Indian Phytopathology*, *72*(1), 79–88. DOI: 10.1007/s42360-018-0081-4

. Karimi, E., & Mohseni Fard, E. (2017). Nanomaterial effects on soil microorganisms. *Nanoscience and plant–soil systems*, 137-200.

Kasel, D., Bradford, S. A., Simunek, J., Putz, T., Vereecken, H., & Klumpp, E. (2013). Limited transport of functionalized multi-walled carbon nanotubes in two natural soils. *Environmental Pollution*, *180*, 152–158. DOI: 10.1016/j.envpol.2013.05.031 PMID: 23770315

Keller, A. A., Wang, H., Zhou, D., Lenihan, H. S., Cherr, G., Cardinale, B. J., Miller, R., & Ji, Z. (2010). Stability and aggregation of metal oxide nanoparticles in natural aqueous matrices. *Environmental Science & Technology*, *44*(6), 1962–1967. DOI: 10.1021/es902987d PMID: 20151631

Keller, A. A., Zheng, Y., Praetorius, A., Quik, J. T., & Nowack, B. (2024). Predicting environmental concentrations of nanomaterials for exposure assessment-a review. *NanoImpact*, *33*, 100496. DOI: 10.1016/j.impact.2024.100496 PMID: 38266914

Khan, S., Naushad, M., Al-Gheethi, A., & Iqbal, J. (2021). Engineered nanoparticles for removal of pollutants from wastewater: Current status and future prospects of nanotechnology for remediation strategies. *Journal of Environmental Chemical Engineering*, *9*(5), 106160. DOI: 10.1016/j.jece.2021.106160

Khan, S. B., Faisal, M., Rahman, M. M., & Jamal, A. (2011). Exploration of CeO2nanoparticles as a chemi-sensor and photo-catalyst for environ-mental applications. *The Science of the Total Environment*, *409*(15), 2987–2992. DOI: 10.1016/j.scitotenv.2011.04.019 PMID: 21570707

Khanna, K., Kohli, S. K., Handa, N., Kaur, H., Ohri, P., Bhardwaj, R., Yousaf, B., Rinklebe, J., & Ahmad, P. (2021). Enthralling the impact of engineered nanoparticles on soil microbiome: A concentric approach towards environmental risks and cogitation. *Ecotoxicology and Environmental Safety*, *222*, 112459. DOI: 10.1016/j.ecoenv.2021.112459 PMID: 34217114

Kirkegaard, M. L., Kines, P., Jeschke, K. C., & Jensen, K. A. (2020). Risk perceptions and safety cultures in the handling of nanomaterials in academia and industry. *Annals of Work Exposures and Health*, *64*(5), 479–489. DOI: 10.1093/annweh/wxaa022 PMID: 32155241

Klaine, S. J., Alvarez, P. J., Batley, G. E., Fernandes, T. F., Handy, R. D., Lyon, D. Y., Mahendra, S., McLaughlin, L. J., & Lead, J. R. (2008). Nanomaterials in the environment: Behavior, fate, bioavailability, and effects. *Environmental Toxicology and Chemistry*, *27*(9), 1825–1851. DOI: 10.1897/08-090.1 PMID: 19086204

Klitzke, S., Metreveli, G., Peters, A., Schaumann, G. E., & Lang, F. (2015). The fate of silver nanoparticles in soil solution—Sorption of solutes and aggregation. *The Science of the Total Environment*, *535*, 54–60. DOI: 10.1016/j.scitotenv.2014.10.108 PMID: 25434472

Kumar, A., Gupta, K., Dixit, S., Mishra, K., & Srivastava, S. (2018). A review on positive and negative impacts of nanotechnology in agriculture. *International Journal of Environmental Science and Technology*, *16*(4), 2175–2184. DOI: 10.1007/s13762-018-2119-7

Kumar, C. V., Karthick, V., Kumar, V. G., Inbakandan, D., Rene, E. R., Suganya, K. U., & Sowmiya, P. (2022). The impact of engineered nanomaterials on the environment: Release mechanism, toxicity, transformation, and remediation. *Environmental Research*, *212*, 113202. DOI: 10.1016/j.envres.2022.113202 PMID: 35398077

. Kumar R. (2020). Application of nanotechnology incrop disease management. Universe Interna-tional Journal of Interdisciplinary Research 1.

. Lambers, H., Mougel, C., Jaillard, B., & Hinsinger, P. (2009). Plant-microbe-soil interactions in the rhizosphere: an evolutionary perspective. https://doi.org/DOI: 10.1007/s11104-009-0042-x

Lasagna-Reeves, C., Gonzalez-Romero, D., Barria, M. A., Olmedo, I., Clos, A., Ramanujam, V. S., Urayama, A., Vergara, L., Kogan, M. J., & Soto, C. (2010). Bioaccumulation and toxicity of gold nano-particles after repeated administration in mice. *Biochemical and Biophysical Research Communications*, *393*(4), 649–655. DOI: 10.1016/j.bbrc.2010.02.046 PMID: 20153731

Lead, J. R., Batley, G. E., Alvarez, P. J. J., Croteau, M.-N., Handy, R. D., McLaughlin, M. J., Judy, J. D., & Schirmer, K. (2018). Nanomaterials in the environment: Behavior, fate, bioavailability and effects—An updated review. *Environmental Toxicology and Chemistry*, *37*(8), 2029–2063. DOI: 10.1002/etc.4147 PMID: 29633323

Lecoanet, H. F., & Wiesner, M. R. (2004). Velocity effects on fullerene and oxide nanoparticle deposition in porous media. *Environmental Science & Technology*, *38*(16), 4377–4382. DOI: 10.1021/es035354f PMID: 15382867

Li, S., Liu, X. Q., Wang, F. Y., & Miao, Y. F. (2015). Effects of ZnO Nanoparticles, $ZnSO_4$ and Arbuscular Mycorrhizal Fungus on the Growth of Maize. *Huan Jing ke Xue=. Huan Jing Ke Xue*, *36*(12), 4615–4622. PMID: 27012001

Liu, T., Qian, K., Fan, J., Qin, Z., Yi, L., Wang, J., & Dou, W. (2024). Nano-based smart pesticide formulations utilizing starch nanocrystals: Achieving high stability and extended effective duration. *Industrial Crops and Products*, *214*, 118470. DOI: 10.1016/j.indcrop.2024.118470

Lombi, E., Donner, E., Tavakkoli, E., Turney, T. W., Naidu, R., Miller, B. M., & Scheckel, K. G. (2012). Fate of zinc oxide nanoparticles during anaerobic digestion of wastewater and post-treatment processing of sewage sludge. *Environmental Science & Technology*, *46*(16), 9089–9096. DOI: 10.1021/es301487s PMID: 22816872

Lowry, G. V., Gregory, K. B., Apte, S. C., & Lead, J. R. (2012). Transformations of nanomaterials in the environment. *Environmental Science & Technology*, *46*(13), 6893–6899. DOI: 10.1021/es300839e PMID: 22582927

Luo, P. G., Yang, F., Yang, S. T., Sonkar, S. K., Yang, L., Broglie, J. J., Liu, Y., & Sun, Y. P. (2014). Carbon-based quantum dots for fluorescence imag-ing of cells and tissues. *RSC Advances*, *4*(21), 10791–10807. DOI: 10.1039/c3ra47683a

Ma, Q., Zhang, Q., Yang, S., Yilihamu, A., Shi, M., Ouyang, B., Guan, X., & Yang, S. T. (2020). Toxicity of nanodiamonds to white rot fungi Phanerochaete chrysosporium through oxidative stress. *Colloids and Surfaces. B, Biointerfaces*, *187*, 110658. DOI: 10.1016/j.colsurfb.2019.110658 PMID: 31810567

Ma, R., Levard, C., Judy, J. D., Unrine, J. M., Durenkamp, M., Martin, B., Jefferson, B., & Lowry, G. V. (2013). Fate of zinc oxide and silver nanoparticles in a pilot wastewater treatment plant and in processed biosolids. *Environmental Science & Technology*, *48*(1), 104–112. DOI: 10.1021/es403646x PMID: 24266610

Mazari, S. A., Ali, E., Abro, R., Khan, F. S. A., Ahmed, I., Ahmed, M., Nizamuddin, S., Siddiqui, T. H., Hossain, N., Mubarak, N. M., & Shah, A. (2021). Nanomaterials: Applications, waste-handling, environmental toxicities, and future challenges–A review. *Journal of Environmental Chemical Engineering*, *9*(2), 105028. DOI: 10.1016/j.jece.2021.105028

McBeath, T. M., & McLaughlin, M. J. (2014). Efficacy of zinc oxides as fertilisers. *Plant and Soil*, *374*(1-2), 843–855. DOI: 10.1007/s11104-013-1919-2

McGee, C. F., Clipson, N., & Doyle, E. (2020). Exploring the influence of raising soil pH on the ecotoxicological effects of silver nanoparticles and micron particles on soil microbial communities. *Water, Air, and Soil Pollution, 231*(4), 1–8. DOI: 10.1007/s11270-020-04540-y

Medici, S., Peana, M., Pelucelli, A., & Zoroddu, M. A. (2021, November). An updated overview on metal nanoparticles toxicity. [). Academic Press.]. *Seminars in Cancer Biology, 76*, 17–26. DOI: 10.1016/j.semcancer.2021.06.020 PMID: 34182143

Medina-Pérez, G., Fernández-Luqueño, F., Vazquez-Nuñez, E., López-Valdez, F., Prieto-Mendez, J., Madariaga-Navarrete, A., & Miranda-Arámbula, M. (2019). Remediating polluted soils using nanotechnologies: Environmental benefits and risks. *Polish Journal of Environmental Studies, 28*(3), 1013–1030. DOI: 10.15244/pjoes/87099

Meier, M. J., Dodge, A. E., Samarajeewa, A. D., & Beaudette, L. A. (2020). Soil exposed to silver nanoparticles reveals significant changes in community structure and altered microbial transcriptional profiles. *Environmental Pollution, 258*, 113816. DOI: 10.1016/j.envpol.2019.113816 PMID: 31864930

Mishra, S., Keswani, C., Abhilash, P. C., Fraceto, L. F., & Singh, H. B. (2017). Integrated approach of agri-nanotechnology: Challenges and futuretrends. *Frontiers in Plant Science, 8*. Advance online publication. DOI: 10.3389/fpls.2017.00471 PMID: 28421100

Moll, J., Klingenfuss, F., Widmer, F., Gogos, A., Bucheli, T. D., Hartmann, M., & van der Heijden, M. G. (2017). Effects of titanium dioxide nanoparticles on soil microbial communities and wheat biomass. *Soil Biology & Biochemistry, 111*, 85–93. DOI: 10.1016/j.soilbio.2017.03.019

Mondal, K. K., & Mani, C. (2012). Investigation of the antibacterial properties of nanocopper against *Xanthomonas axonopodis* pv. punicae, the incitant of pomegranate bacterial blight. *Annals of Microbiology, 62*(2), 889–893. DOI: 10.1007/s13213-011-0382-7

Mueller, N. C., & Nowack, B. (2010). Nanoparticles for remediation: Solving big problems with little particles. *Elements (Quebec), 6*(6), 395–400. DOI: 10.2113/gselements.6.6.395

Mukhopadhyay, R., Sarkar, B., Khan, E., Alessi, D. S., Biswas, J. K., Manjaiah, K. M., Eguchi, M., Wu, K. C. W., Yamauchi, Y., & Ok, Y. S. (2022). Nanomaterials for sustainable remediation of chemical contaminants in water and soil. *Critical Reviews in Environmental Science and Technology, 52*(15), 2611–2660. DOI: 10.1080/10643389.2021.1886891

Muneer, N. (2023). Emerging Materials and Technologies for the Next Generation of Engineering Solutions. *Liberal Journal of Language and Literature Review*, *1*(01), 39–46.

Mura, S., Seddaiu, G., Bacchini, F., Roggero, P. P., & Greppi, G. F. (2013). Advances of nanotechnology in agro-environmental studies. *Italian Journal of Agronomy*, *8*(3), e18–e18. DOI: 10.4081/ija.2013.e18

Navarro, D. A., Kookana, R. S., McLaughlin, M. J., & Kirby, J. K. (2016). Fullerol as a potential pathway for mineralization of Fullerene nanoparticles in biosolid-amended soils. *Environmental Science & Technology Letters*, *3*(1), 7–12. DOI: 10.1021/acs.estlett.5b00292

Niño-Martínez, N., Salas Orozco, M. F., Martínez-Castañón, G. A., Torres Méndez, F., & Ruiz, F. (2019). Molecular mechanisms of bacterial resistance to metal and metal oxide nanoparticles. *International Journal of Molecular Sciences*, *20*(11), 2808. DOI: 10.3390/ijms20112808 PMID: 31181755

Omanovic-miklicanin, E., & Science, F. (2017). Green Inter-net of Things and Green Nanotechnology Rolein agriculture. VIII international scientific agri-culture symposium "AGROSYM 2017

Osmond, M. J., & Mccall, M. J. (2010). Zinc oxide nanoparticles in modernsun-screens: An analysis of potential exposure and hazard. *Nanotoxicology*, *4*(1), 15–41. DOI: 10.3109/17435390903502028 PMID: 20795900

Ouyang, K., Yu, X. Y., Zhu, Y., Gao, C., Huang, Q., & Cai, P. (2017). Effects of humic acid on the interactions between zinc oxide nanoparticles and bacterial biofilms. *Environmental Pollution*, *231*, 1104–1111. DOI: 10.1016/j.envpol.2017.07.003 PMID: 28851497

Paul, S. K., Mazumder, S., & Naidu, R. (2024). Herbicidal weed management practices: History and future prospects of nanotechnology in an eco-friendly crop production system. *Heliyon*, *10*(5), e26527. DOI: 10.1016/j.heliyon.2024.e26527 PMID: 38444464

Peyrot, C., Wilkinson, K. J., Desrosiers, M., & Sauvé, S. (2014). Effects of silver nanoparticles on soil enzyme activities with and without added organic matter. *Environmental Toxicology and Chemistry*, *33*(1), 115–125. DOI: 10.1002/etc.2398 PMID: 24115203

Pradhan, A., Seena, S., Pascoal, C., & Cássio, F. (2011). Can metal nanoparticles be a threat to microbial decomposers of plant litter in streams? *Microbial Ecology*, *62*(1), 58–68. DOI: 10.1007/s00248-011-9861-4 PMID: 21553058

Praetorius, A., Labille, J., Scheringer, M., Thill, A., Hungerbuhler, K., & Bottero, J. Y. (2014). Heteroaggregation of titanium dioxide nanoparticles with model natural colloids under environmentally relevant conditions. *Environmental Science & Technology*, *48*(18), 10690–10698. DOI: 10.1021/es501655v PMID: 25127331

Prakash, P., & S, S. C. (2023). Nano-phytoremediation of heavy metals from soil: A critical review. *Pollutants*, *3*(3), 360–380. DOI: 10.3390/pollutants3030025

Prasad, R., Kumar, M., & Kumar, V. (Eds.). (2017). *Nanotechnology*. Springer Singapore., DOI: 10.1007/978-981-10-4573-8

Pulido-Reyes, G., Rodea-Palomares, I., Das, S., Sakthivel, T. S., Leganes, F., Rosal, R., Seal, S., & Fernández-Piñas, F. (2015). Untangling the bio-logical effects of cerium oxide nanoparticles: The role of surfacevalence states. *Scientific Reports*, *5*(1), 15613. DOI: 10.1038/srep15613 PMID: 26489858

Raja, M. A., & Husen, A. (2020). Role of nanomaterials in soil and water quality management. In *Nanomaterials for agriculture and forestry applications* (pp. 491–503). Elsevier. DOI: 10.1016/B978-0-12-817852-2.00020-2

. Rajput V. D., Singh A., Singh V. K., Minkina T., Sushkova S. N. (2020). Impact of Nanoparticles on Soil Resource. In book: Nanomaterials for Soil RemediationEdition: 1st EditionPublisher: Elsevier. 65-78. https://doi.org/.DOI: 10.1016/B978-0-12-822891-3.00004-9

Rajput, V. D., Singh, A., Ghazaryan, K., Minkina, T. M., & Al-Tawaha, A. R. M. (Eds.). (2024). *Sustainable agriculture: nanotechnology and biotechnology for crop production and protection*. Walter de Gruyter GmbH & Co KG.

Reidy, B., Haase, A., Luch, A., Dawson, K. A., & Lynch, I. (2013). Mechanisms of silver nanoparticle release, transformation and toxicity: A critical review of current knowledge and recommendations for future studies and applications. *Materials (Basel)*, *6*(6), 2295–2350. DOI: 10.3390/ma6062295 PMID: 28809275

Riding, M. J., Martin, F. L., Jones, K. C., & Semple, K. T. (2015). Carbon nanomaterials in clean and contaminated soils: Environmental implications and applications. *Soil (Göttingen)*, *1*(1), 1–21. DOI: 10.5194/soil-1-1-2015

Rudrappa, T., Biedrzycki, M. L., & Bais, H. P. (2008). Causes and consequences of plant-associated biofilms. *FEMS Microbiology Ecology*, *64*(2), 153–166. DOI: 10.1111/j.1574-6941.2008.00465.x PMID: 18355294

Sajid, M. (2022). Nanomaterials: Types, properties, recent advances, and toxicity concerns. *Current Opinion in Environmental Science & Health*, *25*, 100319. DOI: 10.1016/j.coesh.2021.100319

Schlich, K., & Hund-Rinke, K. (2015). Influence of soil properties on the effect of silver nanomaterials on microbial activity in five soils. *Environmental Pollution*, *196*, 321–330. DOI: 10.1016/j.envpol.2014.10.021 PMID: 25463729

Sedlak, H. R., Hnilova, M., Grosh, C., Fong, H., Baneyx, F., Schwartz, D., Sarikaya, M., Tamerler, C., & Traxler, B. (2012). Engineered Escherichia coli silver-binding periplasmic protein that promotes silver tolerance. *Applied and Environmental Microbiology*, *78*(7), 2289–2296. DOI: 10.1128/AEM.06823-11 PMID: 22286990

Seneviratne, M., Gunaratne, S., Bandara, T., Weerasundara, L., Rajakaruna, N., Seneviratne, G., & Vithanage, M. (2016). Plant growth promotion by Bradyrhizobium japonicum under heavy metal stress. *South African Journal of Botany*, *105*, 19–24. DOI: 10.1016/j.sajb.2016.02.206

Shen, Z., Chen, Z., Hou, Z., Li, T., & Lu, X. (2015). Ecotoxicological effect of zinc oxide nanoparticles on soil microorganisms. *Frontiers of Environmental Science & Engineering*, *9*(5), 912–918. DOI: 10.1007/s11783-015-0789-7

Shi, H., Magaye, R., Castranova, V., & Zhao, J. (2013). Titanium dioxide nanoparticles: A review of current toxicological data. *Particle and Fibre Toxicology*, *10*(1), 15. DOI: 10.1186/1743-8977-10-15 PMID: 23587290

Singh, M., Goswami, S. P., Sachan, P., Sahu, D. K., Beese, S., & Pandey, S. K. (2024). Nanotech for Fertilizers and Nutrients-Improving Nutrient use Efficiency with Nano-Enabled Fertilizers. *Journal of Experimental Agriculture International*, *46*(5), 220–247. DOI: 10.9734/jeai/2024/v46i52372

Singh, N., Joshi, E., Singh, D., & Sasode, N. C. (2020). Application of nanotechnology in agriculture. *Research Today*, *2*, 163–165.

Smolders, E., & Degryse, F. (2002). Fate and effect of zinc from tire debris in soil. *Environmental Science & Technology*, *36*(17), 3706–3710. DOI: 10.1021/es025567p PMID: 12322741

Son, K. H., Hong, J. H., & Lee, J. W. (2016). Carbon nanotubes as cancer thera-peutic carriers and mediators. *International Journal of Nanomedicine*, *11*, 5163–5185. DOI: 10.2147/IJN.S112660 PMID: 27785021

Sondi, I., & Salopek-Sondi, B. (2004). Silver nanoparticles as antimicrobial agent: A case study on E. coli as a model for Gram-negative bacteria. *Journal of Colloid and Interface Science*, *275*(1), 177–182. DOI: 10.1016/j.jcis.2004.02.012 PMID: 15158396

Soni, D., Gandhi, D., Tarale, P., Bafana, A., Pandey, R. A., & Sivanesan, S. (2017). Oxidative stress and genotoxicity of zinc oxide nanoparticles to Pseudomonas species, human promyelocytic leukemic (HL-60), and blood cells. *Biological Trace Element Research, 178*(2), 218–227. DOI: 10.1007/s12011-016-0921-y PMID: 28058665

Stiegler, L. M., Luchs, T., & Hirsch, A. (2020). Shell-by-Shell Functionalization of Inorganic Nanoparticles. *Chemistry (Weinheim an der Bergstrasse, Germany), 26*(39), 8483–8498. DOI: 10.1002/chem.202000195 PMID: 32167598

Subhan, M. A., & Subhan, T. (2022). Safety and global regulations for application of nanomaterials. In *Nanomaterials Recycling* (pp. 83–107). Elsevier. DOI: 10.1016/B978-0-323-90982-2.00005-6

Subhan, M. A., Subhan, T., Choudhury, K. P., & Neogi, N. (2024). Safety measures, regulations, ethical, and legal issues for nanomaterials. In *Handbook of Nanomaterials* (Vol. 2, pp. 791–828). Elsevier. DOI: 10.1016/B978-0-323-95513-3.00006-X

Sukhanova, A., Bozrova, S., Sokolov, P., Berestovoy, M., Karaulov, A., & Nabiev, I. (2018). Dependence of nanoparticle toxicity on their physical and chemical properties. *Nanoscale Research Letters, 13*(1), 1–21. DOI: 10.1186/s11671-018-2457-x PMID: 29417375

Sun, M., Yu, Q., Hu, M., Hao, Z., Zhang, C., & Li, M. (2014). Lead sulfide nanoparticles increase cell wall chitin content and induce apoptosis in Saccharomyces cerevisiae. *Journal of Hazardous Materials, 273*, 7–16. DOI: 10.1016/j.jhazmat.2014.03.008 PMID: 24704549

Tarafdar, J. C. (2015). *Tapan Adhikari. Nanotechnology in Soil Science*. Book., https://www.researchgate.net/publication/292539409

Thio, B. J., Montes, M. O., Mahmoud, M. A., Lee, D. W., Zhou, D., & Keller, A. A. (2012). Mobility of capped silver nanoparticles under environmen-tally relevant conditions. *Environmental Science & Technology, 46*(13), 6985–6991. DOI: 10.1021/es203596w PMID: 22133047

Tran, T. K., Nguyen, M. K., Lin, C., Hoang, T. D., Nguyen, T. C., Lone, A. M., & Nguyen, D. D. (2023). Review on fate, transport, toxicity and health risk of nanoparticles in natural ecosystems: Emerging challenges in the modern age and solutions toward a sustainable environment. *The Science of the Total Environment*, •••, 169331. PMID: 38103619

. Tripathi, D. K., Tripathi, A., Shweta, Singh, S., Singh, Y., Vishwakarma, K., Yadav, G., Sharma, S., Singh, V. K., Mishra, R. K., Upadhyay, R. G., Dubey, N. K., Lee, Y. & Chauhan, D. K. (2017). Uptake, accumulation and toxicity of silver nanoparticle in autotrophic plants, and heterotrophic microbes: a concentric review. *Frontiers in microbiology, 8*, 07. https://doi.org/DOI: 10.3389/fmicb.2017.00007

Vijayakumar, P. S., Abhilash, O. U., Khan, B. M., & Prasad, B. L. V. (2010). Nanogold-loadedsharp-edged carbon bullets as plant-gene carriers. *Advanced Functional Materials*, *20*(15), 2416–2423. DOI: 10.1002/adfm.200901883

Vimbela, G. V., Ngo, S. M., Fraze, C., Yang, L., & Stout, D. A. (2017). Antibacterial properties and toxicity from metallic nanomaterials. *International Journal of Nanomedicine*, *12*, 3941–3965. DOI: 10.2147/IJN.S134526 PMID: 28579779

Waalewijn-Kool, P. L., Rupp, S., Lofts, S., Svendsen, C., & van Gestel, C. A. (2014). Effect of soil organic matter content and pH on the toxicity of ZnO nanoparticles to Folsomia candida. *Ecotoxicology and Environmental Safety*, *108*, 9–15. DOI: 10.1016/j.ecoenv.2014.06.031 PMID: 25038266

Wang, L., Hu, C., & Shao, L. (2017). The antimicrobial activity of nanoparticles: Present situation and prospects for the future. *International Journal of Nanomedicine*, *12*, 1227–1249. DOI: 10.2147/IJN.S121956 PMID: 28243086

Wang, P., Menzies, N. W., Lombi, E., McKenna, B. A., Johannessen, B., Glover, C. J., Kappen, P., & Kopittke, P. M. (2013). Fate of ZnO nanoparticles in soils and cowpea (*Vigna unguiculata*). *Environmental Science & Technology*, *47*(23), 13822–13830. DOI: 10.1021/es403466p PMID: 24195448

Wilson, M. A., Tran, N. H., Milev, A. S., Kannangara, G. S. K., Volk, H., & Lu, G. H. M. (2008). Nanomaterials in soils. *Geoderma*, *146*(1-2), 291–302. DOI: 10.1016/j.geoderma.2008.06.004

Wong, S. W., Leung, K. M., & Djurišić, A. B. (2013). A comprehensive review on the aquatic toxicity of engineered nanomaterials. *Reviews in Nanoscience and Nanotechnology*, *2*(2), 79–105. DOI: 10.1166/rnn.2013.1025

Xie, P., Yang, S. T., He, T., Yang, S., & Tang, X. H. (2017). Bioaccumulationand toxicity of carbon nanoparticles suspension injection in intra-venously exposed mice. *International Journal of Molecular Sciences*, *18*(12), 2562. DOI: 10.3390/ijms18122562 PMID: 29186019

Xin, X., Zhao, F., Zhao, H., Goodrich, S. L., Hill, M. R., Sumerlin, B. S., Stoffella, P. J., Wright, A. L., & He, Z. (2020). Comparative assessment of polymeric and other nanoparticles impacts on soil microbial and biochemical properties. *Geoderma, 367*, 114278. DOI: 10.1016/j.geoderma.2020.114278

Xing, B., Vecitis, C., & Senesi, N. (2016). *Engineered nanoparticles and theenvironment: physicochemical processes and toxicity. IUPACseries on biophysicochemical processes in environmental systems* (Vol. 4). Wiley. DOI: 10.1002/9781119275855

Xu, C., Peng, C., Sun, L., Zhang, S., Huang, H., Chen, Y., & Shi, J. (2015). Distinctive effects of TiO2 and CuO nanoparticles on soil microbes and their community structures in flooded paddy soil. *Soil Biology & Biochemistry, 86*, 24–33. DOI: 10.1016/j.soilbio.2015.03.011

Yadav, A., Yadav, K., Ahmad, R., & Abd-Elsalam, K. A. (2023). Emerging frontiers in nanotechnology for precision agriculture: Advancements, hurdles and prospects. *Agrochemicals, 2*(2), 220–256. DOI: 10.3390/agrochemicals2020016

Yashveer, S., Singh, V., Kaswan, V., Kaushik, A., & Tokas, J. (2014). Green biotechnology, nan-otechnology and bio-fortification: Perspec-tives on novel environment-friendly crop im-provement strategies. *Biotechnology & Genetic Engineering Reviews, 30*(2), 113–126. DOI: 10.1080/02648725.2014.992622 PMID: 25598358

. Yavuz, C. T., Mayo, J. T., Yu, W. W., Prakash, A., Falkner, J. C., Yean, S., Cong, L., Shipley, H. J., Kan, A., Tomson, M., Natelson, D. & Colvin, V. L. (2006). Low-field magnetic separation of monodisperse Fe3O4 nanocrystals. *science, 314*(5801), 964-967. https://doi.org/DOI: 10.1126/science.1131475

Yeo, M. K., & Pak, S. W. (2008). Exposing zebrafish to silver nanoparticlesduring caudal fin regeneration disrupts caudal fin growth and p53signaling. *Molecular & Cellular Toxicology, 4*(4), 311–317.

Zahedi, S. M., Abdelrahman, M., Hosseini, M. S., Hoveizeh, N. F., & Tran, L. S. P. (2019). Alleviation of the effect of salinity on growth and yield of strawberry by foliar spray of selenium-nanoparticles. *Environmental Pollution, 253*, 246–258. DOI: 10.1016/j.envpol.2019.04.078 PMID: 31319241

Zaki, S. A., Ouf, S. A., Abd-Elsalam, K. A., Asran, A. A., Hassan, M. M., Kalia, A., & Albarakaty, F. M. (2022). Trichogenic silver-based nanoparticles for suppression of fungi involved in damping-off of cotton seedlings. *Microorganisms, 10*(2), 344. DOI: 10.3390/microorganisms10020344 PMID: 35208801

Zhang, L., Wu, L., Si, Y., & Shu, K. (2018). Size-dependent cytotoxicity of silver nanoparticles to Azotobacter vinelandii: Growth inhibition, cell injury, oxidative stress and internalization. *PLoS One, 13*(12), e0209020. DOI: 10.1371/journal.pone.0209020 PMID: 30566461

Zhang, P., Guo, Z., Zhang, Z., Fu, H., White, J. C., & Lynch, I. (2020). Nanomaterial transformation in the soil–plant system: Implications for food safety and application in agriculture. *Small, 16*(21), 2000705. DOI: 10.1002/smll.202000705 PMID: 32462786

Zhang, W., Pu, Z., Du, S., Chen, Y., & Jiang, L. (2016). Fate of engineeredcerium oxide nanoparticles in an aquatic environment andtheir toxicity toward 14 ciliated protist species. *Environmental Pollution, 212*, 584–591. DOI: 10.1016/j.envpol.2016.03.011 PMID: 26986089

Zhao, X., Yu, M., Xu, D., Liu, A., Hou, X., Hao, F., Long, Y., Zhou, Q., & Jiang, G. (2017). Distribution, bioaccumulation, trophic transfer, and influ-ences of CeO2 nanoparticles in a constructed aquatic food web. *Environmental Science & Technology, 51*(9), 5205–5214. DOI: 10.1021/acs.est.6b05875 PMID: 28383254

Chapter 6
Harnessing Nano-Enabled Nutrient Delivery Systems for Enhanced Plant Growth

Tharindu D. Nuwarapaksha
http://orcid.org/0000-0002-7554-5289
Coconut Research Institute, Sri Lanka

Shashi S. Udumann
http://orcid.org/0000-0002-5717-7419
Coconut Research Institute, Sri Lanka

Nuwandhya S. Dissanayaka
http://orcid.org/0000-0002-2062-1133
Coconut Research Institute, Sri Lanka

Naveen Dilshan
http://orcid.org/0009-0009-8449-9374
Wayamba University of Sri Lanka, Sri Lanka

Bandara Gajanayake
Wayamba University of Sri Lanka, Sri Lanka

Anjana J. Atapattu
http://orcid.org/0000-0003-4806-4647
Coconut Research Institute, Sri Lanka

ABSTRACT

This chapter focuses on nano-enabled nutrient delivery systems as emerging opportunities in agricultural technology that can address some of the current issues of nutrient use in agriculture. Different forms of nanomaterials such as nanofertilizers, nanosensors, nanopesticides, nanoemulsions, and nanoclays are described about their uses and how they work. The chapter focuses on the mode of utilization of nano-nutrient by the plant through root absorption, foliar absorption, and cellular uptake pathway of nano-particles. These systems show better uptake of the nutrients compared with the basic fertilizers hence reducing pollution while increasing the yields of crops. Case studies shows how nano-enabled systems are a better solution

DOI: 10.4018/979-8-3693-7006-3.ch006

than traditional ones. This chapter lays out the opportunities of nano-enabled nutrient delivery systems in agriculture sector, but also the need to further investigate the valorization of these systems.

1. INTRODUCTION

1.1 Challenge of Efficient Nutrient Management in Agriculture

Effective nutrient management in agricultural processes is one of the significant challenges of the modern approaches to farming (Chao et al., 2006; Dissanayaka et al., 2023a). Due to the ever-rising global population, people are demanding food crops and this puts a lot of pressure on farmers and agronomists (Kumar et al., 2022). This is a complex process as the plants require nutrients to be delivered and taken up at optimal rates with limited losses (Rahman & Zhang, 2018). This not only leads to poor plants growth of crops but may also lead to some detrimental effects to the environment such as leaching, soil erosion, and water pollution. This task is not easy given the constantly changing processes within the soil chemistry as well as plant physiology. The proposed crop requires different nutrients at different growth stages and at different times because of factors such as weather, microbial activity, and the previous cropping history of the site (Tripathi et al., 2022). For optimal nutrient delivery, it is essential to understand these interactions and adjust fertilization patterns as appropriate.

New solutions for dealing with this challenge have emerged due to the development of technologies (Atapattu et al., 2024b). Nano fertilizer application enhances nutrient management through improved absorption efficiency, while GPS mapping, soil probes, and data analysis from precision agriculture enable targeted nutrient placement, collectively optimizing plant nutrition and reducing waste (Hedley, 2015; Kanjana, 2017). The adoption of these technologies is not easy as it entails considerable capital investment and technical knowledge, which the majority of farmers are unable to afford especially in the developing countries. Still another aspect of the challenge is cyclical or long-term sources of nutrients. There are rising concerns about the environmental effects of synthetic fertilizers and because some nutrients like phosphorus are non-renewable there is a push to find other effective nutrient management strategies (Bergström et al., 2005). This involves considering the use of organic fertilizers, a more efficient cycling of nutrients within agricultural production systems, as well as the preparation of new types of fertilizers that would optimize nutrient utilization (Nuwarapaksha et al., 2024; Nuwarapaksha et al., 2022). There is an added challenge of deciding when to prioritize economic gains and when to safeguard the environment. Although, more efficient nutrient delivery results in lower costs and higher yields over time, initial capital and potential short-term yield downsides can be daunting for low margin agricultural producers (Luna Juncal et al., 2023).

1.2 Overview of Nanotechnology in Agriculture

Nanotechnology is a relatively new scientific discipline that involves controlling matter at a scale of 1 to 100 nm and this science is gradually gaining major importance in agriculture (Sekhon, 2014). Providing new approaches to the long-standing issues of crop management, food safety, and preservation of the natural environment, this field significantly contributes to the enhancement of the human condition. Nanotechnology has a vast area of applications from agricultural field to another in the following manner (Acharya & Pal, 2020). Figure 1 below gives types of nanomaterials used in agriculture, revealing how different nanomaterials and techniques can be used to enhance plant growth, enhance soil conditions and protect crops. The nanoformulations in these miracle crops may increase nutrient utilization rate and give precise pest control. With this characteristic, it enhances the specific release and release destination enables reduced densities of chemicals used while promoting profitability in crop production. Farm management and agricultural practices are particularly enhanced by nanosensors (Bhagat et al., 2023). They can be as small grain of sand and will be able to measure chemical content in the soil, the health of a plant, or the climate when placed in a specific location. Collected data enables the farmer to decide on the appropriate time to water or fertilize the crops and control pests and diseases to avoid wastage of resources and hunger. It also important to note that nanotechnology is also advancing food packaging and preservation techniques (Pradhan et al., 2015).

Nanocomposites and nanocoatings can preserve food for longer periods and can also detect presence of pathogens in food and can also tell when a food item was last fresh (Alfei et al., 2020). This not only reduces the risks associated with the consumption of contaminated foods but also pushes down the efficiency on wasted food across distribution channels. Nanoparticles have been considered as potential vectors in the field of crop improvement, particularly the delivery of genetic material (Ahmar et al., 2021). This could in some way reduce the time it takes to breed plants and come up with new ways of making the crops genetically different thus leading to better crops that are capable of withstanding certain harsh conditions like drought or even having better nutritional value. Irrespective of whether nanotechnology has proven to have potential in water management, which is a crucial factor in agriculture, is another application area (Kuhn et al., 2022). The areas of application include water treatment and separation, and purification of water containing impurities and contaminated soil. Moving on to nanotechnology application includes use of nanocoating which makes agricultural machinery more effective and durable. However, the technology holds the immense potential in the field of agriculture though there are concerns related to safety and environmental impact measures as well. Additional research should thus be carried out to tackle these concerns to guarantee that the impact of nanotech applications is not only efficiency oriented without compromising the safety of people and the environment. Ultimately, it contributes to achieving Sustainable Development Goals (SDGs) by enhancing soil fertility and crop yields, thereby ensuring food security in the global scenario (Atapattu et al., 2024a; Dissanayaka et al., 2023b; Nuwarapaksha et al., 2023).

Figure 1. Types of nanomaterials use in agriculture

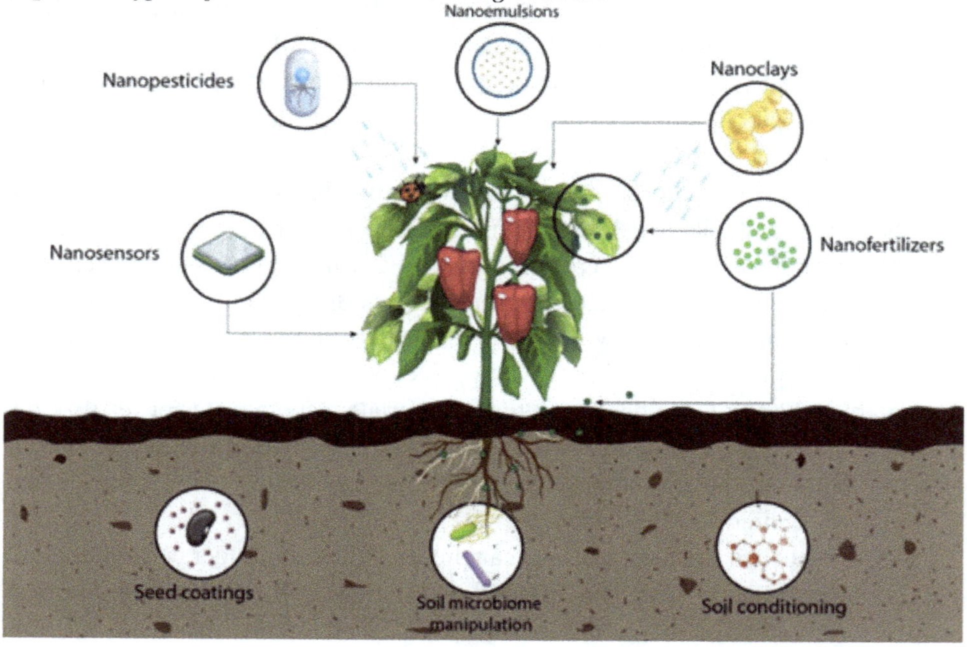

1.3 Potential of Nano-Enabled Nutrient Delivery Systems in Agriculture

The use of nanotechnology in nutrient delivery systems is one of the most emerging fields of agriculture that can potentially bring about drastic change in the way nutrients are delivered to plant, animals, and consequently increase production (Elemike et al., 2019). These systems make use of nanotechnology to enhance the process of feeding plants by boosting nutrient delivery. The basis of such systems is nanoparticles or nanostructures responsible for loading and releasing plant nutrients at the required site (Monreal et al., 2016). These can include macro nutrients that include nitrogen, phosphorus, and potassium or the micro nutrients that include iron, zinc, manganese amongst others. One advantage of the use of these carriers is that they are much smaller than those used in the conventional methods hence resulting in improved penetration into tissue of plants as well as precise delivery of nutrients (Vega-Vásquez et al., 2020). Another advantage of nano enabled nutrient delivery systems is that it is possible to get controlled and sustained release of nutrients. It is unfortunate that most of the conventional fertilizers face rapid rate of release and nutrients are either leached or lost through volatilization. Thus, nanocarriers can be designed in such a way that they release nutrients only when stimulated by certain signs, such as pH, temperature, or enzymatic activity (Karimi et al., 2016). It also helps to regulate the delivery of the nutrients over the lifetime of the plant hence requiring infrequent applications of the fertilizers.

The efficiency optimization of nano-enabled systems may cause the decrease of fertilizer usage by multiple times (Guo et al., 2018). These systems also help reduce waste and threat to the environment by supplying nutrients at the specific part of the plant where they are required. This is especially important where we see negative externalities from the agricultural process such as water pollution by chemical fertilizers which leads to eutrophication of water bodies. Nano-enabled nutrient delivery systems also provide the possibility to administer nutrients in a more optimized manner (Wang et al., 2024). When these systems are integrated with nano sensors and precision agriculture, they can help farmers deliver nutrients according to the requirement of plants or sub-division of fields. Such precision could double or even triple the crop yields while at the same time reducing the resources used in the farming process (Bongiovanni & Lowenberg-Deboer, 2004). These systems can be made to overcome some of the problems associated with nutrient acquisition. As with any novel approach for delivering nutrients to plants, there remains much work still to be done in exploring the full impact of nanotechnology on soil organisms and plant health over the long term (Kumari et al., 2023). Some of the other concerns that also warrant research include safety issues such as the buildup of these nanoparticles in the environment and the food chain (Hardy et al., 2018). Thus, there are expectations that this technology will enhance sustainable and efficient use of resources in agriculture practice and become an essential solution for food shortage problems caused by climate change and increased population around the world. The objective of this chapter is to provide a comprehensive review of nano-enabled nutrient delivery systems in agriculture, exploring their mechanisms, benefits, challenges, and future potential for enhancing crop production and sustainability.

2. TYPES OF NANOMATERIALS USED IN AGRICULTURE

2.1 Nanofertilizers

Nanofertilizers are fertralizers that are engineered in a way to have a size of between 1 to 100 nm (Yadav et al., 2023a). Such materials are developed to supplement low nutrient availability problems in plants as opposed to other common fertilizers. Because of these physical properties, which are nano-scaled, these fertilizers can be made to go where they are needed at a given time, thus helping the plants absorb the nutrient where they are required. Generally, they use slow-release systems that provide the crops with the nutrients as they grow in the season (Zareabyaneh & Bayatvarkeshi, 2015). This controlled release also helps to prevent soil nutrient loss through leaching and nutrient loss through volatilization hence reducing the effects of the environment. The nanofertilizers can be formulated to provide specific amount of macro and micronutrients to the plants and so it can correct the deficiencies in plants more efficiently (Elemike et al., 2019). Examples of those are nano-zinc, nano-iron, and nano-copper containing preparations. These materials get through the tissues of the plant with more ease thereby increasing the assimilation of nutrients in the plant (Rani et al., 2022). Some nanofertilizers also contain the biostimulants or what some call microorganisms that enhance plant growth and development (Mahapatra et al., 2022). Based on the related literature, nanofertilizers have exhibited potential in raising crop productivity as well as the nutritive value of produce and abiotic stress tolerance. Nevertheless, further investigation is required to characterize their impact on the long-term stability of the soil biological components and to fine-tune the preparation's composition depending on the type of crop to be cultivated and the prevailing environmental conditions.

2.2 Nanosensors

Nanosensors in agriculture are micro-devices developed to sense and analyze different parameters such as environmental and biomolecular at the nanoscale (Akyildiz & Jornet, 2010). Speaking of these enhanced tools, the former takes advantage of distinct characteristics of nanomaterials to provide high sensitivity and specificity of the measurements. Using nanosensors, the soil moisture, pH and nutrient content of the soil can be detected and these results fed real time into precision farming (Beegum & Das, 2022). They are also employed to identify plant pathogens, pests and diseases when they are still in incidence stage thus management can be easily implemented. There are nanosensors that are able to detect plant stress hormones or metabolites, which help to determine the health state of plants, and possible yield (Ang & Lew, 2022). In livestock farming, nanosensors are used in health and disease diagnosis, and identification of cycles of reproduction (Rani et al., 2022). Another important use can be identified as environmental control, where nanosensors can detect the presence or such toxic substances as pollutants, pesticide residues or heavy metals in soil and water. The information that is gathered by these sensors can be utilized in the smart farming management systems and also can initiate the actions and decisions (Cambra Baseca et al., 2019). As the technology progresses, nanosensors are mobile, robust, and wire-less, thus are easily incorporated in gigantic agricultural firms. However, several obstacles include cost, the ability to effectively and afford to sustain the reusable materials in more strictly field conditions, and the collaboration of standard practices of the data analysis and what action should be taken on the information acquired.

2.3 Nanopesticides

Nanopesticides are a new age strategy of pest control as nanotechnology promotes the efficiency and safety of pesticidal techniques (Hajji-Hedfi & Chhipa, 2021). These formulations are generally of active ingredients either incorporated within or associated with the nanoparticles with a size of 1-100 nm. Due to the fact that most of these pesticides are in the nano-scale, better penetration into the plant tissues or insect bodies can be achieved, meaning that more pests can be controlled using a smaller number of pesticides. Some of the materials prevalent in nanopesticide research are nanoemulsions, nanoencapsulations, and polymer based nanocarriers (Hayles et al., 2017). It is possible to provide pest control formulations with these longer acting characteristics that can provide protection for a longer time while at the same time minimizing the bioavailability of the toxic constituent in the environment. Certain nanopesticides have been described to comprise targeting strategies so that only the targeted pests are affected and no other beneficial insects (Pradhan & Mailapalli, 2020). Nanomaterials also help in improving the shelf life of the active ingredients protecting them from breaking down factors such as UV radiation or high and low temperatures (Liu et al., 2020). Nanopesticides based research also focusing on developing possibilities of using nanoparticles that have good pesticidal activity within them like nanosilver or nanocopper into the pesticides (Rai et al., 2018). Although research carried out on nanopesticides reveal the advantages of lowering environmental effects and enhancing pest control efficacy, further researches are needed to be able to identify the effects of nanopesticides in a long run-on ecosystems and non-target organism.

2.4 Nanoemulsions

Nanoemulsions in agriculture are stable dispersion of two-phase liquids, where at least one being non miscible with the other and the droplet size of the dispersed phase ranges between 10 and 200 nm (Murugan & Abd-Elsalam, 2022). These advanced formulations are made by high-energy process or low energy phase inversion process and product forms a thermodynamically unstable but being kinetically stable. Nanoemulsions also find usage in agricultural activities in the pesticide, herbicide, and plant growth regulators administrations (Tyagi et al., 2023). These emulsions contain tiny droplets that can enhance the effectiveness and solubility of the active ingredients to make the content more effective with less quantity hence positively affecting the environment. It allows increasing the water dissolution of low-water soluble agrochemicals and their better absorption by the plants. They also have better stabilities and shelf lives as compared to typical emulsions (Ahari & Naeimabadi, 2021). At certain times, nanoemulsions are applied as an encapsulation agent for essential oils or other natural agents that are being used as a replenishment for synthetic pesticides (Tiwari et al., 2022). It should be noted that these formulations can be developed in foliar or in soil application either through drenching. Even though nanoemulsions have enormous applications in agriculture, further studies are directed at refining the formulations for a particular crop-pest complex and evaluating the environmental effects of nanoemulsions over the years.

2.5 Nanoclays

Nanoclay is defined as nanoscale particles of layered mineral silicates that are usually sourced from the natural minerals like montmorillonite, bentonite or kaolin (Kausar et al., 2022). These materials possess a developed layered structure with at least one dimension of the layer being in the nanometer range mostly below 100 nm thick (Kausar et al., 2017). Nanoclays also find application in the agriculture sector by virtue of the involvement of a large surface area, capacity for adsorption, as well as their feature that enables formation of intercalated and exfoliated nanocomposites. A major potential use is in the modification of soil properties whereby nanoclays may increase water available to plants, reduce loss of fertilizers, and increase the aggregation of soils (Kalia et al., 2020). In case of its use in formulation of fertilizers, nanoclays are capable of working as nutrient carriers and they display controlled release properties thus enhancing efficiency of use of the nutrient. In crop protection nanoclays are used to deliver pesticides and herbicides maybe decreasing their hazardous effect to the environment (Kalia et al., 2020). It has been established that some nanoclays have embedding antimicrobial properties that in theory can be utilized for making plants disease proof. In the case of seed coating, nanoclays are useful in enhancing seed vigor besides offering a shield against pathogen attack. Nanoclays are also used in agriculture especially in packaging material to improve the barrier properties of bio-degradable films and this may help in prolonging the shelf life of fresh produce (Pandey et al., 2022). Even though nanoclays have many advantages, further research is required to make the assessment of its effects on the soil ecosystem permanent and to improve the effectiveness of its application in the sphere of agriculture.

3. UNIQUE PROPERTIES OF NANOMATERIALS RELEVANT TO PLANT NUTRIENT DELIVERY

Nanomaterials have characteristics that make them ideal for plant nutrient delivery systems, as will be discussed below. Most of them are in the size range between 1 to 100 nm and their small size enables them to penetrate into cell and sub cellular structures to improve their bio distribution and nutrients delivery across barriers (Mittal et al., 2020; Pillay et al., 2015). The higher surface area to volume ratio renders nanomaterials as more reactive and holds more and releases more nutrients than conventional carriers. They also use this property to enhance the dispersion and solubility of nutrients that have low solubility in water hence enhancing bioavailability. The synthesis and processing of nanomaterials enable one to endow them with specific surface chemistries and functions such as targeting specific tissues or cells in the plants, providing plant nutrients to specific cells in a controlled manner over time, and to protect sensitive compounds from degradation by the hostile environment in the plants (Sembada & Lenggoro, 2024). The sharp control in size, shape, and composition of nanocarriers helps to achieve proper physicochemical characteristics and create intelligent delivery systems based on stimuli sensitivity to pH, temperature, and enzymes (Lombardo et al., 2019). This responsiveness can be used to control the release of nutrients and release them where they are wanted or at certain time when they are needed. Furthermore, some nanomaterials possess optical, magnetic or electrical characteristics that can be used to visualize the nutrient delivery process in real-time and provide information regarding the distribution and elimination of the nutrients in plants (Harish et al., 2022). Moreover, the portability of the nanomaterials can also enable the delivery of two or more plant nutrients at a time or nutrients combined with other bioactive compounds which expands the opportunities for multifaceted nutritional requirements (Ansari, 2023). With future advances in this field, the novel characteristics of nanomaterials explain new methodologies of plant nutrient delivery, which would further enhance the aims of combating nutrient deficiencies and enhancing plant health.

4. MECHANISMS OF NANO-NUTRIENT UPTAKE

4.1 Root Absorption

One of the key ways through which the plants are able to assimilate nano-nutrients is by absorbing them through the root (Farhangi-Abriz & Ghassemi-Golezani, 2023). Particulate matter can penetrate plant root tissues through three major routes. The apoplastic pathway is conducted through cell walls as well as through the spaces between cells (Novaković et al., 2018). This route is particularly relevant for large nano particles that fail to cross cellular membranes and enter the cytosol. The simplistic pathway enables the transport of nanoparticles through contiguous plant cells from one cell to the next through the channels known as plasmodesmata (Schwab et al., 2016). It is important to maintain the transportation of nutrients at the systemic level all across the plant body. Transmembrane transport is a process whereby small particles or nanoparticles are transported across the cell membrane through specific receptors or ports (Barbu et al., 2009). The ability of roots to uptake nanoparticles is therefore a function of properties of the nanoparticles such as size and chemistry, and the type of plant and root architecture (Ma et al., 2010). There is a trend towards the smaller particles in terms of uptake and translocation. It has been identified that the rhizosphere, the first one to two millimeters of soil, which is directly influenced by root secretions and microorganisms, is responsible for altering nanoparticle properties and governing the uptake process. Root exudates are capable of changes the chemical nature of nanoparticle surfaces, which may then affect their absorption (Ahmed et al., 2021). Also, mycorrhizal fungi attached to the roots of plants are capable of helping in nanoparticle uptake, that is they help in increasing the absorptive length of the plant (Etesami et al., 2021).

4.2 Foliar Absorption

This is another avenue through which the nano nutrient can be taken up by the plants through the foliage phytotoxicity. This method could be useful when it comes to supplying nutrients to the leaves since this is where the nutrients may be required more commonly (Kannan, 2010). Stomata the minute windows on the surface of the leaves that allow the exchange of gases are also a portal for nanoparticles (Lew et al., 2020). These openings provide pathways through which nanoparticles gain access into the interior of the leaf for either absorption into the cells or transfer to other tissues of the plant. The second mechanism of uptake is based on the fact that plants have a waxy layer on the leaf surface and through this layer, nanoparticles can penetrate (Wang et al., 2023). This process depends on the size, charge and lipophilicity of this nanoparticle. Since the cuticle has aqueous pores, hydrophilic nanoparticles may be absorbed through these pores when the nanoparticles are lipophilic in nature, then they diffuse through the layers of wax (Banerjee et al., 2019). Which are hydroscopic structures present particularly at leaf margins and function in water secretion, offer the other inlet to admit nanoparticles. Foliar absorption efficiency can be further utilized through surfactants or other adjuvants that aid in the adhesion of the nanoparticles to the external layers of the foliage and penetration (Ramezani et al., 2019). This method of nano-nutrient delivery is capable of providing nutrient deficiencies in a short time, and in cases where uptake by roots is hindered due to soil conditions.

5. CELLULAR UPTAKE OF ABSORBED NANOPARTICLES

Endocytosis is the most common uptake mechanism, where the cell membrane engulfs nanoparticles forming vesicles that internalize them (Francia et al., 2020). This process can be further divided into several subtypes, including clathrin-mediated endocytosis, caveolin-mediated endocytosis, clathrin and caveolin-mediated endocytosis, phagocytosis and micropinocytosis (Figure 2) (Kaźmierczak et al., 2020). Clathrin-mediated endocytosis involves the formation of clathrin-coated pits on the plasma membrane. These pits invaginate and pinch off to form vesicles, which transport cargo into the cell. Clathrin-mediated endocytosis is receptor-dependent and typically internalizes particles smaller than 200 nm (Deng et al., 2018). Caveolin-mediated endocytosis occurs through flask-shaped invaginations called caveolae, rich in cholesterol and sphingolipids. Caveolin proteins are essential for this process, which usually internalizes particles around 50-80 nm in size (Cohen et al., 2004). Some endocytic events can involve both clathrin and caveolin, demonstrating the complexity and overlap of these mechanisms (Mayor & Pagano, 2007). Phagocytosis is primarily used by specialized cells like macrophages to engulf large particles (>500 nm). It involves actin-driven membrane extensions that surround and internalize the particle, forming a phagosome (Kou et al., 2013). Macropinocytosis is a non-selective process where the cell membrane ruffles and folds back on itself, creating large vesicles that engulf extracellular fluid and particles. This mechanism can internalize a wide range of particle sizes (Zaki & Tirelli, 2010). The choice of endocytic pathway depends on various factors, including nanoparticle size, shape, surface charge, and cell type. Understanding these mechanisms is crucial for designing effective nanoparticle-based nutrient delivery systems and predicting cellular interactions with nanomaterials.

Figure 2. Cellular uptake mechanisms are absorbed nanoparticles; (a) Clathrin-mediated endocytosis (b) Caveolin-mediated endocytosis (c) Clathrin and caveolin-mediated endocytosis; (d) phagocytosis; (e) micropinocytosis

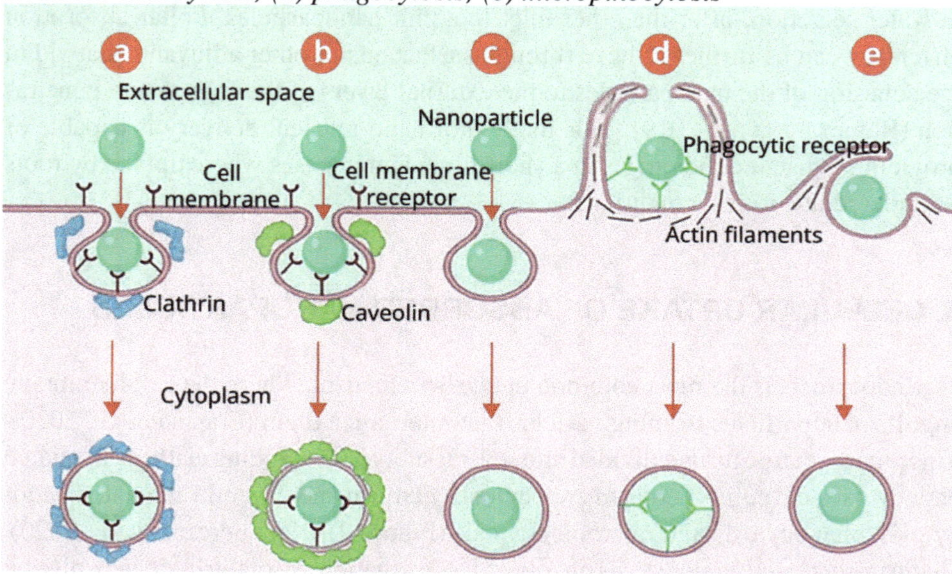

6. ENHANCED NUTRIENT USE EFFICIENCY WITH CASE STUDIES

Nano fertilizers are now a modern innovative solution in modern farming practices as they counter many drawbacks of the typical fertilizers (Nongbet et al., 2022). These consumer products incorporate the particle size of nanoparticles to improve the mobility and absorption of nutrients in plants. The ways in which nano fertilizers enhance efficiency are complex and interrelated with some of the main mechanisms discussed below. The particles are miniscule and this hulks up the surface area to volume ratio, thereby enhancing the solubility and bioavailability of nutrients (Huang et al., 2010). This leads to increasing solubility that will help the plants to be able to absorb nutrients more effectively hence there will be no need for a lot of fertilizer for plants to grow to their best potential. Also, it is important to notice that one of the predominant advantages of nano fertilizers is the ability to deliver nutrients to particular plant organs. They can easily enter plant tissues due to their small size which makes them to be directly delivered to the areas that require the nutrients such as the growing tips or flowering structures (Sharma &

Kumar, 2024). Such an application of the nutrients ensures that a high proportion of the applied fertilizers is utilized in promoting the growth of crops. Slow-release and sustained nutrient delivery are some of the other important benefits provided by many nano fertilizers (Haydar et al., 2024).

By the use of sustained release mechanisms like encapsulation, coating techniques, these products offer sustained releases that match plant cycling and its various metabolic requirements. It can also be efficient in nutrient delivery from planting time to harvest time and reduces the chances of nutrient leaching that otherwise could lead to toxicity of the nutrients or pollution of the environment. Among all the benefits associated with the use of nano fertilizers, one of the most important in environmental terms is certainly the reduction in nutrient leaching and volatilization (Avila-Quezada et al., 2022). Nano fertilizers focus on increasing comparatively the uptake efficiency and decreasing the release rates, which means that more nutrients are retained in the root area where plant can use them rather than being washed away by water or evaporating in air (Haydar et al., 2024). Not only does it add economic sense to the way fertilizer is utilized but also reduces the possible negative effects that come with high nutrient leaching or emission of nitrogen gases. However, in the summary of the work nano fertilizers have more enhancement when compared to conventional fertilizers (Figure 3).

The effectiveness of nanofertilizers in terms of absorption, distribution, and accumulation is influenced by various factors such as soil pH, organic matter content, and soil texture, as well as the inherent properties of the nanoparticles, like size and coating (Toksha et al., 2021). Nano fertilizers can be absorbed through both roots and leaves, affecting their behavior, bioavailability, and absorption within the plant. Studies have demonstrated that nano fertilizers are significantly more effective than conventional fertilizers. Macronutrient nano fertilizers enhance plant growth by 19%, micronutrient nano fertilizers by 18%, and nano fertilizers carrying macronutrients by 29% compared to regular fertilizers (Kah et al., 2018). Additionally, nanochitosan nitrogen based nano fertilizers containing nitrogen, phosphorus, and potassium have been shown to increase sugar content and improve the properties of wheat (Abdel-Aziz et al., 2016). In another study, the application of silver nanoparticles to wheat resulted in increased shoot and root length, leaf area, and chlorophyll, carbohydrate, and protein contents (Salama, 2012). Furthermore, nano fertilizers provide a slow nutrient release over 40-50 days, compared to the 4-10 days of conventional fertilizers (Avila-Quezada et al., 2022).

Figure 3. Comparison of conventional and nano fertilizers: Impact on plant growth and soil health

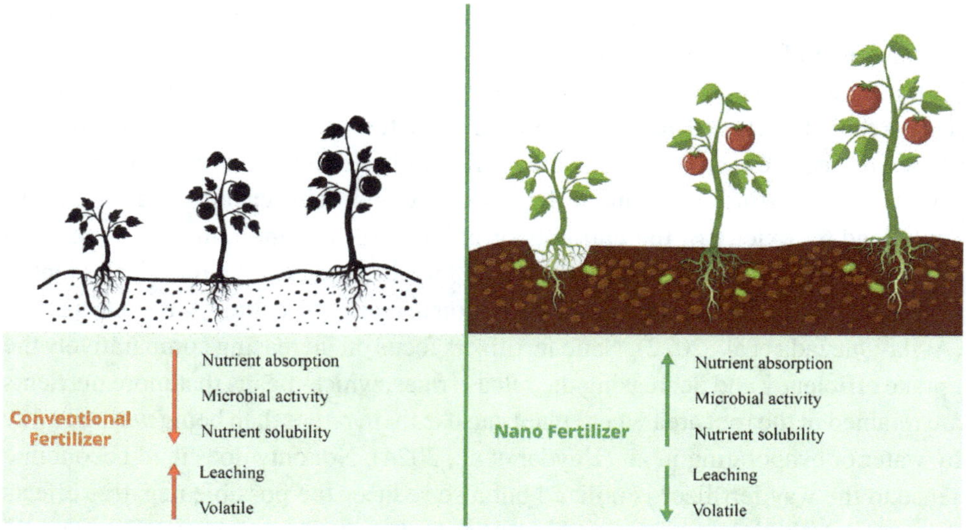

7. ENVIRONMENTAL AND SAFETY CONSIDERATIONS OF NANO NUTRIENT MANAGEMENT

Nano fertilizers are being used increasingly in the process of farming and this has raised some concerns to the years on the impact it has and if it's safe for the environment (Jakhar et al., 2022). The main concern is mainly on the behaviour of nanoparticles in the soil and water system, the effect they have on the organisms that live in the soils, the issue of biomagnification among living organisms and unavailability of laws to control the use of nanoparticles. For its part, there are quite a lot of uncertainties about the manner in which nanoparticles may behave in the environment (Grieger et al., 2009). After getting dispersed in the soil, these particles can undergo processes like aggregation, dissolution, or interacting with other particle like organic substances and minerals. The mobility within the soil and water environment is also governed by factors like particle size, charge, and state (Pan &Xing, 2012). This mobility has the potential of contaminating the groundwater or the surface water point sources. Some research highlighted those certain classes of nanoparticles had the potential to induce microbial proliferation and growth, while others alleged that nanoparticles could reduce microbial abundance and efficiency adversely (Raffi & Husen, 2019).

The impacts of successive uses of nano fertilizer on soil ecosystems are also the consequences that influence subsequent generations, and are still unpredictable. There are high chances that nanoparticles will bio accumulate by the food chain, which poses a danger to human and other forms of life (Uddin et al., 2020). More research should be carried out regarding the mechanisms of nanoparticles entering into the plants and their subsequent build up as well as transport to the food chains. Some particles can be metabolized with the organisms by breaking down into new substances which may be more toxic or soluble. Because of these potential risks, it is crucial to enact beneficial laws regarding the proper application of default message and perform crucial essentiality assessment procedures (Yadav et al., 2023b). It is necessary to set down the standard and comprehensive approaches for the risk assessment while considering the manufacturing and application of nano fertilizers and repercussions on nature (Rahman et al., 2021). This includes the norms to utilize in testing, exposure measures, and monitoring that can enhance the safety of the deployment of these advanced agribusiness inputs.

8. CHALLENGES AND LIMITATIONS FOR NANO NUTRIENT MANAGEMENT

8.1 Limitation of Scalability and High Cost

Translating successful nano-enabled nutrient delivery systems from lab experiments and pilot to mainstream agriculture practices is a daunting task (Raza et al., 2023). The creation of nanomaterials can be intricate and requires costly equipment and substances leading to a situation where the cost outcomes may exceed the advantages in most crops or farming conditions. When the scale of production of nanoparticles is considered, the challenges of the uniformity of the quality and properties of nanoparticles increases significantly (Operti et al., 2021). Application methods that have been adopted in these controlled environment studies may not be used in large field application without considerable alteration and maybe accompanied by specialized equipment. The feasibility of these systems is significantly influenced by the economic return from the crop highly valued crops generally attract such investment (Malézieux, 2012). This indicates that staple crops producing lower returns, may well be disadvantageous in terms of the cost benefit ratio if the cost of production cannot be substantially lowered. Moreover, the stability and reusability of nano-delivery systems in soil contexts are still an issue, which can impact the general economy of nano-application (Wahab et al., 2024). These challenges in relation to scalability and costs will require further cross disciplinary inputs from materials scientists, agronomists and agricultural engineers to advance production technologies as well as formulate new methods of applying the nano fertilizers in varying agricultural environments.

8.2 High Variability in Efficacy Across Different Crop Conditions

Due to diverse responses to nano-enabled nutrient delivery systems in different crops are highly variable, soil types, and other growing conditions, scaling up their use is not a simple matter (Marchiol et al., 2020). There is large variation of nanoparticle fate and performance in different soil environments, and factors such as soil pH, organic matter content, and microbial activity affects stability of nanoparticles, their mobility and rate of nutrient release. The same way the uptake efficiency and potential phytotoxicity of nanoparticles varies with different plant species and their type of interaction (Ruttkay-Nedecky et al., 2017). Since lab or greenhouse tests can be conducted under carefully controlled conditions, they bear a poor relation to effective field efficacy when scaled up. Other factors such as change of weather, temperature, humidity and light conditions can also affect the efficiency of nano-delivery systems. However, there are still questions about the impacts of repeated application of nanoparticles on the health of the soil, its microorganisms, and the function of ecosystems in general (Yayehrad et al., 2021). Due to this variability and uncertainty, the efficacy of these products under different agriculture systems and environments needs to be established through rigorous field trials to generate accurate and effective application protocols. But it also points toward the requirement for more sophisticated nano-carrier systems that can perform stably across various milieus, or crop-targeted formulations for the best results.

8.3 Public Perception and Acceptance Barriers

Perception by the public and acceptance are some of the major barriers that face the application of nano-enabled nutrient delivery systems in agriculture (Hofmann et al., 2020). Public concern on how these nanoparticles may get into the food chain and the effects that they will have on human health and the environment may cause public resistance. This is because like GMOs, the application of nanotechnology may cause initial apprehension among the consumers which may lead to changes in buying behavior and market acceptance of produce from nano enhanced crops (Kuang et al., 2020). Such issues demand explanation of the advantages and disadvantages of nano-delivery systems, as well as the protective measures that are in place. However, there are no existing standardized safety assessment methodologies and legislation for agricultural nanomaterials as of the current time (Kumari et al., 2023). Governments and regulatory agencies around the world have not yet figured out how to best assess and regulate them, therefore causing more confusion in the industry as well as in the consumers. To overcome these challenges, it requires a multiple approach that includes labeling, public health education, collaborating with stakeholders, and creating science-based regulations. Communication and sustainable innovation will remain crucial in making nano-related nutrient delivery systems acceptable enough to bring about the intended benefits for sustainable agriculture (Coles & Frewer, 2013).

8.4 Technical Challenges in Formulation and Application

The design and use of nano-based nutrient delivery systems create many technical difficulties that should be solved to increase the use of these systems in agriculture (Yadav et al., 2023c). Making nutrition payable in a slow and sustained manner in soil conditions poses a challenge in this aspect since a number of factors determines the particles' composition, size and surface properties. Dispersion of nanoparticles and avoiding clumping is one of the biggest challenges as it affects the properties of the nanoparticles and also affects the distribution of the nanoparticles in the soil (Yamini et al., 2023). Another major challenge is identifying potential correct application methods for the specific crop systems and different growth stages because the existing irrigation methods for applying fertilizers may not be effective with nanoparticles (Vasuki et al., 2023). Readily quantitative, and even distribution of the nanoparticles in the root zone without significant replenishing of the aerial parts or off-target deposition demands advanced application methods. Also, positions of nanoparticles in relation to pesticides, herbicides, and other agrochemicals must be taken into consideration to have no undesired side effects or poor efficacy (Vurro et al., 2019). Solving these technical issues requires the cross-disciplinary work of materials science, soil chemistry and agricultural engineering to create effective nano delivery systems and methods for field applications to be effective and stable. Continuous research on nanotech and precision agriculture will provide future possible solutions to these bottlenecks and opportunities of nano-enabled nutrient delivery to increase crop yield and resource utilization efficiency.

9. FUTURE PROSPECTS FOR NANO NUTRIENT MANAGEMENT

9.1 Integration With Precision Agriculture Technologies

Further development of nano-enabled nutrient delivery systems is in harmonizing them with the precision agriculture technologies (Zain et al., 2024). Integrating nanotechnology in smart farming systems enhances the concepts of precise input management in crop nutrition. Sensors, drones and Satellite technologies can assist in providing accurate information on the state of the soil, the health of crops and conditions of the environment (Inoue, 2020). When the data is run through artificial intelligence and machine learning algorithms, it can lead to defining incredibly specific and targeted uses of nano-nutrients. The IoT devices can thus prompt relevant nano-fertilizer dispensing systems that respond to alteration in crop demands and adverse situations as they happen (Dhanaraju et al., 2022). It ensures optimal use of nutrients while at the same time reducing on the use of excess fertilizers that pollute the environment. Blockchain technology can be used to establish the traceability of the consumption of nano-nutrients satisfying the regulation and consumers' concerns (Bosona & Gebresenbet, 2023). Taking into account these technologies' interaction, one may describe the future of nano-enabled nutrient delivery as a subsystem of advanced efficacy-oriented and data-driven agricultural systems with enhanced crop yield and resource-saving effects.

9.2 Development of Biodegradable Nano-Carriers

The prospect of advanced biodegradable nano-carriers holds great promise in the future of sustainable nano-based nutrient delivery systems (Dasauni & Nailwal, 2022). The advanced work will be to develop nanoparticles that are capable of releasing needed nutrients in plants and disintegrating in the soil into harmless particles. This way does not raise questions about the accumulation of nanoparticles and potential long-term consequences for eccentricities. There is the enormous interest of the scientists in the synthesis of nano-carriers using bio-based materials such as chitosan, alginate, cellulose derivatives and plant-based polymers (Serri et al., 2023). These materials provide the familiar benefits of being naturally available, renewable and in many cases biocompatible. Further researches would be directed on determining the mechanisms for controlling the degradation rate of these materials in different types of soil environments. The characterization and identification of the breakdown products should also be investigated to minimize the chances of compromising the soil health or interfere with the microbial populations (Bertola et al., 2021). In this case, the successful development of biodegradable nano-carriers might be the real breakthrough that will allow mean more frequent and precise nutrient application without negative impacts from long-term nanoparticle presence. This spurred research that could lead to the development of fully sustainable, closed-loop nutrient cycling in agriculture.

9.3 Develop Nano-Enabled Nutrient Sensors for Real-Time Monitoring

Expedition towards nano enabled nutrient sensors for live tracking could be viewed as the future direction of precision agriculture (Yadav et al., 2023b). Such next-generation nanosensors may thus totally transform the knowledge of nutrient flow within the agricultural crops and soil, thus hugely enhancing the levels of precision in nutrient delivery. Scientists are in the process of developing the soil based nanosensors that will be able to detect the amount of certain nutrients in the soil with large accuracy. These sensors could give constant and real time information on nutrient availability, PH, and other parameters of the soil (Yin et al., 2021). At the same time, the initiatives to create integrated nanosensors into plant tissues will provide direction insights into the nutrient status of plants or crops and their intake and utilization. These nano-enabled sensors potentially complemented with wireless communication technologies and data analytics platforms could give feedback systems in dynamic nutrient management (Nadporozhskaya et al., 2022). Peasants and machines could get notifications of their scarcity or excess consequently to help farmers adjust them at the right time. The ability to monitor the fields in real time would greatly decrease fertilizer losses and increase fertilizer use efficiency and potentially reduce unfavorable effects on the environment (Shanmugavel et al., 2023). Moreover, the information recorded by these nanosensors may potentially help develop even more realistic forecasts of how crops would develop and what nutrients they'd require to be given in the future, thus gradually increasing the optimality of agricultural processes as a whole over the course of time.

9.4 Combining Nano-Nutrients With Other Beneficial Nanoparticles

The future Nano agriculture technology includes target of using single nanoparticles to deliver nutrients and other agro-chemicals of benefit (Mishra et al., 2017). Scientists are looking at methods by which nutrients and pesticides can be encapsulated in a nanoparticle, thereby lessening the quantity of chemicals used and its effect on the ecosystem. A nanoparticle could be built to carry nutrients and gradually release them while at the same time controlling pests (Wakeil et al., 2017). Another potential area of research is the use of nano-nutrient chelated with growth promoters or plant hormone because this combination can improve nutrient availability and increased plant growth and stress tolerance. Nano-systems that involve the use of nano-enabled seed coatings containing multiple agrochemicals are also being pursued (Zargar et al., 2023). Such coatings could deliver seeds all nutrients, growth promoters, and protectants which are required by plants during germination and initial growth to minimize the requirement of further treatment. However, self-assemblies of these combined nanoparticles and plant systems are quite challenging and extensive research has to be undertaken in order to determine the efficacy and safety of the system (Yadav et al., 2020). They will include further research on the ratios and temporal release profiles of various components, and the possible interaction or cross-inhibition that may occur. In its progression, it may result to the development of multifunctional technologies suitable for enhancing crop production in an environmentally sustainable manner.

10. CONCLUSION

Nano technologies in particular have showcased some potential solutions to the issues with nutrient delivery and producing crops sustainably. As a result of incorporating nanomaterial characteristics into these systems, the developed systems can therefore boast of better nutrient use efficiency, targeted delivery, and controlled release when compared to customary fertilizers. Nano-nutrient uptake through root absorption, foliar absorption and enhanced cellular internalization are the pathways that deliver various nutrients to the plants effectively. Despite the great opportunities, several obstacles are still to be discussed; these include the ability to expand the use of these technologies, the variations in productivity and effectiveness in varying crops and climate conditions, and overall concerns about environmental impacts. Another form of challenge includes the social attitudes as well as technical issues that arise in formulation and application of coatings also present challenges to widespread implementation. There are a number of future prospects for developing the nano-enabled nutrient delivery systems in conjunction with precision agriculture technologies. There is also more potential to improve sustainability and efficiency through the creation of biodegradable nano-carriers and the ability to monitor nutrient levels in real-time. For this reason, when integrating nano-nutrients with other functional nanoparticles, there may be a range of synergistic benefits for crops. When doing research, one has to be mindful of which approaches yield both good crop productivity and environmental sustainability. More research needs to be carried out on the impacts of nanomaterials on the soil and food security after the long-term release of nanomaterials. Based on these findings, nano-optimized nutrient delivery methods could go a long way in providing food security in the face of a growing global population and with a succinctly low effect on the environment, represents the future of sustainable agriculture.

REFERENCES

Abdel-Aziz, H. M. M., Hasaneen, M. N. A., & Omer, A. M. (2016). Nano chitosan-NPK fertilizer enhances the growth and productivity of wheat plants grown in sandy soil. *Spanish Journal of Agricultural Research, 14*(1), e0902. DOI: 10.5424/sjar/2016141-8205

Acharya, A., & Pal, P. K. (2020). Agriculture nanotechnology: Translating research outcome to field applications by influencing environmental sustainability. *NanoImpact, 19*, 100232. DOI: 10.1016/j.impact.2020.100232

Ahari, H., & Naeimabadi, M. (2021). Employing Nanoemulsions in Food Packaging: Shelf Life Enhancement. *Food Engineering Reviews, 13*(4), 858–883. DOI: 10.1007/s12393-021-09282-z

Ahmar, S., Mahmood, T., Fiaz, S., Mora-Poblete, F., Shafique, M. S., Chattha, M. S., & Jung, K.-H. (2021). Advantage of Nanotechnology-Based Genome Editing System and Its Application in Crop Improvement. *Frontiers in Plant Science, 12*, 663849. Advance online publication. DOI: 10.3389/fpls.2021.663849 PMID: 34122485

Ahmed, B., Rizvi, A., Ali, K., Lee, J., Zaidi, A., Khan, M. S., & Musarrat, J. (2021). Nanoparticles in the soil–plant system: A review. *Environmental Chemistry Letters, 19*(2), 1545–1609. DOI: 10.1007/s10311-020-01138-y

Akyildiz, I. F., & Jornet, J. M. (2010). Electromagnetic wireless nanosensor networks. *Nano Communication Networks, 1*(1), 3–19. DOI: 10.1016/j.nancom.2010.04.001

Alfei, S., Marengo, B., & Zuccari, G. (2020). Nanotechnology application in food packaging: A plethora of opportunities versus pending risks assessment and public concerns. *Food Research International, 137*, 109664. DOI: 10.1016/j.foodres.2020.109664 PMID: 33233243

Ang, M. C.-Y., & Lew, T. T. S. (2022). Non-destructive Technologies for Plant Health Diagnosis. *Frontiers in Plant Science, 13*, 884454. Advance online publication. DOI: 10.3389/fpls.2022.884454 PMID: 35712566

Ansari, M. A. (2023). Nanotechnology in Food and Plant Science: Challenges and Future Prospects. *Plants, 12*(13), 2565. DOI: 10.3390/plants12132565 PMID: 37447126

Atapattu, A. J., Ranasinghe, C. S., Nuwarapaksha, T. D., Udumann, S. S., & Dissanayaka, N. S. (2024a). Sustainable Agriculture and Sustainable Development Goals (SDGs). In *Emerging Technologies and Marketing Strategies for Sustainable Agriculture* (pp. 1-27). IGI Global. https://doi.org/DOI: 10.4018/979-8-3693-4864-2.ch001

Atapattu, A. J., Udumann, S. S., Nuwarapaksha, T. D., & Dissanayaka, N. S. (2024b). Upcycling Coconut Husk By-Products: Transitioning from Traditional Applications to Emerging High-Value Usages. In Neelancherry, R., Gao, B., & Wisniewski, A.Jr., (Eds.), *Agricultural Waste to Value-Added Products*. Springer., DOI: 10.1007/978-981-97-2535-9_12

Avila-Quezada, G. D., Ingle, A. P., Golińska, P., & Rai, M. (2022). Strategic applications of nano-fertilizers for sustainable agriculture: Benefits and bottlenecks. *Nanotechnology Reviews*, *11*(1), 2123–2140. DOI: 10.1515/ntrev-2022-0126

Banerjee, K., Pramanik, P., Maity, A., Joshi, D. C., Wani, S. H., & Krishnan, P. (2019). Methods of Using Nanomaterials to Plant Systems and Their Delivery to Plants (Mode of Entry, Uptake, Translocation, Accumulation, Biotransformation and Barriers). In *Advances in Phytonanotechnology* (pp. 123–152). Elsevier. DOI: 10.1016/B978-0-12-815322-2.00005-5

Barbu, E., Molnàr, É., Tsibouklis, J., & Górecki, D. C. (2009). The potential for nanoparticle-based drug delivery to the brain: Overcoming the blood–brain barrier. *Expert Opinion on Drug Delivery*, *6*(6), 553–565. DOI: 10.1517/17425240902939143 PMID: 19435406

Beegum, S., & Das, S. (2022). Nanosensors in agriculture. In *Agricultural Nanobiotechnology* (pp. 465–478). Elsevier., DOI: 10.1016/B978-0-323-91908-1.00012-2

Bergström, L., Bowman, B. T., & Sims, J. T. (2005). Definition of sustainable and unsustainable issues in nutrient management of modern agriculture. *Soil Use and Management*, *21*(1), 76–81. DOI: 10.1111/j.1475-2743.2005.tb00111.x

Bertola, M., Ferrarini, A., & Visioli, G. (2021). Improvement of Soil Microbial Diversity through Sustainable Agricultural Practices and Its Evaluation by -Omics Approaches: A Perspective for the Environment, Food Quality and Human Safety. *Microorganisms*, *9*(7), 1400. DOI: 10.3390/microorganisms9071400 PMID: 34203506

Bhagat, R., Ingle, A. P., & Chen, H. (2023). Nanosensors and nanobiosensors for sustainable agriculture. In *Nanotechnology in Agriculture and Agroecosystems* (pp. 93–112). Elsevier., DOI: 10.1016/B978-0-323-99446-0.00014-3

Bongiovanni, R., & Lowenberg-Deboer, J. (2004). Precision Agriculture and Sustainability. *Precision Agriculture*, *5*(4), 359–387. DOI: 10.1023/B:PRAG.0000040806.39604.aa

Bosona, T., & Gebresenbet, G. (2023). The Role of Blockchain Technology in Promoting Traceability Systems in Agri-Food Production and Supply Chains. *Sensors (Basel)*, *23*(11), 5342. DOI: 10.3390/s23115342 PMID: 37300069

Cambra Baseca, C., Sendra, S., Lloret, J., & Tomas, J. (2019). A Smart Decision System for Digital Farming. *Agronomy (Basel)*, *9*(5), 216. DOI: 10.3390/agronomy9050216

Chao, G. A. O., Sun, B., & Zhang, T. L. (2006). Sustainable Nutrient Management in Chinese Agriculture: Challenges and Perspective. *Pedosphere*, *16*(2), 253–263. DOI: 10.1016/S1002-0160(06)60051-9

Cohen, A. W., Hnasko, R., Schubert, W., & Lisanti, M. P. (2004). Role of Caveolae and Caveolins in Health and Disease. *Physiological Reviews*, *84*(4), 1341–1379. DOI: 10.1152/physrev.00046.2003 PMID: 15383654

Coles, D., & Frewer, L. J. (2013). Nanotechnology applied to European food production – A review of ethical and regulatory issues. *Trends in Food Science & Technology*, *34*(1), 32–43. DOI: 10.1016/j.tifs.2013.08.006

Dasauni, K., & Nailwal, T. K. (2022). Nano–delivery system: In the agriculture sector. In *Nano-enabled Agrochemicals in Agriculture* (pp. 467–484). Elsevier., DOI: 10.1016/B978-0-323-91009-5.00014-8

Deng, H., Dutta, P., & Liu, J. (2018). Stochastic simulations of nanoparticle internalization through transferrin receptor dependent clathrin-mediated endocytosis. *Biochimica et Biophysica Acta (BBA) -. Biochimica et Biophysica Acta. G, General Subjects*, *1862*(9), 2104–2111. DOI: 10.1016/j.bbagen.2018.06.018 PMID: 29959983

Dhanaraju, M., Chenniappan, P., Ramalingam, K., Pazhanivelan, S., & Kaliaperumal, R. (2022). Smart Farming: Internet of Things (IoT)-Based Sustainable Agriculture. *Agriculture*, *12*(10), 1745. DOI: 10.3390/agriculture12101745

Dissanayaka, D. M. N. S., Dissanayake, D. K. R. P. L., Udumann, S. S., Nuwarapaksha, T. D., & Atapattu, A. J. (2023a). Agroforestry a key tool in the climate-smart agriculture context: A review on coconut cultivation in Sri Lanka. *Frontiers in Agronomy*, *5*, 1162750. DOI: 10.3389/fagro.2023.1162750

Dissanayaka, N. S., Udumann, S. S., Nuwarapaksha, T. D., & Atapattu, A. J. (2023b). Agroforestry: An Avenue for Resilient and Productive Farming through Integrated Crops and Livestock Production. In Kiba, D. I. (Ed.), *Transitioning to zero hunger* (pp. 115–136). MDPI Books., DOI: 10.3390/books978-3-03897-863-3-5

El Wakeil, N., Alkahtani, S., & Gaafar, N. (2017). Is nanotechnology a promising field for insect pest control in IPM programs? In *New Pesticides and Soil Sensors* (pp. 273–309). Elsevier., DOI: 10.1016/B978-0-12-804299-1.00008-4

Elemike, E. E., Uzoh, I. M., Onwudiwe, D. C., & Babalola, O. O. (2019). The Role of Nanotechnology in the Fortification of Plant Nutrients and Improvement of Crop Production. *Applied Sciences (Basel, Switzerland)*, *9*(3), 499. DOI: 10.3390/app9030499

Etesami, H., Jeong, B. R., & Glick, B. R. (2021). Contribution of Arbuscular Mycorrhizal Fungi, Phosphate–Solubilizing Bacteria, and Silicon to P Uptake by Plant. *Frontiers in Plant Science*, *12*, 699618. Advance online publication. DOI: 10.3389/fpls.2021.699618 PMID: 34276750

Farhangi-Abriz, S., & Ghassemi-Golezani, K. (2023). Nutrient Availability Regulates Root System Behavior. In *Plant Ionomics* (pp. 96–119). Wiley., DOI: 10.1002/9781119803041.ch6

Francia, V., Montizaan, D., & Salvati, A. (2020). Interactions at the cell membrane and pathways of internalization of nano-sized materials for nanomedicine. *Beilstein Journal of Nanotechnology*, *11*, 338–353. DOI: 10.3762/bjnano.11.25 PMID: 32117671

Grieger, K. D., Hansen, S. F., & Baun, A. (2009). The known unknowns of nanomaterials: Describing and characterizing uncertainty within environmental, health and safety risks. *Nanotoxicology*, *3*(3), 222–233. DOI: 10.1080/17435390902944069

Guo, H., White, J. C., Wang, Z., & Xing, B. (2018). Nano-enabled fertilizers to control the release and use efficiency of nutrients. *Current Opinion in Environmental Science & Health*, *6*, 77–83. DOI: 10.1016/j.coesh.2018.07.009

Hajji-Hedfi, L., & Chhipa, H. (2021). Nano-based pesticides: Challenges for pest and disease management. *Euro-Mediterranean Journal for Environmental Integration*, *6*(3), 69. DOI: 10.1007/s41207-021-00279-y

Hardy, A., Benford, D., Halldorsson, T., Jeger, M. J., Knutsen, H. K., More, S., Naegeli, H., Noteborn, H., Ockleford, C., Ricci, A., Rychen, G., Schlatter, J. R., Silano, V., Solecki, R., Turck, D., Younes, M., Chaudhry, Q., Cubadda, F., Gott, D., & Mortensen, A. (2018). Guidance on risk assessment of the application of nanoscience and nanotechnologies in the food and feed chain: Part 1, human and animal health. *EFSA Journal*, *16*(7). Advance online publication. DOI: 10.2903/j.efsa.2018.5327 PMID: 32625968

Harish, V., Tewari, D., Gaur, M., Yadav, A. B., Swaroop, S., Bechelany, M., & Barhoum, A. (2022). Review on Nanoparticles and Nanostructured Materials: Bioimaging, Biosensing, Drug Delivery, Tissue Engineering, Antimicrobial, and Agro-Food Applications. *Nanomaterials (Basel, Switzerland)*, *12*(3), 457. DOI: 10.3390/nano12030457 PMID: 35159802

Haydar, M. S., Ghosh, D., & Roy, S. (2024). Slow and controlled release nanofertilizers as an efficient tool for sustainable agriculture: Recent understanding and concerns. *Plant Nano Biology*, *7*, 100058. DOI: 10.1016/j.plana.2024.100058

Hayles, J., Johnson, L., Worthley, C., & Losic, D. (2017). Nanopesticides: a review of current research and perspectives. In *New Pesticides and Soil Sensors* (pp. 193–225). Elsevier. DOI: 10.1016/B978-0-12-804299-1.00006-0

Hedley, C. (2015). The role of precision agriculture for improved nutrient management on farms. *Journal of the Science of Food and Agriculture*, *95*(1), 12–19. DOI: 10.1002/jsfa.6734 PMID: 24816925

Hofmann, T., Lowry, G. V., Ghoshal, S., Tufenkji, N., Brambilla, D., Dutcher, J. R., Gilbertson, L. M., Giraldo, J. P., Kinsella, J. M., Landry, M. P., Lovell, W., Naccache, R., Paret, M., Pedersen, J. A., Unrine, J. M., White, J. C., & Wilkinson, K. J. (2020). Technology readiness and overcoming barriers to sustainably implement nanotechnology-enabled plant agriculture. *Nature Food*, *1*(7), 416–425. DOI: 10.1038/s43016-020-0110-1

Huang, Q., Yu, H., & Ru, Q. (2010). Bioavailability and Delivery of Nutraceuticals Using Nanotechnology. *Journal of Food Science*, *75*(1). Advance online publication. DOI: 10.1111/j.1750-3841.2009.01457.x PMID: 20492195

Inoue, Y. (2020). Satellite- and drone-based remote sensing of crops and soils for smart farming – a review. *Soil Science and Plant Nutrition*, *66*(6), 798–810. DOI: 10.1080/00380768.2020.1738899

Jakhar, A. M., Aziz, I., Kaleri, A. R., Hasnain, M., Haider, G., Ma, J., & Abideen, Z. (2022). Nano-fertilizers: A sustainable technology for improving crop nutrition and food security. *NanoImpact*, *27*, 100411. DOI: 10.1016/j.impact.2022.100411 PMID: 35803478

Kah, M., Kookana, R. S., Gogos, A., & Bucheli, T. D. (2018). A critical evaluation of nanopesticides and nanofertilizers against their conventional analogues. *Nature Nanotechnology*, *13*(8), 677–684. DOI: 10.1038/s41565-018-0131-1 PMID: 29736032

Kalia, A., Sharma, S. P., Kaur, H., & Kaur, H. (2020). Novel nanocomposite-based controlled-release fertilizer and pesticide formulations: Prospects and challenges. In *Multifunctional Hybrid Nanomaterials for Sustainable Agri-Food and Ecosystems* (pp. 99–134). Elsevier., DOI: 10.1016/B978-0-12-821354-4.00005-4

Kanjana, D. (2017). Advancement of Nanotechnology Applications on Plant Nutrients Management and Soil Improvement. In *Nanotechnology* (pp. 209–234). Springer Singapore., DOI: 10.1007/978-981-10-4678-0_12

Kannan, S. (2010). *Foliar Fertilization for Sustainable Crop Production.*, DOI: 10.1007/978-90-481-8741-6_13

Karimi, M., Eslami, M., Sahandi-Zangabad, P., Mirab, F., Farajisafiloo, N., Shafaei, Z., Ghosh, D., Bozorgomid, M., Dashkhaneh, F., & Hamblin, M. R. (2016). <scp>pH</scp> -Sensitive stimulus-responsive nanocarriers for targeted delivery of therapeutic agents. *Wiley Interdisciplinary Reviews. Nanomedicine and Nanobiotechnology*, *8*(5), 696–716. DOI: 10.1002/wnan.1389 PMID: 26762467

Kausar, A., Ahmad, I., Maaza, M., & Eisa, M. H. (2022). State-of-the-Art Nanoclay Reinforcement in Green Polymeric Nanocomposite: From Design to New Opportunities. *Minerals (Basel)*, *12*(12), 1495. DOI: 10.3390/min12121495

Kausar, A., Rafique, I., & Muhammad, B. (2017). Aerospace Application of Polymer Nanocomposite with Carbon Nanotube, Graphite, Graphene Oxide, and Nanoclay. *Polymer-Plastics Technology and Engineering*, *56*(13), 1438–1456. DOI: 10.1080/03602559.2016.1276594

Kaźmierczak, Z., Szostak-Paluch, K., Przybyło, M., Langner, M., Witkiewicz, W., Jędruchniewicz, N., & Dąbrowska, K. (2020). Endocytosis in cellular uptake of drug delivery vectors: Molecular aspects in drug development. *Bioorganic & Medicinal Chemistry*, *28*(18), 115556. DOI: 10.1016/j.bmc.2020.115556 PMID: 32828419

Kou, L., Sun, J., Zhai, Y., & He, Z. (2013). The endocytosis and intracellular fate of nanomedicines: Implication for rational design. *Asian Journal of Pharmaceutical Sciences*, *8*(1), 1–10. DOI: 10.1016/j.ajps.2013.07.001

Kuang, L., Burgess, B., Cuite, C. L., Tepper, B. J., & Hallman, W. K. (2020). Sensory acceptability and willingness to buy foods presented as having benefits achieved through the use of nanotechnology. *Food Quality and Preference*, *83*, 103922. DOI: 10.1016/j.foodqual.2020.103922

Kuhn, R., Bryant, I. M., Jensch, R., & Böllmann, J. (2022). Applications of Environmental Nanotechnologies in Remediation, Wastewater Treatment, Drinking Water Treatment, and Agriculture. *Applied Nanoscience*, *3*(1), 54–90. DOI: 10.3390/applnano3010005

Kumar, L., Chhogyel, N., Gopalakrishnan, T., Hasan, M. K., Jayasinghe, S. L., Kariyawasam, C. S., Kogo, B. K., & Ratnayake, S. (2022). Climate change and future of agri-food production. In *Future Foods* (pp. 49–79). Elsevier., DOI: 10.1016/B978-0-323-91001-9.00009-8

Kumari, A., Rana, V., Yadav, S. K., & Kumar, V. (2023). Nanotechnology as a powerful tool in plant sciences: Recent developments, challenges and perspectives. *Plant Nano Biology*, *5*, 100046. DOI: 10.1016/j.plana.2023.100046

Kumari, R., Suman, K., Karmakar, S., Mishra, V., Lakra, S. G., Saurav, G. K., & Mahto, B. K. (2023). Regulation and safety measures for nanotechnology-based agri-products. *Frontiers in Genome Editing*, *5*, 1200987. Advance online publication. DOI: 10.3389/fgeed.2023.1200987 PMID: 37415849

Lew, T. T. S., Koman, V. B., Gordiichuk, P., Park, M., & Strano, M. S. (2020). The Emergence of Plant Nanobionics and Living Plants as Technology. *Advanced Materials Technologies*, *5*(3), 1900657. Advance online publication. DOI: 10.1002/admt.201900657

Liu, W., Zhang, M., & Bhandari, B. (2020). Nanotechnology – A shelf life extension strategy for fruits and vegetables. *Critical Reviews in Food Science and Nutrition*, *60*(10), 1706–1721. DOI: 10.1080/10408398.2019.1589415 PMID: 30896257

Lombardo, D., Kiselev, M. A., & Caccamo, M. T. (2019). Smart Nanoparticles for Drug Delivery Application: Development of Versatile Nanocarrier Platforms in Biotechnology and Nanomedicine. *Journal of Nanomaterials*, *2019*, 1–26. DOI: 10.1155/2019/3702518

Luna Juncal, M. J., Masino, P., Bertone, E., & Stewart, R. A. (2023). Towards nutrient neutrality: A review of agricultural runoff mitigation strategies and the development of a decision-making framework. *The Science of the Total Environment*, *874*, 162408. DOI: 10.1016/j.scitotenv.2023.162408 PMID: 36828071

Ma, X., Geiser-Lee, J., Deng, Y., & Kolmakov, A. (2010). Interactions between engineered nanoparticles (ENPs) and plants: Phytotoxicity, uptake and accumulation. *The Science of the Total Environment*, *408*(16), 3053–3061. DOI: 10.1016/j.scitotenv.2010.03.031 PMID: 20435342

Mahapatra, D. M., Satapathy, K. C., & Panda, B. (2022). Biofertilizers and nanofertilizers for sustainable agriculture: Phycoprospects and challenges. *The Science of the Total Environment*, *803*, 149990. DOI: 10.1016/j.scitotenv.2021.149990 PMID: 34492488

Malézieux, E. (2012). Designing cropping systems from nature. *Agronomy for Sustainable Development*, *32*(1), 15–29. DOI: 10.1007/s13593-011-0027-z

Marchiol, L., Iafisco, M., Fellet, G., & Adamiano, A. (2020). *Nanotechnology support the next agricultural revolution: Perspectives to enhancement of nutrient use efficiency* (pp. 27–116). DOI: 10.1016/bs.agron.2019.12.001

Mayor, S., & Pagano, R. E. (2007). Pathways of clathrin-independent endocytosis. *Nature Reviews. Molecular Cell Biology*, *8*(8), 603–612. DOI: 10.1038/nrm2216 PMID: 17609668

Mishra, S., Keswani, C., Abhilash, P. C., Fraceto, L. F., & Singh, H. B. (2017). Integrated Approach of Agri-nanotechnology: Challenges and Future Trends. *Frontiers in Plant Science*, *8*. Advance online publication. DOI: 10.3389/fpls.2017.00471 PMID: 28421100

Mittal, D., Kaur, G., Singh, P., Yadav, K., & Ali, S. A. (2020). Nanoparticle-Based Sustainable Agriculture and Food Science: Recent Advances and Future Outlook. *Frontiers in Nanotechnology*, *2*, 579954. Advance online publication. DOI: 10.3389/fnano.2020.579954

Monreal, C. M., DeRosa, M., Mallubhotla, S. C., Bindraban, P. S., & Dimkpa, C. (2016). Nanotechnologies for increasing the crop use efficiency of fertilizer-micronutrients. *Biology and Fertility of Soils*, *52*(3), 423–437. DOI: 10.1007/s00374-015-1073-5

Murugan, K., & Abd-Elsalam, K. A. (2022). Sustainable nanoemulsions for agri-food applications: Today and future trends. In *Bio-Based Nanoemulsions for Agri-Food Applications* (pp. 1–11). Elsevier., DOI: 10.1016/B978-0-323-89846-1.00012-7

Nadporozhskaya, M., Kovsh, N., Paolesse, R., & Lvova, L. (2022). Recent Advances in Chemical Sensors for Soil Analysis: A Review. *Chemosensors (Basel, Switzerland)*, *10*(1), 35. DOI: 10.3390/chemosensors10010035

Nongbet, A., Mishra, A. K., Mohanta, Y. K., Mahanta, S., Ray, M. K., Khan, M., Baek, K.-H., & Chakrabartty, I. (2022). Nanofertilizers: A Smart and Sustainable Attribute to Modern Agriculture. *Plants*, *11*(19), 2587. DOI: 10.3390/plants11192587 PMID: 36235454

Novaković, L., Guo, T., Bacic, A., Sampathkumar, A., & Johnson, K. L. (2018). Hitting the Wall—Sensing and Signaling Pathways Involved in Plant Cell Wall Remodeling in Response to Abiotic Stress. *Plants*, *7*(4), 89. DOI: 10.3390/plants7040089 PMID: 30360552

Nuwarapaksha, T. D., Dissanayake, W. K., Gunathilaka, W. S., Udumann, S. S., Dissanayaka, N. S., & Atapattu, A. J. (2024). Assessing the Optimum Harvesting Stage of *Tithonia diversifolia* as Climate Smart Soil Amendment for Coconut Plantations. *Biol. Life Sci. Forum*, *30*, 1. DOI: 10.3390/IOCAG2023-15967

Nuwarapaksha, T. D., Udumann, S. S., Dissanayaka, D. M. N. S., Dissanayake, D. K. R. P. L., & Atapattu, A. J. (2022). Coconut based multiple cropping systems: An analytical review in Sri Lankan coconut cultivations. *Circular Agricultural Systems, 2*(1), 8. DOI: 10.48130/CAS-2022-0008

Nuwarapaksha, T. D., Udumann, S. S., Dissanayaka, N. S., & Atapattu, A. J. (2023). Coconut-Based Livestock Farming: A Sustainable Approach to Enhancing Food Security in Sri Lanka. In Kiba, D. I. (Ed.), *Transitioning to Zero Hunger* (pp. 197–213). MDPI Books., DOI: 10.3390/books978-3-03897-863-3-8

Operti, M. C., Bernhardt, A., Grimm, S., Engel, A., Figdor, C. G., & Tagit, O. (2021). PLGA-based nanomedicines manufacturing: Technologies overview and challenges in industrial scale-up. *International Journal of Pharmaceutics, 605*, 120807. DOI: 10.1016/j.ijpharm.2021.120807 PMID: 34144133

Pan, B., & Xing, B. (2012). Applications and implications of manufactured nanoparticles in soils: A review. *European Journal of Soil Science, 63*(4), 437–456. DOI: 10.1111/j.1365-2389.2012.01475.x

Pandey, S., Sharma, K., & Gundabala, V. (2022). Antimicrobial bio-inspired active packaging materials for shelf life and safety development: A review. *Food Bioscience, 48*, 101730. DOI: 10.1016/j.fbio.2022.101730

Pillay, V., Murugan, K., Choonara, Y. E., Kumar, P., Bijukumar, D., & du Toit, L. C. (2015). Parameters and characteristics governing cellular internalization and trans-barrier trafficking of nanostructures. *International Journal of Nanomedicine, 2191*, 2191. Advance online publication. DOI: 10.2147/IJN.S75615

Pradhan, N., Singh, S., Ojha, N., Shrivastava, A., Barla, A., Rai, V., & Bose, S. (2015). Facets of Nanotechnology as Seen in Food Processing, Packaging, and Preservation Industry. *BioMed Research International, 2015*, 1–17. DOI: 10.1155/2015/365672 PMID: 26613082

Pradhan, S., & Mailapalli, D. R. (2020). *Nanopesticides for Pest Control.*, DOI: 10.1007/978-3-030-33281-5_2

Raffi, M. M., & Husen, A. (2019). Impact of Fabricated Nanoparticles on the Rhizospheric Microorganisms and Soil Environment. In *Nanomaterials and Plant Potential* (pp. 529–552). Springer International Publishing., DOI: 10.1007/978-3-030-05569-1_21

Rahman, K., & Zhang, D. (2018). Effects of Fertilizer Broadcasting on the Excessive Use of Inorganic Fertilizers and Environmental Sustainability. *Sustainability (Basel), 10*(3), 759. DOI: 10.3390/su10030759

Rahman, M., Haque, K. M. S., & Khan, M. Z. H. (2021). A review on application of controlled released fertilizers influencing the sustainable agricultural production: A Cleaner production process. *Environmental Technology & Innovation*, *23*, 101697. DOI: 10.1016/j.eti.2021.101697

Rai, M., Ingle, A. P., Pandit, R., Paralikar, P., Shende, S., Gupta, I., Biswas, J. K., & da Silva, S. S. (2018). Copper and copper nanoparticles: Role in management of insect-pests and pathogenic microbes. *Nanotechnology Reviews*, *7*(4), 303–315. DOI: 10.1515/ntrev-2018-0031

Ramezani, M., Ramezani, F., & Gerami, M. (2019). Nanoparticles in Pest Incidences and Plant Disease Control. In *Nanotechnology for Agriculture: Crop Production & Protection* (pp. 233–272). Springer Singapore. DOI: 10.1007/978-981-32-9374-8_12

Rani, N., Kumari, K., Sangwan, P., Barala, P., Yadav, J., Vijeta, , Rahul, , & Hooda, V. (2022). Nano-Iron and Nano-Zinc Induced Growth and Metabolic Changes in Vigna radiata. *Sustainability (Basel)*, *14*(14), 8251. DOI: 10.3390/su14148251

Raza, A., Charagh, S., Salehi, H., Abbas, S., Saeed, F., Poinern, G. E. J., Siddique, K. H. M., & Varshney, R. K. (2023). Nano-enabled stress-smart agriculture: Can nanotechnology deliver drought and salinity-smart crops? *Journal of Sustainable Agriculture and Environment*, *2*(3), 189–214. DOI: 10.1002/sae2.12061

Ruttkay-Nedecky, B., Krystofova, O., Nejdl, L., & Adam, V. (2017). Nanoparticles based on essential metals and their phytotoxicity. *Journal of Nanobiotechnology*, *15*(1), 33. DOI: 10.1186/s12951-017-0268-3 PMID: 28446250

Salama, H. M. H. (2012). Effects of silver nanoparticles in some crop plants, Common bean (*Phaseolus vulgaris* L.) and corn *(Zea mays L.)*. In *International. Research Journal of Biotechnology*, *3*, 10.

Schwab, F., Zhai, G., Kern, M., Turner, A., Schnoor, J. L., & Wiesner, M. R. (2016). Barriers, pathways and processes for uptake, translocation and accumulation of nanomaterials in plants – Critical review. *Nanotoxicology*, *10*(3), 257–278. DOI: 10.3109/17435390.2015.1048326 PMID: 26067571

Sekhon, B. (2014). Nanotechnology in agri-food production: An overview. *Nanotechnology, Science and Applications*, *31*, 31. Advance online publication. DOI: 10.2147/NSA.S39406 PMID: 24966671

Sembada, A. A., & Lenggoro, I. W. (2024). Transport of Nanoparticles into Plants and Their Detection Methods. *Nanomaterials (Basel, Switzerland)*, *14*(2), 131. DOI: 10.3390/nano14020131 PMID: 38251096

Serri, C., Cruz-Maya, I., Bonadies, I., Rassu, G., Giunchedi, P., Gavini, E., & Guarino, V. (2023). Green Routes for Bio-Fabrication in Biomedical and Pharmaceutical Applications. *Pharmaceutics, 15*(6), 1744. DOI: 10.3390/pharmaceutics15061744 PMID: 37376192

Shanmugavel, D., Rusyn, I., Solorza-Feria, O., & Kamaraj, S.-K. (2023). Sustainable SMART fertilizers in agriculture systems: A review on fundamentals to in-field applications. *The Science of the Total Environment, 904*, 166729. DOI: 10.1016/j.scitotenv.2023.166729 PMID: 37678530

Sharma, R., & Kumar, V. (2024). Nano enabled agriculture for sustainable soil. *Waste Management Bulletin, 2*(1), 152–161. DOI: 10.1016/j.wmb.2024.01.002

Tiwari, S., Upadhyay, N., Singh, B. K., Singh, V. K., & Dubey, N. K. (2022). Facile Fabrication of Nanoformulated Cinnamomum glaucescens Essential Oil as a Novel Green Strategy to Boost Potency Against Food Borne Fungi, Aflatoxin Synthesis, and Lipid Oxidation. *Food and Bioprocess Technology, 15*(2), 319–337. DOI: 10.1007/s11947-021-02739-3

Toksha, B., Sonawale, V. A. M., Vanarase, A., Bornare, D., Tonde, S., Hazra, C., Kundu, D., Satdive, A., Tayde, S., & Chatterjee, A. (2021). Nanofertilizers: A review on synthesis and impact of their use on crop yield and environment. *Environmental Technology & Innovation, 24*, 101986. DOI: 10.1016/j.eti.2021.101986

Tripathi, S., Bahuguna, R. N., Shrivastava, N., Singh, S., Chatterjee, A., Varma, A., & Jagadish, S. K. (2022). Microbial biofortification: A sustainable route to grow nutrient-rich crops under changing climate. *Field Crops Research, 287*, 108662. DOI: 10.1016/j.fcr.2022.108662

Tyagi, S., Rajput, P., Pramanik, A., Dixit, V., Sinha, A., Pal, A. K., Chaubey, K. K., Manjunath, B. T., Verma, D. K., Bachheti, R. K., & Bachheti, A. (2023). *Plant Growth-Promoting Rhizobacteria Nanoemulsion and Their Applications.*, DOI: 10.1007/978-981-99-5398-1_7

Uddin, M., Desai, F., & Asmatulu, E. (2020). Engineered nanomaterials in the environment: Bioaccumulation, biomagnification and biotransformation. *Environmental Chemistry Letters, 18*(4), 1073–1083. DOI: 10.1007/s10311-019-00947-0

Vasuki, A., Paulpandi, V. K., Singh, R. D., Gurusamy, A., Mahendran, P. P., Sivakumar, T., & Prabakaran, K. (2023). Influence of Irrigation Methods and Nanofertilizers Application on the Yield of Transplanted Lowland Rice (Oryza sativa L.) in Periyar Vaigai Command Area of Madurai. *Agricultural Science Digest - A Research Journal, Of.* DOI: 10.18805/ag.D-5840

Vega-Vásquez, P., Mosier, N. S., & Irudayaraj, J. (2020). Nanoscale Drug Delivery Systems: From Medicine to Agriculture. *Frontiers in Bioengineering and Biotechnology*, *8*, 79. Advance online publication. DOI: 10.3389/fbioe.2020.00079 PMID: 32133353

Vurro, M., Miguel-Rojas, C., & Pérez-de-Luque, A. (2019). Safe nanotechnologies for increasing the effectiveness of environmentally friendly natural agrochemicals. *Pest Management Science*, *75*(9), 2403–2412. DOI: 10.1002/ps.5348 PMID: 30672106

Wahab, A., Muhammad, M., Ullah, S., Abdi, G., Shah, G. M., Zaman, W., & Ayaz, A. (2024). Agriculture and environmental management through nanotechnology: Eco-friendly nanomaterial synthesis for soil-plant systems, food safety, and sustainability. *The Science of the Total Environment*, *926*, 171862. DOI: 10.1016/j.scitotenv.2024.171862 PMID: 38527538

Wang, X., Xie, H., Wang, P., & Yin, H. (2023). Nanoparticles in Plants: Uptake, Transport and Physiological Activity in Leaf and Root. *Materials (Basel)*, *16*(8), 3097. DOI: 10.3390/ma16083097 PMID: 37109933

Wang, Z., Fu, S., Guo, Y., Han, Y., Ma, C., Li, R., & Yang, X. (2024). Classification and design strategies of polysaccharide-based nano-nutrient delivery systems for enhanced bioactivity and targeted delivery: A review. *International Journal of Biological Macromolecules*, *256*, 128440. DOI: 10.1016/j.ijbiomac.2023.128440 PMID: 38016614

Yadav, A., Yadav, K., & Abd-Elsalam, K. A. (2023a). Nanofertilizers: Types, Delivery and Advantages in Agricultural Sustainability. *Agrochemicals*, *2*(2), 296–336. DOI: 10.3390/agrochemicals2020019

Yadav, A., Yadav, K., Ahmad, R., & Abd-Elsalam, K. A. (2023b). Emerging Frontiers in Nanotechnology for Precision Agriculture: Advancements, Hurdles and Prospects. *Agrochemicals*, *2*(2), 220–256. DOI: 10.3390/agrochemicals2020016

Yadav, N., Garg, V. K., Chhillar, A. K., & Rana, J. S. (2023c). Recent advances in nanotechnology for the improvement of conventional agricultural systems: A review. *Plant Nano Biology*, *4*, 100032. DOI: 10.1016/j.plana.2023.100032

Yadav, S., Sharma, A. K., & Kumar, P. (2020). Nanoscale Self-Assembly for Therapeutic Delivery. *Frontiers in Bioengineering and Biotechnology*, *8*, 127. Advance online publication. DOI: 10.3389/fbioe.2020.00127 PMID: 32158749

Yamini, V., Shanmugam, V., Rameshpathy, M., Venkatraman, G., Ramanathan, G., & Garalleh, A. L. (2023). Environmental effects and interaction of nanoparticles on beneficial soil and aquatic microorganisms. *Environmental Research, 236*, 116776. DOI: 10.1016/j.envres.2023.116776 PMID: 37517486

Yayehrad, A. T., Siraj, E. A., Wondie, G. B., Alemie, A. A., Derseh, M. T., & Ambaye, A. S. (2021). Could Nanotechnology Help to End the Fight Against COVID-19? Review of Current Findings, Challenges and Future Perspectives. *International Journal of Nanomedicine, 16*, 5713–5743. DOI: 10.2147/IJN.S327334 PMID: 34465991

Yin, H., Cao, Y., Marelli, B., Zeng, X., Mason, A. J., & Cao, C. (2021). Soil Sensors and Plant Wearables for Smart and Precision Agriculture. *Advanced Materials, 33*(20), 2007764. Advance online publication. DOI: 10.1002/adma.202007764 PMID: 33829545

Zain, M., Ma, H., Ur Rahman, S., Nuruzzaman, M., Chaudhary, S., Azeem, I., Mehmood, F., Duan, A., & Sun, C. (2024). Nanotechnology in precision agriculture: Advancing towards sustainable crop production. *Plant Physiology and Biochemistry, 206*, 108244. DOI: 10.1016/j.plaphy.2023.108244 PMID: 38071802

Zaki, N. M., & Tirelli, N. (2010). Gateways for the intracellular access of nanocarriers: A review of receptor-mediated endocytosis mechanisms and of strategies in receptor targeting. *Expert Opinion on Drug Delivery, 7*(8), 895–913. DOI: 10.1517/17425247.2010.501792 PMID: 20629604

Zareabyaneh, H., & Bayatvarkeshi, M. (2015). Effects of slow-release fertilizers on nitrate leaching, its distribution in soil profile, N-use efficiency, and yield in potato crop. *Environmental Earth Sciences, 74*(4), 3385–3393. DOI: 10.1007/s12665-015-4374-y

Zargar, M., Bayat, M., Saquee, F. S., Diakite, S., Ramzanovich, N. M., & Akhmadovich, K. A. S. (2023). New Advances in Nano-Enabled Weed Management Using Poly(Epsilon-Caprolactone)-Based Nanoherbicides. *Agriculture, 13*(10), 2031. DOI: 10.3390/agriculture13102031

Chapter 7

Nano-Biochar in Sustainable Agriculture: Applications, Benefits, and Future Perspectives

Omkar Singh
http://orcid.org/0000-0003-1971-185X
Department of Agriculture, KVCOS, Swami Vivekanand Subharti University, India

Shivangi Singh
KVK Bijnor, Sardar Vallabhbhai Patel University of Agriculture and Technology, Meerut, India

Uday Pratap Shahi
http://orcid.org/0000-0002-7951-5087
Sardar Vallabhbhai Patel University of Agriculture and Technology, Meerut, India

Vaishali Singh
Sardar Vallabhbhai Patel University of Agriculture and Technology, Meerut, India

Krishna Kumar Singh
KVK Bijnor, Sardar Vallabhbhai Patel University of Agriculture and Technology, Meerut, India

Sakshi Singh
Faculty of Biology, Yerevan State University, Armenia

Tatevik Derdzyan
Faculty of Biology, Yerevan State University, Armenia

João Ricardo Sousa
http://orcid.org/0000-0003-4365-9086
Center for the Research and Technology of Agroenvironmental and Biological Sciences, Inov4Agro, Universidadede Trás-os-Montes e Alto Douro, Quinta de Prados, Vila Real, Portugal

ABSTRACT

Innovative soil amendment research is driven by the need for sustainable agriculture to provide global food security and reduce environmental damage. Biochar, made from organic feedstocks that promote soil health and sequester carbon. Bulk

DOI: 10.4018/979-8-3693-7006-3.ch007

characteristics may limit its efficacy. Nanotechnology can modify biochar's functioning. Sub-micrometer particle size, large surface area, and unique physicochemical properties make nano-biochar (NBC) a significant breakthrough. Nano-biochar in sustainable agriculture is covered in this chapter. NBC production and characterization are explained, along with its properties compared to regular biochar. In this chapter, we discuss NBC's many uses as a soil conditioner, remediator, agrochemical transporter, and plant stress reducer. A heavy metal contaminated soil remediation case study shows NBC's practicality and efficacy. Finally, the chapter discusses nano-biochar's role in next-generation sustainable agriculture systems.

1. INTRODUCTION

Feeding nine billion people by 2050 while maintaining the ecological integrity of our planet is a dual challenge in 21st century (Singh O, 2023). Traditional farming methods, which mostly depend on artificial fertilizers, pesticides, and intense tillage, have increased output significantly but come at a high environmental cost. These costs include widespread soil degradation, water pollution, biodiversity loss, and substantial greenhouse gas (GHG) emissions which collectively undermine the long-term sustainability of our food systems (Tilman et al., 2002; Saikanth et al, 2023; Singh et al., 2023). Over the past 20 years, biochar has garnered significant scientific attention in the pursuit of sustainable soil management solutions. According to Lehmann and Joseph (2015), biochar is a solid, carbon-rich substance created when organic biomass is thermally broken down (pyrolyzed) in an oxygen-limited environment. 'Terra Preta de Índio' (Amazonian dark earths), extremely fertile anthropogenic soils that have remained productive for generations, is where its conceptual origins lie (Glaser et al., 2001). Recent studies on biochar have demonstrated that it can enhance soil structure, increase nutrient and water retention, and stimulate beneficial microbial activity which contribute to increased soil fertility. A possible method for mitigating climate change through long-term carbon storage in soil is provided by its refractory carbon structure which also enables it to function as a substantial carbon sink (Woolf et al., 2010).

Despite these advantages, the physical and chemical properties of conventional, or macro-scale, biochar may constrain its effectiveness. The large particle size, typically greater than 1 mm, can result in non-uniform distribution within the soil matrix, limited diffusion rates, and reduced interaction with soil colloids. Consequently, researchers have explored methods to enhance biochar functionality, leading to the integration of nanotechnology with biochar technology.

Nanotechnology, defined as the manipulation of matter at the atomic and molecular levels (1-100 nm), has revolutionized various industries, including electronics and medicine (Siddiqui et al., 2025; Singh et al., 2025a-b). In agriculture, this approach, known as agri-nanotechnology, provides targeted solutions for environmental remediation, pest control, and nutrient management (Prasad et al., 2017; Singh et al., 2023; Singh et al., 2024a). Nano-biochar (NBC) is produced by reducing biochar particle size to the nano-scale, typically less than 100 nm or occasionally sub-micrometer levels. This reduction significantly increases specific surface area, pore volume, and the density of functional groups, resulting in enhanced reactivity compared to bulk biochar (Ok et al., 2021). Nano-biochar exhibits unique functional properties that distinguish it from conventional biochar. Its colloidal form enables improved dispersion in soil and water, more effective interactions with microorganisms and plant roots, and greater sorption capacity for pollutants and nutrients. These characteristics position NBC as a promising soil amendment capable of addressing limitations of traditional biochar and providing more targeted and efficient solutions for sustainable agriculture.

This chapter presents a comprehensive and critical analysis of the function of nano-biochar in this context. The synthesis and characterization procedures are first reviewed, with emphasis on the distinguishing features of nano-biochar compared to traditional biochar. Subsequently, the chapter examines its diverse applications, including enhancement of agricultural productivity, improvement of soil health, and remediation of contaminated environments. Recent scientific evidence supports a detailed discussion of the mechanisms underlying these effects. A relevant case study is included to illustrate real-world effectiveness. The chapter then critically evaluates the current challenges and limitations, including concerns about potential toxicity and environmental fate, before discussing future perspectives and research directions that may shape the development of this technology. This analysis aims to provide a comprehensive understanding of nano-biochar's potential to contribute to a more resilient and sustainable agricultural future.

2. SYNTHESIS AND CHARACTERIZATION OF NANO-BIOCHAR

The transition from bulk biochar to nano-biochar involves a fundamental shift in material properties, driven by a reduction in particle size to the sub-micrometer or nanometer scale. This transformation is achieved through specific synthesis strategies and post-production modifications. The resulting material possesses a unique suite of characteristics that must be meticulously evaluated to understand its behaviour and potential applications in agroecosystems.

2.1. Synthesis and Production Methods

Nano-biochar production can be broadly divided into two methods: (1) direct synthesis of nano-sized carbonaceous materials from biomass (bottom-up technique) and (2) post-production size reduction of conventional biochar (top-down approach).

2.1.1. Top-Down Approach: Size Reduction of Pre-formed Biochar

Currently, the most widely used and straightforward method for producing nanostructured biochar (NBC) involves the pyrolysis of various biomass feedstocks, including wood chips, manure, sewage sludge, and agricultural residues such as rice husk and straw. The properties of the resulting biochar, such as carbon content, aromaticity, and surface chemistry, are primarily influenced by pyrolysis conditions, particularly temperature and heating rate (Tomczyk et al., 2020). Higher pyrolysis temperatures, typically above 600°C, yield biochar with increased surface area and enhanced carbon stability, both of which are critical for subsequent nanosizing processes.

Once the bulk biochar is produced, mechanical grinding is employed to reduce its particle size.

- ***Ball milling:*** This method is the most popular. A revolving chamber filled with grinding media—such as steel or ceramic balls—is filled with bulk biochar. The biochar particles are broken down by the tumbling balls' repetitive impact and shearing forces. To attain the required particle size distribution, the size/type of grinding media, rotation speed, and duration can all be adjusted (Lyu et al., 2018). It is possible to mill both wet and dry. By adding a liquid medium, such as water or ethanol, wet milling can aid to reduce heat generation and avoid particle agglomeration, which frequently results in a more uniform and smaller particle size.
- ***Ultrasonication:*** In a liquid biochar suspension, cavitation bubbles are produced using high-frequency sound waves (>20 kHz). These bubbles quickly develop and burst, producing powerful localized shockwaves and microjets that efficiently break up biochar particles and agglomerates. After initial coarse milling, this technique works especially well for de-agglomeration and attaining a finer particle size distribution (Uchimiya et al., 2011).

The top-down approach is beneficial because it builds on the infrastructure already in place for the production of biochar and is comparatively simple and scalable. It is an energy-intensive process, though, and the strong mechanical stresses might occasionally change the material's crystalline structure or surface chemistry.

2.1.2. Bottom-Up Approach: Direct Synthesis

This approach involves synthesizing nano-carbon materials directly from biomass precursors, often under controlled chemical conditions.

- **Hydrothermal Carbonization (HTC):** Hydrothermal Carbonization (HTC) utilizes hot, compressed subcritical water at temperatures ranging from 180°C to 250°C to convert biomass. This process closely resembles the natural formation of coal. When specific temperature, pressure, and reaction time parameters are applied, nano-sized spherical carbon particles known as hydrochar are generated (Sevilla & Fuertes, 2009). Hydrochar typically exhibits a high concentration of oxygen-containing functional groups on its surface, enhancing its suitability for targeted applications.
- **Template-Assisted Pyrolysis:** Template-assisted pyrolysis involves impregnating a porous template, such as zeolites or mesoporous silica, with a solution of biomass-derived precursors, including lignocellulosic extracts or sugars. The impregnated template is then subjected to pyrolysis, which carbonizes the precursor within the nanopores. Subsequent removal of the template, for example by dissolving silica with hydrofluoric acid, produces a nano-structured carbon replica. Although this approach is more complex and costly, it allows for precise control over particle size and porosity (Titirici et al., 2007).

Compared to top-down methods, the bottom-up strategy is currently less established and more costly for large-scale agricultural applications. However, it allows for greater control over the final product's morphology and surface chemistry. Table 1 provides an overview of production processes.

Table 1. Comparison of nano-biochar production methods

Method Category	Specific Technique	Principle	Advantages	Disadvantages
Top-Down	Ball Milling	Mechanical attrition and shearing forces break down bulk biochar particles.	Simple, scalable, utilizes existing biochar production.	High energy consumption, potential for contamination, broad size distribution.
	Ultrasonication	Acoustic cavitation disintegrates particles in a liquid medium.	Effective for de-agglomeration, narrow size distribution.	Lower throughput, high energy cost for large volumes.
Bottom-Up	Hydrothermal Carbonization	Biomass conversion in hot, compressed water to form carbonaceous spheres.	High yield of carbon, retention of oxygen functional groups.	Produces hydrochar (less aromatic than pyrochar), complex wastewater.
	Template-Assisted Pyrolysis	Pyrolysis of biomass precursors within a nano-structured template.	Precise control over size, shape, and porosity.	Complex, multi-step process, high cost, use of hazardous chemicals.

2.2. Characterization of Nano-Biochar

A detailed description of NBC's physicochemical characteristics is necessary to comprehend its operation and forecast its environmental behaviour. For this purpose, a variety of analytical methods are employed.

2.2.1. Size, Morphology, and Surface Properties

- **Electron Microscopy:** Two essential instruments for imaging NBC are Transmission Electron Microscopy (TEM) and Scanning Electron Microscopy (SEM). The shape of the particles and the porosity structure inherited from the parent biomass are shown by SEM's high-resolution photos of the surface topography and morphology. Because of its better resolution, TEM makes it possible to see individual nanoparticles up close, allowing for accurate size measurements as well as information on their internal structure and crystallinity (Wang et al., 2020).
- **Dynamic Light Scattering (DLS):** The hydrodynamic diameter of NBC particles suspended in a liquid can be determined using the Dynamic Light Scattering (DLS) technique. In addition to measuring the zeta potential, which represents the surface charge of the particles and their colloidal stability in suspension, it offers a quick evaluation of the particle size distribution.

Good dispersion and stability are usually indicated by a significantly negative zeta potential.

- **Brunauer-Emmett-Teller (BET) Analysis**: This technique uses cryogenic temperatures to physisorb a gas, usually nitrogen, onto the material's surface. The specific surface area (SSA) is a crucial factor that determines NBC's reactive capability, is ascertained using it. The overall pore volume and pore size distribution (micropores <2 nm, mesopores 2-50 nm, macropores >50 nm) are also revealed by the measurement of the gas adsorption-desorption isotherm. The SSA of NBC is generally much higher than that of its bulk counterpart.

Figure 1. Synthesis and characterization of nano-biochar

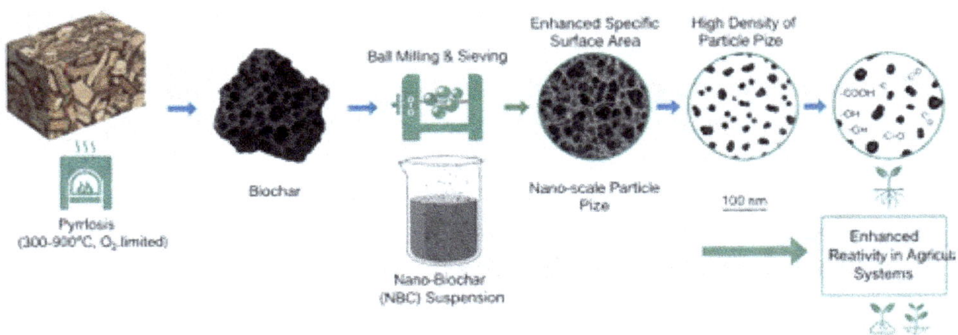

2.2.2. Chemical Composition and Functional Groups

- **Elemental Analysis (CHNS/O):** The elemental analysis (CHNS/O) calculates the weight percentage of oxygen, sulfur, nitrogen, hydrogen, and carbon. Atomic ratios like H/C and O/C are computed using the results, and they reveal information on the NBC's degree of carbonization, aromaticity (lower H/C), and polarity (higher O/C).
- **Fourier-Transform Infrared Spectroscopy (FTIR):** • Fourier-Transform Infrared Spectroscopy (FTIR): This technique determines whether different functional groups are present on the NBC surface. The vibrational frequencies of chemical bonds are correlated with the absorption of infrared radiation at particular wavenumbers. Hydroxyl (-OH), carboxyl (-COOH), phenolic, and carbonyl (C=O) groups are frequently found on NBC and are essential

for cation exchange, nutrient binding, and interaction with contaminants (Sun et al., 2017).

- **X-ray Photoelectron Spectroscopy (XPS):** XPS is a surface-sensitive method that gives precise details about the chemical states and elemental makeup of atoms on NBC's outermost surface. With the ability to measure the relative abundance of various carbon and oxygen species (such as C-C, C-O, C=O, and O=C-O), it provides a deeper comprehension of the surface chemistry than FTIR.

2.2.3. Crystalline Structure and Mineralogy

- **X-ray** Diffraction (XRD): • X-ray Diffraction (XRD): This technique is used to examine a material's crystalline structure. A broad peak at 20 = 20-30°, which corresponds to the plane of turbostratic graphitic structures which is commonly seen in the XRD pattern of NBC. This suggests that the carbon matrix is primarily amorphous or weakly crystalline. The presence of crystalline mineral components (such as silica, calcite, and sylvite) formed from the original biomass ash who is indicated by sharp peaks in the pattern. These components can affect the characteristics and interactions of NBC in soil.

2.3. Comparative Properties: Nano-Biochar vs. Conventional Biochar

Significant changes in characteristics occur when bulk biochar is converted to nano-biochar; these are compiled in Table 2. The sharp rise in specific surface area is the most notable alteration. Depending on the feedstock and production process, conventional biochar may have an SSA of 50–400 m^2/g, but NBC can easily surpass 1000 m^2/g (Wang et al., 2020). This increased surface area enables more locations for microbial colonization, catalysis, and adsorption. Additionally, a larger density of reactive surface functional groups and edge sites per unit mass is achieved through the reduction in particle size. Furthermore, unlike macro-particles, which are rather stationary, NBC's colloidal form (<1 μm) enables it to travel more readily through soil pores with water, possibly reaching deeper soil layers and interacting more closely with plant roots and soil microorganisms. Its long-term fate and potential for transmission into aquatic systems, a crucial subject of ongoing research, are also called into doubt by its increased mobility.

Table 2. Comparison of key physicochemical properties of conventional biochar and nano-biochar

Property	Conventional Biochar (Macro)	Nano-Biochar (Nano)	Implication for Agricultural Application
Particle Size	> 1 mm (typically)	< 1000 nm (often < 100 nm)	Enhanced dispersion in soil, better interaction with roots and microbes.
Specific Surface Area (SSA)	Moderate (e.g., 50-400 m^2/g)	Very High (e.g., > 400 m^2/g, can reach >1000 m^2/g)	Increased capacity for water retention, nutrient adsorption, and pollutant immobilization.
Pore Volume	High, dominated by macropores	High, with a greater proportion of micropores/ mesopores	Provides more habitat for microorganisms and sites for molecular sorption.
Surface Functional Groups	Present on surfaces and in pores	Higher density per unit mass, more accessible	Enhanced cation exchange capacity (CEC) and reactivity.
Dispersibility/Mobility	Low, immobile in soil matrix	High, can move with water through soil pores	Potential for deeper soil profile influence but also risk of leaching.
Reactivity	Moderate, diffusion-limited	High, rapid kinetics due to short diffusion paths	Faster nutrient release, quicker immobilization of contaminants.
Application Rate	High (t/ha)	Low (kg/ha)	Potentially more cost-effective and practical for large-scale use.

3. MECHANISMS OF NANO-BIOCHAR IN AGROECOSYSTEMS

The intricate relationships that nano-biochar has with the physical, chemical, and biological elements of soil are what give it its positive impacts in agriculture. Due to its unique nano-scale characteristics, it can influence agroecosystem processes in ways that are significantly more noticeable and effective than those of traditional biochar.

3.1. Improvement of Soil Physical Properties

Healthy soil structure is vital for root penetration, water infiltration, and aeration. NBC contributes to the improvement of soil physical properties primarily by promoting the formation and stabilization of soil aggregates.

- **Aggregate Formation**: Organic matter, microbial exudates, and mineral cements hold together clusters of primary soil particles, such as sand, silt, and clay, to form aggregates. NBC particles are strong binding agents because of their large surface area and surface charge. They are capable of 'gluing' organic materials and mineral particles together to create micro-aggregates by forming bridges between them (Rillig et al., 2010). Because of their small size, they can enter intra-aggregate pore spaces and reinforce the structure from the inside out.
- **Water holding capacity (WHC):** The potential of biochar to improve the soil's water-holding capacity is one of its most important advantages; this is especially important in arid and semi-arid areas. NBC amplifies this effect in several ways. First, water can be held in by its own internal porosity. Second, and perhaps more significantly, it enhances soil aggregation, which increases the soil's overall pore volume, particularly the number of micropores and mesopores that store water available to plants (Blanco-Canqui, 2017). Water molecules are also drawn to and retained by the hydrophilic oxygen-containing functional groups on the NBC surface. This leads to better water usage efficiency and less water stress for crops during dry spells.
- **Bulk Density and Soil Aeration:** The development of stable aggregates produces a network of inter-aggregate macropores that enhance soil drainage and aeration, avoiding anaerobic situations and waterlogging that could damage plant roots. NBC can also aid in lowering soil bulk density, which facilitates root growth, by adding lightweight, porous carbon to the soil matrix and strengthening its structure.

3.2. Modification of Soil Chemical Properties

NBCs can have a significant impact on soil chemistry, specifically pH, cation exchange capacity, and nutrient dynamics, due to their high reactivity and surface area.

- **Liming Effect and pH Buffering:** Many biochars have an alkaline character, particularly those made from feedstocks strong in alkali and alkaline earth metals (such as manure and agricultural wastes) at high temperatures. NBC can raise the pH of acidic soils by acting as a liming agent. This is advantageous since low pH can cause manganese and aluminum toxicity, and decrease the availability of vital minerals such as molybdenum and phosphorus. Beyond merely increasing pH, NBC's high buffering capacity due to its concentration of functional groups (such as carboxyl and phenolic) helps to keep soil pH steady and provide a more stable chemical environment for microbes and plants (Yuan et al., 2011).

- **Cation Exchange Capacity (CEC):** CEC measures the soil's capacity to hold onto positively charged nutrients (cations), such as ammonium (NH_4^+), potassium (K^+), magnesium (Mg^{2+}), and calcium (Ca^{2+}), so that irrigation or rainfall does not wash them away. The negatively charged carboxyl and phenolic functional groups generated on the surface of NBC during pyrolysis and subsequent environmental oxidation are the primary cause of its extremely high CEC. As a "nutrient magnet" that retains vital minerals and releases them gradually for plant uptake, NBC contributes disproportionately to the soil's overall CEC due to its significantly larger surface area and higher functional group density compared to bulk biochar (Clough et al., 2013).
- **Nutrient Retention and Availability:** NBC improves nutrient utilization efficiency by directly reducing the leaching of critical cations by raising CEC. Although this process is more intricate and contingent upon the particular characteristics of the NBC and the soil, its porous structure and surface chemistry also enable it to adsorb anions such as nitrate (NO_3^-) and phosphate (PO_4^{3-}) (Hale et al., 2013). In acidic soils, NBC can decrease the fixation of phosphorus by iron and aluminum oxides, improving plant availability. Additionally, NBC can influence plant-mycorrhizal symbioses by acting as a sorbent for signaling molecules, such as strigolactones, which are involved in nutrient acquisition.

Figure 2. Mechanisms of nano-biochar in soil

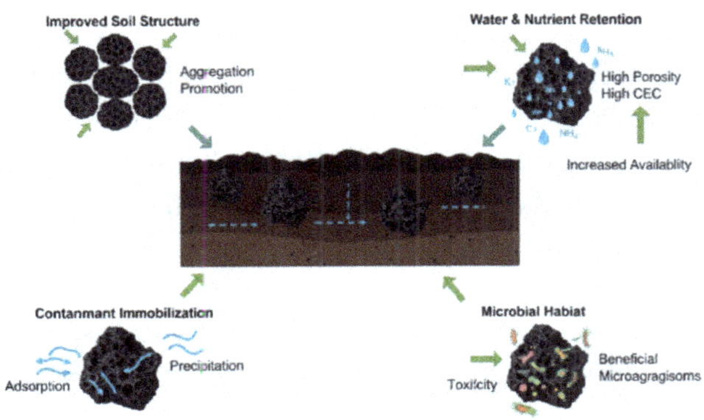

3.3. Influence on Soil Biological Properties and Plant-Microbe Interactions

The soil is a living ecosystem teeming with microorganisms that drive crucial biogeochemical cycles. NBC can profoundly alter this microbial world, generally in a positive manner.

- **Microbial Habitat and Refuge:** Soil microorganisms, such as bacteria and fungi, find the porous structure of NBC, with its broad variety of pore diameters, to be a perfect physical habitat. According to Quilliam et al. (2013), these pores provide defense against environmental stressors such as desiccation and pH variations as well as predators like protozoa. The overall biomass and diversity of microorganisms may increase as a result of this "refuge effect."
- **Microbial Activity Stimulation:** NBC has the ability to actively promote microbial activity in addition to serving as a passive environment. Although its fundamental structure is quite resistant, some microorganisms can use it as a source of energy and labile carbon. More significantly, it can improve the environment for microbial development by adsorbing microbial inhibitory chemicals, such as poisonous phenols. A more active microbial community is also indirectly supported by the increased water and nutrient availability linked to NBC (Lehmann et al., 2011).
- **Modification of Microbial Community Structure:** The soil microbial community's makeup may change as a result of NBC treatment. It has been demonstrated to promote the growth of advantageous microorganisms like arbuscular mycorrhizal fungus (AMF), phosphate-solubilizing bacteria, and nitrogen-fixing bacteria (like Rhizobium). Most agricultural plants' roots and AMF develop symbiotic partnerships that aid in the plants' uptake of water and nutrients—particularly phosphorus—from the soil. By enhancing soil structure and providing a surface favorable for hyphal growth, NBC can promote AMF colonization (Warnock et al., 2007).
- **Interference with Quorum Sensing:** Many bacteria use quorum sensing, a cell-to-cell communication system, to coordinate group behaviors, such as biofilm formation and the expression of virulence factors in pathogens. According to Gao et al. (2019), NBC's large surface area enables it to adsorb the signaling molecules involved in quorum sensing, which may disrupt the pathogenic activities of soil-borne illnesses and improve the environment for plant health.

4. APPLICATIONS OF NANO-BIOCHAR IN SUSTAINABLE AGRICULTURE

Due to its unique properties, nano-biochar can be utilized in various real-world applications to enhance agricultural sustainability. It is a particularly alluring instrument for contemporary agriculture due to its excellent efficiency at low application rates. Table 3 summarizes the main findings and applications.

4.1. Soil Fertility Improvement and Crop Growth Promotion

NBC is primarily used as a strong soil supplement to increase crop yields and fertility. As explained in the preceding section, its systems work together to produce an excellent growing medium for plants.

- **Improved Nutrient Uptake:** NBC guarantees that plants have a consistent and easily accessible supply of nutrition throughout their growth cycle by decreasing nutrient leaching and raising the availability of important macro- and micronutrients. As a result, fertilizer is used more effectively, which reduces the financial expenses and environmental contamination associated with overuse. Higher levels of nitrogen, phosphorus, and potassium in the tissues of crops cultivated in NBC-amended soils have been associated with improved growth and yield, according to studies (Liu et al., 2021).
- **Improved Root Development:** Deeper and more extensive root proliferation is made possible by the enhanced soil structure, which is characterized by decreased bulk density and higher aeration. Plants with larger root systems can reach deeper into the soil, which improves their access to nutrients and water.
- **Direct Plant Stimulation:** NBC may have direct physiological effects on plants in addition to its indirect effects on soil characteristics. Since they are nanoparticles, they may be absorbed by plant roots and moved to different sections of the plant. According to some research, intracellular NBC may influence metabolic processes by acting as a nano-catalyst or by modulating the expression of genes associated with growth and stress response, although the scope and consequences of this absorption are still being investigated (Shen et al., 2022).

4.2. Remediation of Contaminated Soils

Pollutants like heavy metals (such as cadmium, lead, and arsenic) from mining, industrial processes, and the use of polluted fertilizers and pesticides damage agricultural soils in many regions of the world. Through the food chain, these pollutants can seriously endanger human health and lower crop productivity. For in-situ cleanup, NBC has proven to be an extremely successful agent.

Immobilization of Heavy Metals: NBC serves as an effective sorbent for heavy metal cations because of its large surface area, porous structure, and abundance of functional groups. The primary immobilization mechanisms are as follows:

- o *Adsorption and Complexation on the Surface*: Cations, including Cd^{2+}, Pb^{2+}, and $Cu^{2},^{+}$ bind directly to negatively charged functional groups on the NBC surface, such as carboxyl groups ($-COO^{-}$).
- o *Precipitation*: Heavy metals can form insoluble hydroxides or carbonates due to the localized increase in pH caused by alkaline NBC. Additionally, heavy metals may react with minerals such as phosphates and carbonates present in NBC ash to generate stable, insoluble mineral precipitates, including lead phosphate (Cao et al., 2009).
- o *Redox Reactions:* Certain NBCs can chemically transform metals from highly toxic, mobile forms to less toxic, immobile forms. For example, NBC can reduce the highly toxic and mobile Cr(VI) to the less harmful and immobile Cr(III) (Dong et al., 2011). This process decreases the bioavailability of these pollutants in the soil matrix, thereby reducing their uptake by plants and contributing to safer food production.
- o *Organic Pollutant Degradation:* NBC facilitates the degradation of organic pollutants, including polycyclic aromatic hydrocarbons (PAHs) and pesticides. By adsorbing these contaminants onto its surface, NBC increases their availability for microbial degradation. Furthermore, certain NBCs can directly participate in or accelerate the chemical breakdown of these compounds through catalytic activity (Uchimiya et al., 2011).

4.3. Carrier for Fertilizers and Pesticides

One of the main issues is the inefficiency of traditional agrochemicals. While pesticides might stray from their intended target and harm the environment, a significant portion of applied fertilizers are lost through leaching or volatilization. NBC can be engineered to act as a "smart" delivery system for these inputs.

- *Slow-Release Fertilizers:* Nutrients like nitrogen and potassium can be added to NBC particles by combining them with concentrated fertilizer solutions. The nutrients are absorbed onto the NBC's surface and within its pores. After applying this "nutrient-charged" NBC to the soil, the nutrients are gradually released as the NBC matrix weathers or as they are traded for other ions in the soil solution. This significantly lowers losses and boosts efficiency by coordinating fertilizer supply with crop needs (Ghafari et al., 2022; Singh et al., 2024b).

- *Targeted Pesticide Delivery:* In a similar manner, NBC can be loaded with pesticides. This can minimize worker exposure, reduce the overall amount of pesticide used, and prevent rain from washing the herbicide away. The nano-size of the carrier may also boost the pesticide's penetration into plant tissues or its adhesion to pests. The development of functionalized NBC systems that release their payload only in response to specific triggers—such as pH shifts caused by a fungal infection—is still being researched.

4.4. Mitigation of Abiotic and Biotic Stress

Environmental stressors including disease, salt, and drought frequently affect plants, limiting their ability to grow and produce. NBC can assist plants in overcoming these obstacles.

- *Mitigation of Drought Stress:* As was previously mentioned, NBC's power to greatly expand soil water-holding capacity acts as a buffer for plants during times when rainfall is scarce. It lessens the intensity and duration of water stress by sustaining higher soil moisture levels for longer, enabling plants to continue their physiological processes, such as photosynthesis.
- *Alleviation of Salinity Stress:* Excessive salt concentrations, particularly Na^+, in salty soils can directly irritate plants and induce osmotic stress. By using cation exchange to absorb excess Na+ ions, NBC can lessen their detrimental effects on plants by lowering their concentration in the soil solution (Usman et al., 2016; Singh et al., 2024a).
- *Disease Suppression:* NBC can strengthen the soil's innate ability to inhibit pathogens by encouraging a diverse and healthy soil microbiome. It can promote the development of beneficial microorganisms that produce antimicrobial compounds or outcompete pathogens for resources. It provides the plant with an additional line of defence by interfering with the quorum sensing of pathogenic bacteria.

Table 3. Overview of nano-biochar applications in sustainable agriculture

Application Area	Specific Use	Key Mechanism(s)	Reported Outcomes	Selected Reference
Soil Fertility & Crop Growth	General soil amendment	Improved soil structure, WHC, CEC, and nutrient availability.	Increased crop yields (e.g., 15-50%), improved root growth, higher nutrient uptake.	(Liu et al., 2021)
Soil Remediation	Immobilization of heavy metals (Cd, Pb, As)	Surface adsorption, precipitation, redox reactions.	Reduced heavy metal bioavailability and uptake by crops by over 70%.	(Cao et al., 2009)
	Degradation of organic pollutants (pesticides, PAHs)	Sorption and concentration, catalytic degradation, enhanced microbial degradation.	Increased degradation rates of persistent organic pollutants.	(Uchimiya et al., 2011)
Agrochemical Carrier	Slow-release fertilizer	High sorption capacity for nutrient ions (NH_4^+, K^+, PO_4^{3-}).	Reduced nutrient leaching by up to 80%, synchronized nutrient release with plant demand.	(Ghafari et al., 2022)
	Targeted pesticide delivery	Adsorption of pesticide molecules, controlled release.	Lower pesticide usage, reduced environmental runoff, enhanced efficacy.	(Grillo et al., 2021)
Stress Mitigation	Drought and salinity tolerance	Increased water holding capacity, adsorption of excess Na^+ ions.	Maintained plant growth and yield under water and salt stress conditions.	(Usman et al., 2016)
	Plant disease suppression	Promotion of beneficial microbes, quorum sensing interference.	Reduced incidence of soil-borne diseases like Fusarium wilt.	(Gao et al., 2019)

5. SYNERGISTIC BENEFITS FOR SUSTAINABILITY

Nano-biochar applications generate a cascade of advantages that improve sustainable agriculture goals like climate action, environmental protection, and food security.

5.1. Climate Change Mitigation

Agriculture is a major source of greenhouse gases (GHGs). NBC can contribute to climate change mitigation through several key pathways.

- *Carbon Sequestration:* Long-term carbon storage is the most obvious advantage. The easily broken-down carbon in biomass is converted into a very stable, aromatic form of carbon in NBC by the pyrolysis process. This carbon can stay stable for hundreds to thousands of years when added to soil and is incredibly resistant to microbial destruction (Kuzyakov et al., 2014; Singh et al., 2024d). This creates a strong carbon sink by efficiently removing CO_2 from the atmosphere and storing it in the soil.
- *Lowering Emissions of Nitrous Oxide (N_2O):* With a potential for global warming that is about 300 times greater than that of CO_2, nitrous oxide is a powerful greenhouse gas. The primary microbiological activities that create it in soils are nitrification and denitrification. By enhancing soil aeration (which prevents denitrification), adsorbing the ammonium and nitrate that drive these processes, and possibly reducing the activity of the enzymes involved, NBC can reduce N_2O emissions (Case et al., 2015).
- *Methane (CH_4) Emissions Reduction:* NBC can lower methane emissions in flooded agricultural systems, such as rice paddies. It accomplishes this by increasing the activity of methanotrophs, or bacteria that consume methane, and by enhancing soil aeration and drainage, which changes the soil environment from anaerobic (generating methane) to aerobic (Zhang et al., 2012).

5.2. Enhanced Resource Use Efficiency

Sustainable agriculture must be productive while minimizing resource consumption. NBC is a key enabler of this goal.

- *Water Use Efficiency (WUE):* NBC makes sure that more water from irrigation or rainfall is retained in the root zone and accessible for plant absorption by boosting the soil's ability to hold water. This increases crop output per unit of water used by lowering non-productive water loss via evaporation and deep percolation.
- *Nutrient Use Efficiency (NUE).* NBC guarantees that a significantly higher percentage of applied fertilizers are truly utilized by the crop by significantly lowering the leaching of nitrogen and other nutrients. This saves money and keeps surface and groundwater from becoming contaminated, which causes eutrophication, by enabling farmers to produce the same or even higher yields with a lot less fertilizer (Singh et al., 2023; Singh et al., 2024c)

6. CASE STUDY: REMEDIATION OF CADMIUM-CONTAMINATED PADDY SOIL USING RICE HUSK-DERIVED NANO-BIOCHAR

The viability of agriculture and human health are seriously threatened by cadmium (Cd), a heavy metal that is very poisonous and non-essential. Cancer, bone disease, and severe kidney damage can result from long-term exposure to cadmium. More than half of the world's population depends on rice (Oryza sativa L.) as a staple food, yet it is especially effective at absorbing Cd from polluted soils and transferring it to the grain (Zhao et al., 2010). In some parts of Asia, mining operations, industrial pollutants, and the long-term use of phosphate fertilizers containing cadmium have contaminated large tracts of paddy fields with cadmium. Because rice from these areas frequently exceeds the maximum allowable limits for cadmium (Cd) imposed by food safety regulators, this has resulted in major food safety concerns. Traditional remediation techniques, such as chemical washing or soil excavation, are often unfeasible, detrimental to soil structure, and unsuitable for extensive agricultural regions. Using soil amendments for in-situ immobilization is considered a more practical and sustainable method. This case study evaluates the efficacy of rice husk-derived nano-biochar (RH-NBC) for remediating Cd-contaminated paddy soil by combining results from other research projects (e.g., Wang et al., 2020; Liu et al., 2021).

Methodology: A field trial is established using a randomized complete block design with four treatments: 1. Control (CK): No amendment, 2. Conventional Fertilizer (NPK): Standard local practice, 3. NPK + Conventional Biochar (BC): NPK plus RH-BC applied at a rate of 10 t/ha, 4. NPK + Nano-Biochar (NBC): NPK plus RH-NBC applied at a much lower rate of 1 t/ha. The amendments are incorporated into the top 15 cm of soil before transplanting rice seedlings. Standard water and crop management practices are followed for the duration of the growing season.

Sampling and Analysis: Soil samples are collected before and after the experiment to analyze pH, CEC, and the concentration of available Cd (extracted using DTPA). At harvest, rice plants are separated into roots, straw, and grain. The biomass of each part is measured, and the Cd concentration is determined using Atomic Absorption Spectrometry (AAS) after acid digestion. The results show that the application of biochar amendments, particularly NBC, led to significant improvements in soil properties and a reduction in cadmium (Cd) accumulation in rice grains. The key results are summarized in Table 4 and Figure 3.

Table 4. Effects of biochar amendments on soil properties and rice growth

Treatment	Soil pH (post-harvest)	Soil CEC (cmol+/kg)	Available Soil Cd (mg/kg)	Rice Grain Yield (t/ha)
CK	5.3	8.1	0.95	4.5
NPK	5.2	8.3	0.98	6.2
NPK + BC (10 t/ha)	6.1	11.5	0.42	6.8
NPK + NBC (1 t/ha)	6.3	12.8	0.25	7.1

- *Soil Properties:* In comparison to the control and NPK-only treatments, the pH and CEC of the soil were considerably raised by the BC and NBC treatments. It is noteworthy that the NBC treatment, which was administered at a rate of just 10% of BC, increased pH and CEC more than BC. As a result, the soil's concentration of bioavailable Cd dropped significantly. The BC treatment achieved a 57% reduction in accessible Cd, whereas the NBC treatment reduced it by 74% when compared to the NPK treatment.

Figure 3. Cadmium concentration in rice grain under different treatments

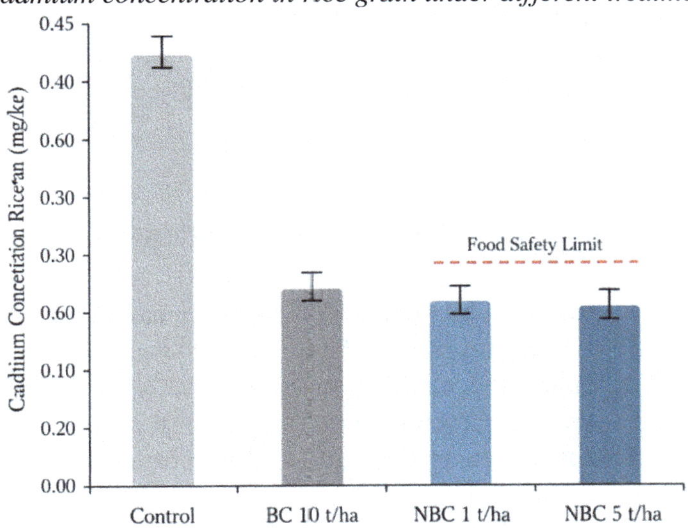

- *Cadmium Acquisition and Rice Yield:* Lower soil Cd bioavailability resulted in decreased rice plant uptake of Cd. The grain from the NPK treatment had a

Cd concentration of 0.60 mg/kg, which was three times the food safety limit, as seen in Figure 3. Grain Cd concentration was successfully lowered by the BC treatment to 0.18 mg/kg, which is somewhat below the limit. In comparison to the NPK treatment, the NBC treatment was even more successful, lowering the grain Cd concentration by 85% to just 0.09 mg/kg. Additionally, both biochar treatments enhanced grain production; however, the NBC treatment produced the highest yield (7.1 t/ha), likely due to enhanced soil fertility and reduced Cd toxicity.

Discussion and Conclusion: Even when sprayed at a much lower rate, this case study shows that nano-biochar is more effective than its traditional counterpart at remediating Cd-contaminated soil. Higher specific surface area, higher functional group density, and improved soil dispersion all contributed to RH-NBC's improved efficacy by increasing the number of active sites available for Cd adsorption and immobilization. Another important factor was the rise in soil pH brought on by NBC, which encouraged the precipitation of Cd as insoluble substances. Practically speaking, there are significant logistical and financial benefits to NBC's lower application rate (1 t/ha) compared to 10 t/ha. By lowering the expenses related to manufacturing, shipping, and using the amendment, it increases farmers' access to remediation technology. Nano-biochar offers a win-win solution that addresses the issues of food safety and food security in contaminated agricultural areas by successfully reducing the Cd content of rice grains to safe levels while also increasing crop yields. This case study provides compelling evidence of nano-biochar's potential as an extremely effective and long-lasting tool for controlling heavy metal contamination in agriculture.

7. CHALLENGES AND RISKS, FUTURE PERSPECTIVES

The widespread use of nano-biochar (NBC) for sustainable agriculture presents several risks and challenges that need to be carefully considered, despite its promising applications. High production costs and scalability problems—especially from energy-intensive manufacturing processes—as well as a notable deficiency in standardized production, characterization, and application methods, are major barriers that prevent consistent use and comparison across studies. Additionally, longer-term field-scale validation is required to verify its efficacy in various settings. The tiny size and high reactivity of NBC also pose ecotoxicological concerns since they may impact beneficial soil species and cast doubt on its environmental mobility, which could result in unintentional movement to water bodies and interactions with co-contaminants. Future research should focus on creating functionalized "smart" nano-biochars with customized properties, conducting thorough life cycle assessments, integrating NBC with precision agriculture technologies for optimal application, and establishing precise regulatory frameworks that ensure environmental monitoring, safety, and quality control to safely and sustainably harness the benefits of NBC. To maximize NBC's potential while minimizing hazards, these issues must be addressed through focused research and policy development.

8. CONCLUSION

Nano-biochar represents a significant advancement at the intersection of environmental engineering materials science, and agriculture. By applying nanotechnology principles, conventional biochar is transformed from a promising yet often underperforming material into an effective, multifunctional tool for sustainable soil management. Nano-biochar exhibits superior precision and effectiveness compared to its bulk counterpart, enabling the remediation of contaminated land, enhancement of soil health, increased agricultural productivity, and mitigation of climate change due to its high surface area, reactivity, and colloidal mobility. Its diverse applications include functioning as a soil conditioner, nutrient adsorbent, pollutant immobilizer, and microbial habitat, collectively supporting key principles of sustainable agriculture such as ecosystem resilience, resource efficiency, and waste-to-resource conversion. The cadmium remediation case study illustrates the potential of nano-biochar to provide practical and cost-effective solutions to pressing real-world challenges, thereby safeguarding food safety and supporting agricultural livelihoods. However, widespread adoption faces obstacles, including high production costs, a lack of standardization, and insufficient understanding of long-term environmental impacts and ecotoxicological risks. Future progress depends on the development of tailored, functionalized materials for specific agricultural challenges, supported by comprehensive life cycle assessments and clear regulatory frameworks.

REFERENCES

Blanco-Canqui, H. (2017). Biochar and soil physical properties. *Soil Science Society of America Journal*, *81*(4), 687–711. DOI: 10.2136/sssaj2017.01.0017

Cao, X., Ma, L., Gao, B., & Harris, W. (2009). Dairy-manure derived biochar effectively sorbs lead and atrazine. *Environmental Science & Technology*, *43*(9), 3285–3291. DOI: 10.1021/es803092k PMID: 19534148

Case, S. D. C., McNamara, N. P., Reay, D. S., & Whitaker, J. (2015). The effect of biochar on nitrous oxide and carbon dioxide emissions from temperate agricultural soils: A meta-analysis. *Soil Biology & Biochemistry*, *89*, 1–11.

Clough, T. J., Condron, L. M., Kammann, C., & Müller, C. (2013). A review of biochar and soil nitrogen dynamics. *Agronomy (Basel)*, *3*(2), 275–293. DOI: 10.3390/agronomy3020275

Dong, X., Ma, L. Q., & Li, Y. (2011). Characteristics and mechanisms of hexavalent chromium removal by biochar from sugar beet tailing. *Journal of Hazardous Materials*, *190*(1-3), 909–915. DOI: 10.1016/j.jhazmat.2011.04.008 PMID: 21550718

Gao, Y., Zhou, H., He, C., Huang, L., Gao, H., & Ruan, R. (2019). Biochar-supported nanoscale iron sulfide composite for enhanced Cr (VI) removal from aqueous solution. *Bioresource Technology*, *281*, 417–424.

Ghafari, N., Narchin, P., & Niazi, A. (2022). Slow-release fertilizer based on nano-biochar: A review on synthesis, characterization, and applications. *Journal of Cleaner Production*, *353*, 131643.

Glaser, B., Haumaier, L., Guggenberger, G., & Zech, W. (2001). The 'Terra Preta' phenomenon: A model for sustainable agriculture in the humid tropics. *Naturwissenschaften*, *88*(1), 37–41. DOI: 10.1007/s001140000193 PMID: 11302125

Grillo, R., Fraceto, L. F., & de Melo, N. F. S. (2021). Biochar as a carrier for pesticides and herbicides: A sustainable and promising approach for agriculture. *Journal of Agricultural and Food Chemistry*, *69*(26), 7241–7243.

Hale, S. E., Alling, V., Martinsen, V., Mulder, J., Breedveld, G. D., & Cornelissen, G. (2013). The sorption of PAHs to seven different biochars as a function of their physicochemical properties. *Chemosphere*, *91*(10), 1412–1419.

Kuzyakov, Y., Subbotina, I., Chen, H., Bogomolova, I., & Xu, X. (2014). Black carbon decomposition and incorporation into soil microbial biomass estimated by ^{14}C labeling. *Soil Biology & Biochemistry*, *73*, 44–51.

Lehmann, J., & Joseph, S. (Eds.). (2015). *Biochar for environmental management: science, technology and implementation*. Routledge. DOI: 10.4324/9780203762264

Lehmann, J., Rillig, M. C., Thies, J., Masiello, C. A., Hockaday, W. C., & Crowley, D. (2011). Biochar effects on soil biota–a review. *Soil Biology & Biochemistry*, *43*(9), 1812–1836. DOI: 10.1016/j.soilbio.2011.04.022

Liu, S., Meng, J., Jiang, L., Yang, X., Lan, Y., Cheng, X., & Chen, W. (2021). Rice husk biochar impacts on rice production and nitrogen and phosphorus uptake in a Chinese paddy soil. *Experimental Results*, •••, 2.

Lyu, H., Gao, B., He, F., Zimmerman, A. R., Ding, C., Huang, H., & Tang, J. (2018). Effects of ball milling on the physicochemical and sorptive properties of biochar: Experimental observations and governing mechanisms. *Environmental Pollution*, *233*, 54–63. DOI: 10.1016/j.envpol.2017.10.037 PMID: 29053998

Ok, Y. S., Chang, S. X., Gao, B., & Wang, H. (2021). Nanobiochar: A new emerging material for environmental remediation and agricultural applications. *Environmental Science. Nano*, *8*(6), 1436–1437.

Prasad, R., Bhattacharyya, A., & Nguyen, Q. D. (2017). Nanotechnology in sustainable agriculture: Recent developments, challenges, and perspectives. *Frontiers in Microbiology*, *8*, 1014. DOI: 10.3389/fmicb.2017.01014 PMID: 28676790

Quilliam, R. S., Glanville, H. C., Wade, S. C., & Jones, D. L. (2013). The Trojan-horse effect: Biochar systems for nutrient co-delivery in soil. *Global Change Biology. Bioenergy*, *5*(2), 202–205.

Rillig, M. C., Wagner, M., Salem, M., Antunes, P. M., George, C., Ramke, H. G., & Zhl, J. (2010). Material derived from hydrothermal carbonization: Effects on plant growth and arbuscular mycorrhiza. *Applied Soil Ecology*, *45*(3), 199–204. DOI: 10.1016/j.apsoil.2010.04.011

Saikanth, D. R. K., Singh, B. V., Rai, A. K., Kumar, U. S., Surender, , Yadav, B., & Singh, O. (2023). Biochar implementation in rice paddies for addressing greenhouse gas emissions and nutrient loss: A review. *International Journal of Plant and Soil Science*, *35*(18), 610–623. DOI: 10.9734/ijpss/2023/v35i183326

Sevilla, M., & Fuertes, A. B. (2009). The production of carbon materials by hydrothermal carbonization of cellulose. *Carbon*, *47*(9), 2281–2289. DOI: 10.1016/j.carbon.2009.04.026

Shen, M., Zhang, H., Ding, Y., & Tang, Z. (2022). Biochar nanoparticles as a potential tool to mitigate abiotic stress in plants: A review. *Environmental Pollution*, *293*, 118536.

Siddiqui, H., Ahmed, K. B. M., Faizan, M., Baran, M. F., Alam, P., & Hayat, S. (2025). Pre-Sowing Seed Treatment with Glucose Boosts the Photosynthetic Efficiency, Metabolic Enzymes and Antioxidant Defense System to increase the Growth of <i>Abelmoschus esculentus</i> (L.) Moench. *Journal of Innovative Solutions for Eco-Environmental Sustainability*, *1*(1), 13854. DOI: 10.46991/JISEES.2025.01.1.13854

Singh, A., Rawat, S., Rajput, V. D., Minkina, T., Mandzhieva, S., Eloyan, A. S., Singh, R. K., Singh, O., El-Ramady, H., & Ghazaryan, K. (2024a). Nanotechnology products in agriculture and environmental protection: Advances and challenges. *Egyptian Journal of Soil Science*, *64*(4), 1355–1378. DOI: 10.21608/ejss.2024.300047.1802

Singh, O. (2023). Nanotechnology for sustainability and food security in agriculture: A nanopriming long story in short. In Singh, A., Rajput, V., Ghazaryan, K., Gupta, S., & Minkina, T. (Eds.), *Nanopriming Approach to Sustainable Agriculture* (pp. 315–339). IGI Global., DOI: 10.4018/978-1-6684-7232-3.ch014

Singh, O., Shahi, U.P., Dhyani, B.P., Kumar, S., Vivek, S., & Sengar, R.S., Shivangi, and Singh, A. (2023). Effect of zinc oxide nanoparticles application on growth and yield of basmati rice (Oryza sativa) in alkaline soil. *Indian Journal of Agronomy*, *68*(3), 1–6.

Singh, O., Singh, S., Shahi, U. P., Singh, R., Singh, P. K., Grigoryan, K. K., Khachatryan, H., Rajput, P., Perelomov, L., Kozmenko, S., & Burachevskaya, M. (2024c). Nanotechnological Innovations for Soil Pollution Remediation and Environmental Conservation. In Rajput, V., Singh, A., Ghazaryan, K., Alexiou, A., & Said Al-Tawaha, A. (Eds.), *Harnessing NanoOmics and Nanozymes for Sustainable Agriculture* (pp. 237–255). IGI Global., DOI: 10.4018/979-8-3693-1890-4.ch012

Singh, O., Singh, S., Singh, P. K., Singh, V. K., Singh, A., Singh, S., Movsesyan, H. S., Minkina, T., Nazarenko, O., Mandzhiev, S., & Latsynnik, E. (2024d). Nanoagriculture: Exploring Environmental Impacts and Sustainable Advantages. In Rajput, V., Singh, A., Ghazaryan, K., Alexiou, A., & Said Al-Tawaha, A. (Eds.), *Harnessing NanoOmics and Nanozymes for Sustainable Agriculture* (pp. 275–296). IGI Global., DOI: 10.4018/979-8-3693-1890-4.ch014

Singh, S., Singh, O., & Shahi, U. P. (2023). Nanotechnology: A New Innovation in Ghazaryan, K., Pandey, D., Singh, S., Varagyan, V., Alexiou, A., Petropoulos, D., Kriemadis, A., Rajput, V., Minkina, T., Singh, R., Sousa, J., Kumar, S., El-Ramady, H., Singh, O., & Singh, A. (2024). Enhancing Crop Production: Unveiling the Role of Nanofertilizers in Sustainable Agriculture and Precision Nutrient Management. *Egyptian Journal of Soil Science*, *64*(3), 981–1007.

Sun, K., Kang, M., Zhang, Z., Jin, J., Wang, Z., Pan, Z., & Xing, B. (2017). A new insight into the adsorption of wastewater-derived organic contaminants on biochar: A review. *Journal of Cleaner Production*, *148*, 590–602.

Tilman, D., Cassman, K. G., Matson, P. A., Naylor, R., & Polasky, S. (2002). Agricultural sustainability and intensive production practices. *Nature*, *418*(6898), 671–677. DOI: 10.1038/nature01014 PMID: 12167873

Titirici, M. M., Thomas, A., & Antonietti, M. (2007). Back in the black: Hydrothermal carbonization of plant material as an efficient chemical process to treat the C_O_2 problem? *New Journal of Chemistry*, *31*(6), 787–789. DOI: 10.1039/b616045j

Tomczyk, A., Sokołowska, Z., & Boguta, P. (2020). Biochar as a soil amendment for improving soil quality. *International Agrophysics*, *34*(2).

Uchimiya, M., Lima, I. M., Klasson, K. T., Chang, S., & Wartelle, L. H. (2011). Immobilization of heavy metal, lead, in contaminated soils by biochar and manuscripts from different biomass. *Chemosphere*, *80*(8), 923–930.

Usman, A. R. A., Abduljabbar, A., Vithanage, M., Ok, Y. S., Ahmad, M., Ahmad, M., & Al-Wabel, M. I. (2016). Biochar production from date palm waste: A review of technologies, characterization, and applications. *Waste and Biomass Valorization*, *7*(5), 1187–1200.

Wang, L., Wang, Y., Ma, F., & Ruan, R. (2020). Synthesis and characterization of a novel magnetic biochar from rice husk for the removal of Cd2+ from aqueous solution. *Bioresource Technology*, *308*, 123287.

Warnock, D. D., Lehmann, J., Kuyper, T. W., & Rillig, M. C. (2007). Mycorrhizal responses to biochar in soil–concepts and mechanisms. *Plant and Soil*, *300*(1), 9–20. DOI: 10.1007/s11104-007-9391-5

Woolf, D., Amonette, J. E., Street-Perrott, F. A., Lehmann, J., & Joseph, S. (2010). Sustainable biochar to mitigate global climate change. *Nature Communications*, *1*(1), 1–9. DOI: 10.1038/ncomms1053 PMID: 20975722

Yuan, J. H., Xu, R. K., & Zhang, H. (2011). The forms of alkalis in the biochar produced from crop residues at different temperatures. *Bioresource Technology*, *102*(9), 5488–5497. DOI: 10.1016/j.biortech.2010.11.018 PMID: 21112777

Zhang, A., Cui, L., Pan, G., Li, L., Hussain, Q., Zhang, X., & Crowley, D. (2012). Effect of biochar amendment on yield and methane and nitrous oxide emissions from a rice paddy from Tai Lake plain, China. *Agriculture, Ecosystems & Environment*, *159*, 1–8.

Zhao, F. J., Ma, J. F., Meharg, A. A., & McGrath, S. P. (2010). Cadmium uptake and accumulation in rice. [). Academic Press.]. *Advances in Agronomy*, *106*, 1–46.

Chapter 8
Monitoring Earth's Vital Signs:
Utilizing Nanosensors for Precision Soil Monitoring and Analysis

Ushaa Eswaran
http://orcid.org/0000-0002-5116-3403
Indira Institute of Technology and Sciences, Jawaharlal Nehru Technological University, Kakinada, India

Vivek Eswaran
http://orcid.org/0009-0002-7475-2398
Medallia, India

Keerthna Murali
http://orcid.org/0009-0009-1419-4268
Dell, India

Vishal Eswaran
http://orcid.org/0009-0000-2187-3108
CVS Health, India

E. Kannan
Vel Tech Rangarajan Dr. Sagunthala R&D Institute of Science and Technology, India

ABSTRACT

This chapter explores the innovative use of nanosensors for precision monitoring and analysis of soil health and environmental conditions. Nanotechnology offers

DOI: 10.4018/979-8-3693-7006-3.ch008

unprecedented opportunities to develop sensors capable of detecting and quantifying various soil parameters with high sensitivity and specificity. By leveraging nanomaterials and nanotechnology principles, these sensors can provide real-time data on soil moisture, pH levels, nutrient concentrations, contaminant presence, and microbial activity, among other critical factors. This chapter discusses the principles underlying nanosensor design, fabrication, and deployment in soil environments. It explores the diverse applications of nanosensors in precision agriculture, environmental monitoring, and soil remediation efforts. Furthermore, the chapter examines the challenges and future prospects of integrating nanosensor technologies into soil science practices, emphasizing their potential to revolutionize soil management strategies and promote sustainable agriculture and environmental stewardship.

1. INTRODUCTION TO NANOSENSORS IN SOIL SCIENCE

The advent of nanotechnology has ushered in a new era of unprecedented opportunities for various scientific disciplines, including soil science. Nanosensors, devices that leverage nanomaterials and nanostructures for sensing and detection purposes, have emerged as a game-changing technology for precision monitoring and analysis of soil health and environmental conditions. These tiny sensors, often measuring just a few nanometers in size, possess remarkable capabilities for detecting and quantifying various soil parameters with high sensitivity and specificity.

Soil, the foundation of terrestrial ecosystems, plays a crucial role in supporting plant growth, maintaining water cycles, and regulating global biogeochemical processes. However, traditional soil monitoring methods often lack the required resolution and real-time data acquisition capabilities to effectively address the complex challenges faced in precision agriculture, environmental monitoring, and soil remediation efforts. Nanosensors offer a solution to this dilemma by providing unprecedented levels of accuracy, sensitivity, and temporal resolution, enabling a comprehensive understanding of soil dynamics and informing sustainable soil management strategies.(Eswaran U., et.al (2023a))

This chapter delves into the innovative applications of nanosensors in soil science, exploring their principles, design, fabrication, and deployment in soil environments. It highlights the diverse applications of nanosensors in precision agriculture, environmental monitoring, and soil remediation efforts, showcasing their potential to revolutionize soil management practices and promote sustainable agricultural and environmental stewardship.

The mind map shown in Figure 1 illustrates the diverse applications of nanosensors in soil science. From precision agriculture to environmental monitoring and soil remediation, nanosensors offer real-time monitoring with high sensitivity and selectivity, enhancing our understanding and management of soil health.

Figure 1. Applications of nanosensors in soil science

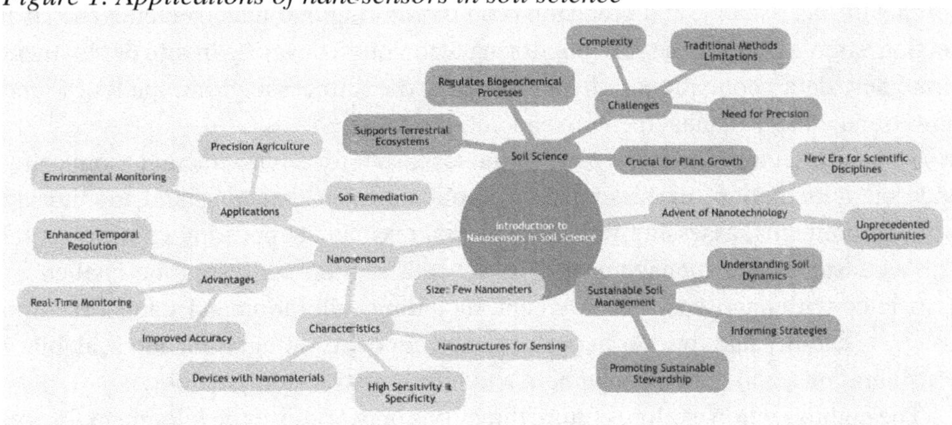

Objectives of the chapter:

1. To explore the principles and techniques involved in nanosensor design, fabrication, and deployment for soil monitoring applications.
2. To examine the diverse types of nanosensors and their specific applications in precision agriculture, environmental monitoring, and soil remediation.
3. To discuss the challenges and future prospects of integrating nanosensor technologies into soil science practices.
4. To highlight the potential benefits and implications of nanosensors for sustainable soil management and environmental stewardship.

Organization of the Chapter

The chapter begins with an introduction that sets the context for the role of nanosensors in soil science, highlighting their potential to revolutionize soil management practices and promote sustainable agriculture and environmental stewardship. The second section delves into the principles of nanosensor design and fabrication, covering topics such as nanomaterial selection, surface functionalization, and sensor platform integration. The third section explores the different types of nanosensors for soil monitoring, categorized based on the nanomaterials employed and their detection mechanisms, including electrochemical, optical, mechanical, and nanobiosensors.

The fourth section focuses on the applications of nanosensors in precision agriculture, discussing their role in nutrient monitoring, soil moisture management, soil health assessment, and precision crop monitoring and management. The fifth section examines real-time soil monitoring techniques, covering in-situ deployment strategies, data acquisition and transmission, data processing and analytics, and power and energy management considerations.

The sixth section shifts the focus to nanosensors for environmental monitoring and soil remediation, discussing their applications in detecting and monitoring soil contaminants, assessing remediation progress, and supporting environmental risk assessment and management. The seventh section addresses the challenges associated with nanosensor deployment, including calibration and standardization issues, reliability and long-term performance concerns, integration and scalability considerations, and data management and interpretation complexities.

The eighth section explores future directions in nanosensor development for soil science, such as advanced nanomaterial engineering, novel transduction mechanisms, sensor networking and integration, biocompatible and biodegradable nanosensors, energy harvesting, and nanosensor-assisted precision farming. The ninth section presents case studies showcasing real-world applications of nanosensors in soil monitoring and analysis, highlighting their diverse implementations and impact.

The tenth section examines the ethical and regulatory considerations surrounding nanosensor deployment, including environmental impacts, human health and safety, ethical and societal implications, and the need for robust regulatory frameworks and responsible innovation practices. Finally, the chapter concludes by summarizing the potential of nanosensors to revolutionize sustainable soil management and emphasizing the importance of interdisciplinary collaboration and responsible innovation in realizing their full potential.

2. PRINCIPLES OF NANOSENSOR DESIGN AND FABRICATION

Nanosensors are designed and fabricated using a combination of nanomaterials, such as nanoparticles, nanotubes, nanowires, and nanocomposites, which exhibit unique properties at the nanoscale. These nanomaterials possess exceptional surface-to-volume ratios, quantum confinement effects, and tailored surface chemistries, enabling highly sensitive and selective detection of various analytes.

The design and fabrication of nanosensors involve several key principles and techniques:

2.1. Nanomaterial Selection and Synthesis

Choosing the appropriate nanomaterial is crucial for the desired sensing application. Nanoparticles (e.g., gold, silver, iron oxide), carbon nanotubes, graphene, and quantum dots are commonly used due to their unique optical, electrical, and chemical properties. Synthesis methods like chemical vapor deposition, sol-gel, and hydrothermal techniques are employed to produce these nanomaterials with precise control over size, shape, and surface characteristics (Baig et al., 2021).

2.2. Surface Functionalization

The surface of nanomaterials can be modified with various functional groups or recognition elements to enhance selectivity and sensitivity towards specific analytes. Techniques such as self-assembled monolayers, covalent binding, and non-covalent interactions (e.g., electrostatic, hydrophobic, or π-π interactions) are employed to functionalize the nanomaterial surface.

2.3. Sensor Platform Integration

Nanosensors are integrated into various platforms or substrates, such as electrodes, optical fibers, or microfluidic devices, depending on the desired transduction mechanism (e.g., electrochemical, optical, or mechanical). Microfabrication techniques, including lithography, deposition, and etching, are used to create these sensor platforms and integrate the nanomaterials (Wen et al., 2019).

2.4. Signal Transduction and Readout

The interaction between the analyte and the functionalized nanomaterial surface induces changes in the physical or chemical properties of the nanomaterial, such as conductivity, optical properties, or mass. These changes are converted into measurable electrical, optical, or mechanical signals through appropriate transduction mechanisms, which are then processed and analyzed using suitable readout systems.

The design and fabrication of nanosensors involve interdisciplinary collaborations among chemists, materials scientists, engineers, and physicists, leveraging expertise in nanomaterial synthesis, surface chemistry, microfabrication, and signal processing.

3. TYPES OF NANOSENSORS FOR SOIL MONITORING: NANOMATERIALS AND DETECTION MECHANISMS

Nanosensors for soil monitoring and analysis can be broadly classified based on the type of nanomaterials employed and the detection mechanisms utilized. Each type of nanosensor offers unique advantages and capabilities for monitoring specific soil parameters or environmental conditions (Eswaran & Eswaran, 2023).

3.1. Electrochemical Nanosensors

- Carbon nanotubes (CNTs), graphene, and metal nanoparticles (e.g., gold, silver) are commonly used for electrochemical sensing.
- Detection mechanisms involve measuring changes in conductivity, redox reactions, or electrochemical impedance upon analyte interaction.
- Applications include monitoring soil pH, nutrient levels (e.g., nitrates, phosphates), heavy metal contaminants, and redox potential.

3.2. Optical Nanosensors

- Quantum dots, noble metal nanoparticles, and upconverting nanoparticles are utilized for optical sensing.
- Detection mechanisms involve measuring changes in fluorescence, surface plasmon resonance, or Raman scattering upon analyte interaction.
- Applications include monitoring soil organic matter, pesticide residues, and microbial activity through fluorescence or Raman spectroscopy.

3.3. Mechanical Nanosensors

- Nanomaterials like carbon nanotubes, graphene, and piezoelectric nanomaterials are employed for mechanical sensing.
- Detection mechanisms involve measuring changes in mass, stress, or strain upon analyte adsorption or desorption.
- Applications include monitoring soil moisture, density, and compaction levels through mass or strain changes.

3.4. Nanobiosensors

- Nanomaterials are combined with biological recognition elements (e.g., enzymes, antibodies, nucleic acids) for selective analyte detection.

- Detection mechanisms involve monitoring changes in optical, electrochemical, or mass-based signals upon analyte-recognition element interaction.
- Applications include monitoring soil microbial activity, presence of specific pathogens, or biomarkers of soil health.

3.5. Nanosensor Arrays and Multiplexed Systems

- Combinations of different nanosensors are integrated into arrays or multiplexed systems for simultaneous monitoring of multiple analytes.
- These systems provide comprehensive soil analysis by combining various detection mechanisms and nanomaterials.
- Applications include real-time monitoring of multiple soil parameters for precision agriculture and environmental monitoring.

Innovative nanosensors offer diverse detection mechanisms like electrochemical, optical, mechanical, and nanobiosensors for precise soil analysis and monitoring is shown in Table 1.

Table 1. Nanosensor types and their applications in soil monitoring

Type of Nanosensor	Nanomaterials	Detection Mechanism	Applications
Electrochemical	Carbon nanotubes, graphene, metal nanoparticles	Conductivity, redox reactions, electrochemical impedance	Soil pH, nutrient levels, heavy metals, redox potential
Optical	Quantum dots, noble metal nanoparticles, upconverting nanoparticles	Fluorescence, surface plasmon resonance, Raman scattering	Soil organic matter, pesticide residues, microbial activity
Mechanical	Carbon nanotubes, graphene, piezoelectric nanomaterials	Mass, stress, strain	Soil moisture, density, compaction
Nanobiosensors	Nanomaterials combined with biological recognition elements	Optical, electrochemical, mass-based signals	Microbial activity, pathogens, biomarkers of soil health
Nanosensor arrays	Combinations of different nanosensors	Multiple detection mechanisms	Simultaneous monitoring of multiple soil parameters

The choice of nanomaterial and detection mechanism is dictated by the specific soil parameters or environmental conditions to be monitored, as well as the desired sensitivity, selectivity, and operational requirements.

4. APPLICATIONS OF NANOSENSORS IN PRECISION AGRICULTURE

Precision agriculture aims to optimize crop production while minimizing environmental impact by tailoring agricultural practices to site-specific conditions and crop requirements. Nanosensors play a crucial role in enabling precision agriculture by providing real-time, high-resolution data on soil characteristics and environmental conditions.

4.1. Nutrient Monitoring and Management

In the field of managing agriculture, the accurate monitoring and management of nutrients in soil are paramount for ensuring optimal crop growth and minimizing environmental degradation. Electrochemical and optical nanosensors, owing to their high sensitivity and specificity, have emerged as invaluable tools for precisely detecting and quantifying essential nutrients such as nitrogen, phosphorus, and potassium in soil samples (Yang et al., 2024).

Electrochemical nanosensors utilize materials like carbon nanotubes, graphene, and metal nanoparticles to detect changes in conductivity or redox reactions upon interaction with target nutrients. Similarly, optical nanosensors, employing quantum dots, noble metal nanoparticles, or upconverting nanoparticles, detect alterations in fluorescence or surface plasmon resonance signals in response to nutrient presence.

The data obtained from these nanosensors not only provide accurate measurements of nutrient levels but also enable farmers to make informed decisions regarding fertilizer application. By tailoring the application of fertilizers based on real-time soil nutrient data, farmers can optimize crop nutrition while minimizing the risk of over-fertilization, which can lead to environmental issues such as nutrient runoff and water pollution.

Furthermore, the precise management of nutrients facilitated by nanosensor technology contributes to sustainable agricultural practices by reducing the environmental impact associated with excessive fertilizer use. By promoting efficient nutrient utilization in agricultural systems, nanosensors play a pivotal role in enhancing both crop productivity and environmental stewardship in modern agriculture.

4.2. Soil Moisture and Irrigation Management

In the area of soil moisture and irrigation management, mechanical nanosensors represent a cutting-edge solution for precise monitoring of soil moisture levels. Specifically, carbon nanotube-based strain sensors have emerged as a promising technology due to their high sensitivity and accuracy in detecting subtle changes in soil moisture content.

These mechanical nanosensors operate by measuring alterations in strain or deformation within the carbon nanotube structure in response to variations in soil moisture. As soil moisture levels fluctuate, the nanosensors detect these changes and convert them into measurable signals, providing real-time data on soil moisture status (Feng et al., 2024).

The availability of real-time moisture data offers significant benefits for irrigation management practices. By integrating this data into irrigation scheduling systems, farmers can implement precision irrigation techniques tailored to the specific moisture needs of their crops and soil conditions. This optimized approach to irrigation helps minimize water waste by delivering the right amount of water precisely when and where it is needed, thereby conserving water resources and reducing operational costs.

Moreover, efficient irrigation scheduling based on accurate soil moisture data helps prevent over-irrigation, a common issue in traditional irrigation practices. Over-irrigation not only wastes valuable water resources but also contributes to soil erosion, nutrient leaching, and waterlogging, which can adversely impact crop health and productivity. By preventing excessive water application, nanosensor-enabled irrigation management promotes sustainable water usage while maintaining optimal soil conditions for crop growth.

In summary, the utilization of mechanical nanosensors for soil moisture monitoring facilitates precise irrigation management strategies that optimize water usage, enhance crop yields, and promote sustainable agricultural practices. By harnessing the power of nanotechnology, farmers can achieve greater efficiency and effectiveness in irrigation practices, contributing to both environmental conservation and agricultural productivity.

4.3. Soil Health and Quality Assessment

When it comes to assessing soil health and quality, nanobiosensors emerge as invaluable tools for detecting and quantifying various key indicators. Specifically, these advanced sensors are adept at discerning soil microbial activity, organic matter content, and the presence of potential pathogens or contaminants with exceptional precision and sensitivity (Nunes et al., 2024).

Nanobiosensors operate by combining nanomaterials with biological recognition elements, such as enzymes, antibodies, or nucleic acids, that specifically target the molecules of interest. Through selective binding interactions between the biological recognition elements and target molecules present in the soil, nanobiosensors generate detectable signals that can be quantified and analyzed.

One of the primary applications of nanobiosensors in soil health assessment is the measurement of microbial activity. By detecting the presence and metabolic activity of soil microorganisms, nanobiosensors provide insights into the overall microbial community structure, diversity, and functionality. This information is crucial for understanding soil fertility, nutrient cycling processes, and ecosystem functioning, thereby facilitating informed decisions on soil management practices.

Furthermore, nanobiosensors play a vital role in assessing soil organic matter content, a key determinant of soil fertility and productivity. By accurately quantifying organic carbon levels in soil samples, these sensors offer valuable information on soil nutrient availability, water retention capacity, and microbial habitat suitability. This knowledge enables farmers to implement targeted soil amendments and organic matter management strategies to enhance soil health and productivity sustainably.

Additionally, nanobiosensors excel in detecting the presence of potential pathogens or contaminants in soil environments. Whether it be harmful bacteria, fungi, or chemical pollutants, nanobiosensors can rapidly and selectively identify these agents, allowing for timely intervention measures to mitigate risks to soil health and human health.

In summary, the deployment of nanobiosensors for soil health and quality assessment facilitates a comprehensive understanding of soil dynamics and enables proactive management strategies to maintain soil fertility, productivity, and sustainability. By harnessing the capabilities of nanotechnology and biosensing, farmers and land managers can make informed decisions that promote healthy soils, resilient ecosystems, and sustainable agricultural practices.

4.4. Precision Crop Monitoring and Management

In the realm of precision crop monitoring and management, nanosensor arrays and multiplexed systems represent cutting-edge technologies that revolutionize the way farmers optimize agricultural practices. These advanced systems offer the capability to simultaneously monitor multiple critical soil parameters, providing a wealth of comprehensive data essential for site-specific crop management strategies.

By integrating various types of nanosensors into arrays or multiplexed systems, farmers gain access to a diverse range of real-time data regarding soil conditions, nutrient levels, moisture content, and other key factors influencing crop growth and productivity. This multidimensional approach to monitoring allows for a nuanced understanding of the dynamic soil-plant-environment interactions, enabling precise adjustments and interventions tailored to the specific needs of individual crops and field conditions.

The comprehensive data generated by nanosensor arrays and multiplexed systems serve as the foundation for data-driven decision-making in crop management. By leveraging this information, farmers can optimize the application of inputs such as fertilizers, pesticides, and irrigation water, ensuring that resources are allocated efficiently and effectively. This targeted approach minimizes wastage, reduces environmental impact, and maximizes yields, thereby enhancing overall agricultural productivity and profitability.

Furthermore, the ability to monitor multiple soil parameters simultaneously empowers farmers to implement proactive measures to address potential challenges or risks to crop health and performance. For example, by detecting early signs of nutrient deficiencies, water stress, or pest infestations, farmers can intervene promptly with targeted interventions, such as nutrient supplementation, irrigation adjustments, or pest management strategies, thereby minimizing yield losses and optimizing crop quality.

In essence, nanosensor arrays and multiplexed systems revolutionize precision crop monitoring and management by providing farmers with actionable insights derived from comprehensive soil data. By embracing these advanced technologies, farmers can embark on a path towards sustainable agriculture practices that optimize resource use, minimize environmental impact, and maximize crop productivity and profitability in a rapidly evolving agricultural landscape.

4.5. Environmental Monitoring and Climate Adaptation

In the context of environmental monitoring and climate adaptation, nanosensors serve as indispensable tools for assessing soil conditions and addressing the challenges posed by climate change. These advanced sensing technologies enable the real-time monitoring of key soil parameters that are sensitive indicators of environmental change, including soil erosion, salinization, and variations in soil organic carbon levels (Aboagye et al., 2024).

Soil erosion, exacerbated by factors such as intensive agricultural practices and extreme weather events, poses a significant threat to soil health and agricultural sustainability. Nanosensors equipped with capabilities to detect changes in soil structure, moisture content, and sediment transport can provide valuable insights into erosion dynamics. By monitoring erosion hotspots and identifying vulnerable areas, farmers can implement targeted erosion control measures, such as contour farming, cover cropping, or terracing, to mitigate soil loss and preserve soil fertility.

Salinization, resulting from factors such as irrigation practices, poor drainage, and climate variability, can lead to the accumulation of salts in the soil, negatively impacting crop growth and productivity. Nanosensors capable of measuring soil salinity levels with high precision enable early detection of salinization trends and facilitate timely interventions to prevent salt accumulation. By implementing measures such as improved irrigation management, drainage systems, and salt-tolerant crop varieties, farmers can effectively manage soil salinity issues and maintain agricultural productivity in saline-affected areas.

Changes in soil organic carbon levels, influenced by factors such as land use change, agricultural practices, and climate variability, have profound implications for soil fertility, carbon sequestration, and greenhouse gas emissions. Nanosensors equipped with capabilities to monitor soil organic carbon dynamics provide valuable data for assessing soil health and carbon storage potential. By tracking changes in soil organic carbon levels over time, farmers can implement practices such as conservation tillage, cover cropping, and organic amendments to enhance soil fertility, sequester carbon, and mitigate climate change impacts.

Overall, nanosensors play a critical role in environmental monitoring and climate adaptation efforts by providing real-time data on soil conditions related to climate change. By leveraging this information, farmers can develop climate-resilient agricultural practices and adaptive management strategies that enhance soil health, preserve natural resources, and ensure the long-term sustainability of agricultural systems in the face of changing environmental conditions.

Line graph in Figure 2 titled "Projected Impact of Nanosensor-based Precision Agriculture" showing the percentage change in crop yields, fertilizer usage, and water usage over the years 2022 to 2026. Crop yields increase steadily from 100% in 2022 to 130% in 2026. Fertilizer usage decreases from 100% in 2022 to 80% in 2026. Water usage decreases from 100% in 2022 to 88% in 2026.

Figure 2. Projected impact of nanosensor-based precision agriculture over time

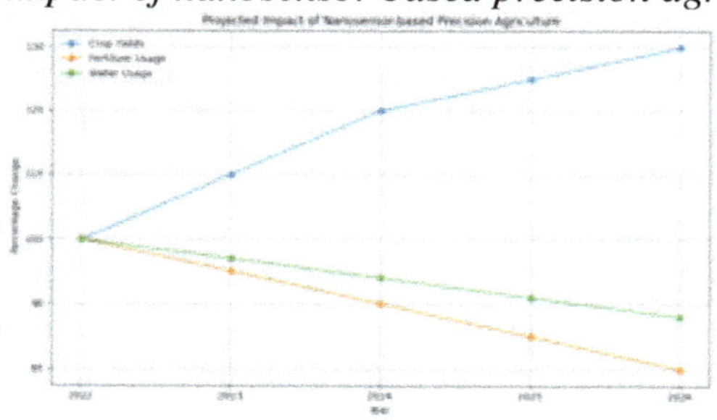

By providing real-time, high-resolution data on soil characteristics and environmental conditions, nanosensors enable farmers and agricultural professionals to make informed decisions, optimize resource utilization, and implement sustainable farming practices tailored to specific field conditions.

5. REAL-TIME SOIL MONITORING: NANOSENSOR DEPLOYMENT AND DATA ACQUISITION

The effective deployment of nanosensors in soil environments and the acquisition of real-time data are essential for leveraging their potential in precision agriculture and environmental monitoring. Several strategies and techniques are employed to ensure successful nanosensor deployment and data acquisition:

5.1. In-Situ Deployment Strategies

Direct soil embedding: One approach to deploying nanosensors involves directly embedding them into the soil matrix at specific depths. This method enables continuous and in-situ monitoring of soil conditions without disturbing the natural soil structure. Nanosensors embedded in the soil can detect changes in key parameters such as moisture content, nutrient levels, and contaminant concentrations over time, providing valuable insights into soil health and environmental dynamics (Anburuvel, 2024).

Sensor probes and arrays: Another deployment strategy involves integrating nanosensors into sensor probes or arrays that are inserted into the soil. These probes can be designed to target specific soil depths or locations of interest, allowing for targeted monitoring of soil properties. Nanosensor arrays enable simultaneous measurement of multiple analytes or soil profiles, providing comprehensive data on soil characteristics and variability. By deploying sensor probes or arrays at strategic locations within the soil profile, researchers and farmers can obtain detailed information on soil conditions and spatial variability, facilitating informed decision-making in agriculture and environmental management.

Microfluidic devices: Nanosensors can also be incorporated into microfluidic devices designed to extract and analyze soil pore water or soil extracts. These devices utilize microfluidic channels to collect samples from the soil matrix and deliver them to nanosensor-based detection systems for analysis. By capturing soil pore water or extracts, microfluidic devices enable real-time monitoring of soil chemistry, including nutrient concentrations, contaminant levels, and microbial activity. This approach offers a minimally invasive and highly sensitive method for assessing soil health and environmental quality, with potential applications in precision agriculture, environmental monitoring, and soil remediation efforts.

5.2. Data Acquisition and Transmission

Wireless sensor networks: Nanosensors can be seamlessly integrated into wireless sensor networks (WSNs), forming a distributed system for monitoring soil conditions over large agricultural or environmental areas. These networks consist of interconnected nanosensor nodes equipped with wireless communication capabilities, allowing them to communicate with each other and with central monitoring stations. Nanosensors deployed in WSNs can continuously collect data on various soil parameters, such as moisture content, nutrient levels, and temperature, providing real-time insights into soil health and environmental dynamics. The use of wireless communication technologies, such as Wi-Fi, Bluetooth, or LoRa, enables remote data acquisition and transmission, eliminating the need for wired connections and facilitating data access from remote or inaccessible locations.

Internet of Things (IoT) integration: Nanosensor data can be seamlessly integrated into IoT platforms, expanding their capabilities for real-time monitoring, data analysis, and decision support systems. IoT platforms provide a scalable and interoperable framework for collecting, storing, and analyzing data from diverse sources, including nanosensors deployed in soil monitoring applications. By integrating nanosensor data into IoT platforms, stakeholders can access comprehensive insights into soil conditions, environmental trends, and agricultural performance through web-based dashboards or mobile applications. IoT-enabled nanosensor systems offer enhanced flexibility, scalability, and accessibility, empowering users to make informed decisions and optimize resource management strategies based on real-time data analytics.

Mobile applications: Field-deployable nanosensor systems can be interfaced with mobile applications, enabling on-site data visualization and analysis for farmers or field technicians. These applications provide user-friendly interfaces for accessing, visualizing, and interpreting nanosensor data in real-time, allowing users to monitor soil conditions, track environmental trends, and make informed decisions on-the-go. Mobile applications can display sensor readings, trend analysis, and alerts related to soil health, nutrient status, or irrigation needs, helping users optimize agricultural practices and resource allocation strategies. Additionally, mobile applications may offer features such as data logging, mapping, and sharing, facilitating collaboration and knowledge exchange among stakeholders in agriculture and environmental management.

5.3. Data Processing and Analytics

Signal processing and calibration: Nanosensor data collected from soil monitoring applications often undergoes signal processing and calibration to enhance accuracy and reliability. Signal processing techniques such as noise filtering, baseline correction, and signal amplification are employed to improve the quality of raw sensor data. Additionally, calibration procedures are conducted to establish the relationship between sensor responses and target analyte concentrations, accounting for factors such as environmental conditions, sensor drift, and matrix effects. By calibrating nanosensor responses against known reference standards or laboratory measurements, accurate quantification of soil parameters such as nutrient levels, contaminant concentrations, or microbial activity can be achieved.

Machine learning and data fusion: Advanced machine learning algorithms and data fusion techniques play a crucial role in analyzing nanosensor data, integrating information from multiple sensors, and extracting meaningful insights. Machine learning models, including supervised learning, unsupervised learning, and deep learning algorithms, can be trained on large datasets to recognize patterns, trends, and anomalies in soil sensor data. Data fusion methods combine data from different nanosensors or sources, such as satellite imagery, weather data, or soil maps, to generate holistic views of soil conditions and environmental dynamics. By leveraging machine learning and data fusion, nanosensor systems can provide predictive modeling capabilities, forecast future soil trends, and support decision-making in precision agriculture, environmental monitoring, and soil remediation efforts.

Decision support systems: Nanosensor data is integrated into decision support systems (DSS), which offer actionable recommendations and insights to stakeholders involved in soil management practices. DSS utilize algorithms, models, and visualization tools to process, analyze, and interpret nanosensor data, facilitating informed decision-making at various levels of soil management. These systems provide real-time alerts, optimization strategies, and scenario-based simulations tailored to specific agricultural or environmental contexts. By harnessing the power of nanosensor data analytics, decision support systems enable users to implement proactive soil management practices, mitigate risks, and optimize resource allocation for sustainable land use and ecosystem preservation.

5.4. Power and Energy Management

Energy harvesting: Nanosensors can be integrated with energy harvesting technologies to enable self-powered operation and extend deployment longevity. Energy harvesting techniques such as solar, thermal, or kinetic energy conversion can be utilized to power nanosensor systems autonomously without the need for external power sources. For example, photovoltaic cells can convert sunlight into electrical energy to power nanosensors in outdoor environments, while thermoelectric generators can harness temperature differentials in the soil to generate electricity. By leveraging energy harvesting, nanosensor systems can operate sustainably for extended periods, reducing the need for frequent maintenance and battery replacements.

Low-power electronics and communication: Nanosensor systems are engineered with low-power electronics and communication protocols to minimize energy consumption and maximize battery life. Integrated circuits, microcontrollers, and sensor interfaces are optimized for energy efficiency, ensuring minimal power consumption during sensor operation and data transmission. Furthermore, energy-efficient communication protocols such as Zigbee, LoRaWAN, or Bluetooth Low Energy (BLE) are employed to facilitate wireless data transmission while conserving power. These low-power design principles enable nanosensor systems to operate reliably in remote field conditions for prolonged durations, enhancing their suitability for long-term environmental monitoring, soil remediation, and precision agriculture applications.

5.5. Sensor Calibration and Maintenance

In-field calibration: Regular calibration of nanosensors is essential to maintain accuracy and reliability in measuring soil parameters. In-field calibration involves adjusting sensor settings or parameters based on reference measurements taken directly from the soil environment. This process helps account for variations in environmental conditions, sensor drift, and other factors that may affect sensor performance over time. By conducting in-field calibration, users can ensure that nanosensor data remains accurate and consistent, enhancing the reliability of soil monitoring and analysis.

Self-calibrating and self-diagnostic systems: To streamline calibration and maintenance procedures, some advanced nanosensor systems are equipped with self-calibrating and self-diagnostic capabilities. These systems utilize built-in algorithms and sensor feedback mechanisms to automatically adjust sensor settings, detect anomalies, and perform diagnostic checks without manual intervention. By implementing self-calibration and self-diagnostic features, nanosensor systems can minimize downtime, reduce the need for specialized technical expertise, and improve overall system robustness. Additionally, self-diagnostic capabilities enable early detection of sensor malfunctions or performance degradation, allowing users to take proactive measures to address issues and ensure continuous operation of soil monitoring networks.

Effective deployment strategies, data acquisition techniques, and integrated data processing and analytics are crucial for harnessing the full potential of nanosensors in real-time soil monitoring and precision agriculture applications.

6. NANOSENSORS FOR ENVIRONMENTAL MONITORING AND SOIL REMEDIATION

In addition to their applications in precision agriculture, nanosensors play a vital role in environmental monitoring and soil remediation efforts. The ability to detect and quantify various contaminants and pollutants in soil is crucial for assessing environmental impacts, identifying potential risks, and guiding remediation strategies.

6.1. Monitoring and Detection of Soil Contaminants

- Heavy metal detection: Electrochemical and optical nanosensors can detect and quantify heavy metal contaminants (e.g., lead, cadmium, mercury) in soil with high sensitivity and selectivity (Amani et al., 2024).
- Organic pollutant monitoring: Nanosensors can detect and monitor the presence of organic pollutants, such as pesticides, polycyclic aromatic hydrocarbons (PAHs), and persistent organic pollutants (POPs), in soil samples.
- Microbial contaminant detection: Nanobiosensors can detect and identify specific microbial contaminants, such as pathogenic bacteria or fungi, in soil environments.

6.2. Soil Remediation Monitoring and Assessment

- Monitoring remediation progress: Nanosensors can track the effectiveness of soil remediation techniques by monitoring changes in contaminant levels, redox potential, and other relevant parameters over time.
- Phytoremediation monitoring: Nanosensors can be used to monitor the uptake and accumulation of contaminants by plants during phytoremediation processes, providing insights into the efficiency and optimization of these techniques.
- Bioremediation monitoring: Nanobiosensors can monitor the activity and proliferation of microorganisms involved in bioremediation processes, enabling optimized management of these remediation strategies.

6.3. Environmental Risk Assessment and Management

- Contaminant transport and fate modeling: Nanosensor data can be integrated into computational models to predict the transport and fate of contaminants in soil and groundwater systems, aiding in risk assessment and management.
- Ecosystem impact assessment: Nanosensors can provide valuable data on the impact of soil contaminants on soil biota, plant life, and overall ecosystem health, informing environmental management decisions.
- Long-term monitoring and early warning systems: Nanosensor networks can be deployed for long-term monitoring of soil and environmental conditions, enabling early detection of potential risks and timely intervention.

The bar chart shown in Figure 3 illustrates the detection limits of nanosensors for various soil contaminants, including heavy metals, organic pollutants, and microbial contaminants. It demonstrates the high sensitivity of nanosensors in detecting these contaminants, with lower detection limits indicating greater sensitivity.

Figure 3. Detection limits of nanosensors for soil contaminants

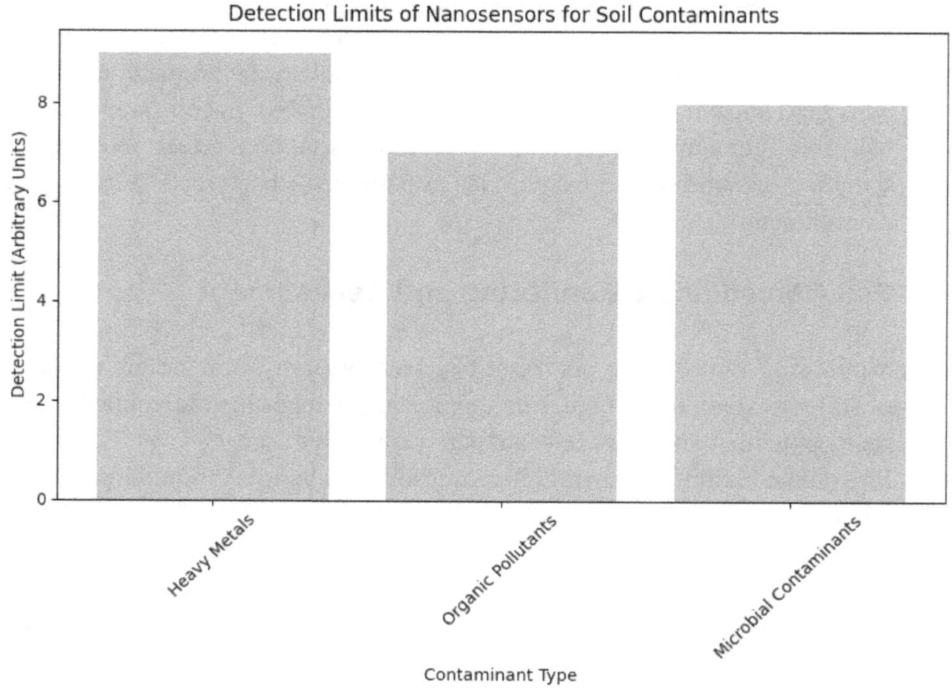

By leveraging the high sensitivity and specificity of nanosensors, environmental monitoring and soil remediation efforts can be significantly enhanced, leading to more effective management of soil contaminants, minimizing environmental impacts, and promoting sustainable land use practices.

7. CHALLENGES IN NANOSENSOR DEPLOYMENT: CALIBRATION, RELIABILITY, AND LONG-TERM PERFORMANCE

While nanosensors offer tremendous potential for soil monitoring and analysis, several challenges must be addressed to ensure their successful deployment and long-term performance in real-world soil environments.

7.1. Calibration and Standardization

- Environmental factors: Soil properties, such as pH, ionic strength, and organic matter content, can influence nanosensor performance, necessitating site-specific calibration and standardization protocols (Amelung et al., 2024).
- Matrix effects: The complex and heterogeneous nature of soil matrices can introduce matrix effects that interfere with nanosensor responses, requiring robust calibration strategies.
- Reference materials and standards: The lack of well-characterized reference materials and standards for nanosensor calibration poses challenges in ensuring accurate and reproducible measurements across different soil types and conditions.

7.2. Reliability and Long-Term Performance

- Fouling and biofouling: Over time, nanosensors can be subject to fouling by organic matter, minerals, or microbial growth, affecting their sensitivity and accuracy.
- Environmental stability: Nanosensors must be designed to withstand harsh soil environments, including temperature fluctuations, moisture changes, and exposure to various chemical species, to ensure reliable long-term performance.
- Sensor drift and degradation: Nanosensor responses can drift or degrade over time due to material degradation, surface modifications, or environmental factors, requiring periodic recalibration or replacement.

7.3. Integration and Scalability

- Multiplexing and sensor arrays: Integrating multiple nanosensors into arrays or multiplexed systems for simultaneous monitoring of various analytes can pose challenges in terms of signal processing, data fusion, and system miniaturization.
- Wireless communication and power management: Enabling wireless communication and efficient power management for nanosensor networks in remote or inaccessible soil environments remains a significant challenge.
- Scalability and cost-effectiveness: Scaling up nanosensor production and deployment for large-scale agricultural or environmental monitoring applications requires cost-effective manufacturing processes and infrastructure.

7.4. Data Management and Interpretation

- Data quality and validation: Ensuring the quality and validity of nanosensor data is crucial for accurate interpretation and decision-making, necessitating robust data processing and validation techniques.
- Data fusion and integration: Integrating data from multiple nanosensors and diverse sources (e.g., remote sensing, weather data) into coherent and meaningful information remains a complex task.
- Interpretation and decision support: Translating nanosensor data into actionable insights and decision support systems requires interdisciplinary collaboration and the development of advanced data analytics tools.

7.5. Regulatory and Ethical Considerations

- Environmental and health impacts: The potential environmental and health impacts of nanosensor deployment, including the release of nanomaterials into soil ecosystems, must be carefully assessed and regulated.
- Ethical implications: The use of nanosensors for soil monitoring and analysis may raise ethical concerns related to privacy, data ownership, and responsible innovation, which need to be addressed through appropriate governance frameworks.

Table 2 outlines key challenges faced in the deployment of nanosensors for environmental monitoring and soil remediation. Overcoming these hurdles is crucial for ensuring the effectiveness and ethical use of nanosensor technology in addressing environmental concerns.

Table 2. Challenges in nanosensor deployment for environmental monitoring and soil remediation

Challenge Category	Key Challenges
Calibration and Standardization	Environmental factors, matrix effects, lack of reference materials
Reliability and Long-term Performance	Fouling, biofouling, environmental stability, sensor drift, degradation
Integration and Scalability	Multiplexing, wireless communication, power management, cost-effectiveness
Data Management and Interpretation	Data quality, validation, fusion, integration, interpretation, decision support
Regulatory and Ethical Considerations	Environmental impacts, health and safety, privacy, equity, public perception

Addressing these challenges through collaborative efforts among researchers, engineers, policymakers, and stakeholders is essential to fully harness the potential of nanosensors for sustainable soil management and environmental stewardship.

8. Future Directions in Nanosensor Development for Soil Science

The field of nanosensors for soil monitoring and analysis is rapidly evolving, with numerous exciting research avenues and future directions that hold the potential to revolutionize soil science practices further.

8.1. Advanced Nanomaterial Engineering

- **Novel nanomaterials:** The development of new classes of nanomaterials with tailored properties, such as enhanced reactivity, selectivity, and stability, can enable the design of more sensitive and robust nanosensors for soil applications (Shah et al., 2024).
- **Hybrid and composite nanomaterials:** Combining different nanomaterials into hybrid or composite structures can synergize their unique properties, enabling multifunctional nanosensors capable of detecting multiple analytes simultaneously.
- **Self-assembled and biomimetic nanostructures:** Exploiting self-assembly principles and biomimetic approaches can lead to the fabrication of highly ordered and functional nanostructures for sensing applications, mimicking the intricate designs found in nature.

8.2. Advanced Transduction Mechanisms and Signal Processing

- **Plasmonic and optical nanosensors:** Leveraging plasmonic and optical properties of nanomaterials can enable highly sensitive and label-free detection of analytes in soil environments through techniques like surface-enhanced Raman spectroscopy (SERS) or localized surface plasmon resonance (LSPR).
- **Nanophotonic and nanoplasmonics integration:** Integrating nanophotonic and nanoplasmonics components into nanosensor platforms can enhance signal transduction, multiplexing capabilities, and miniaturization potential.
- **Artificial intelligence and machine learning:** Incorporating advanced artificial intelligence and machine learning algorithms can improve signal pro-

cessing, data analysis, and interpretation of nanosensor data, enabling predictive modeling and autonomous decision-making.

8.3. Nanosensor Integration and Networking

- **Internet of Bio-Nano Things (IoBNT):** Integrating nanosensors into the emerging paradigm of the Internet of Bio-Nano Things (IoBNT) can enable seamless integration of biological and nanoscale devices, leading to unprecedented monitoring and control of soil ecosystems (Higgs et al., 2024).
- **Soil cyber-physical systems:** Developing soil cyber-physical systems that tightly couple nanosensor networks with computational models, decision support systems, and automated actuation mechanisms can enable real-time, adaptive soil management strategies.
- **Edge computing and fog computing:** Leveraging edge computing and fog computing architectures can facilitate real-time data processing and decision-making at the edge of the nanosensor networks, reducing latency and enabling timely interventions.

8.4. Biocompatible and Biodegradable Nanosensors

- **Green nanomaterials:** Exploring the use of green and sustainable nanomaterials derived from biomass or agricultural waste can lead to the development of environmentally friendly and biocompatible nanosensors for soil applications.
- **Biodegradable nanosensors:** Designing nanosensors that can degrade naturally over time or under specific environmental conditions can minimize the potential environmental impact of nanosensor deployment in soil ecosystems.

8.5. Energy Harvesting and Self-Powered Nanosensors

- **Nanogenerators and energy harvesters:** Integrating nanogenerators and energy harvesters into nanosensor platforms can enable self-powered operation, eliminating the need for external power sources and enabling long-term deployment in remote or inaccessible soil environments.
- **Triboelectric and piezoelectric nanogenerators:** Exploiting triboelectric and piezoelectric effects at the nanoscale can provide a means to harvest energy from soil vibrations, moisture gradients, or other environmental sources.

8.6. Nanosensor-Assisted Precision Farming

- **Precision nutrient and water management:** Nanosensors can be integrated into precision farming systems, providing real-time data on soil nutrient levels, moisture content, and other parameters to enable precise and targeted application of fertilizers and irrigation.
- **Crop stress monitoring and disease detection:** Nanosensors can be deployed to monitor crop health, detect early signs of stress or disease, and inform timely interventions and management strategies.
- **Precision spraying and pest management:** Nanosensors can be used to detect and map the presence of pests or pathogens in fields, enabling targeted and precise application of pesticides or other control measures, reducing unnecessary chemical usage.

8.7. Smart Farming

Integrating nanosensors into precision farming systems represents a crucial aspect of smart farming initiatives. By combining nanosensor data with other technologies such as IoT, big data analytics, and automation, smart farming systems enable data-driven decision-making, autonomous operation, and remote monitoring of agricultural processes. This holistic approach enhances productivity, efficiency, and sustainability across the entire agricultural value chain, from soil management to crop cultivation and distribution. (Eswaran et al., 2024).

The future of nanosensors in soil science lies in the convergence of advanced nanomaterial engineering, transduction mechanisms, sensor networking, and integration with emerging technologies such as artificial intelligence, internet of things, and energy harvesting systems. Interdisciplinary collaborations and responsible innovation practices will be crucial in realizing the full potential of nanosensors for sustainable soil management and precision agriculture.

9. CASE STUDIES: NANOSENSOR APPLICATIONS IN SOIL MONITORING AND ANALYSIS

To illustrate the practical applications of nanosensors in soil monitoring and analysis, several case studies are presented, highlighting the diverse range of nanosensor technologies and their real-world implementations.

9.1. Precision Nutrient Management Using Ion-Selective Nanosensors

- Researchers at a leading agricultural university developed a nanosensor array based on ion-selective electrodes (ISEs) and carbon nanotube field-effect transistors (CNTFETs) for real-time monitoring of soil nutrient levels (Tharini et al., 2024).
- The nanosensor array was deployed in a commercial agricultural field and integrated with a wireless sensor network and decision support system.
- Real-time data on soil nitrate, phosphate, and potassium levels were used to precisely adjust fertilizer application rates, leading to a significant reduction in fertilizer usage and minimizing environmental impacts.

9.2. Optical Nanosensors for Monitoring Soil Organic Matter and Microbial Activity

- A team of environmental scientists developed a fluorescent nanosensor based on quantum dots functionalized with enzymes for monitoring soil organic matter and microbial activity (Parameswari et al., 2024).
- The nanosensors were deployed in various soil types and ecosystems, including agricultural fields, forests, and wetlands.
- By tracking changes in fluorescence intensity and lifetime, the researchers were able to quantify soil organic matter dynamics and microbial activity in response to different environmental conditions and land management practices.

9.3. Phytoremediation Monitoring Using Nanobiosensors

- Researchers at an environmental engineering institute developed a nanobiosensor based on gold nanoparticles and plant-derived antibodies for monitoring the uptake of heavy metal contaminants by plants during phytoremediation processes (Dhanapal et al., 2024).
- The nanobiosensors were integrated into a microfluidic platform for real-time monitoring of plant root exudates and xylem sap.
- By tracking the binding of heavy metals to the plant-derived antibodies, the researchers could quantify the uptake and translocation of contaminants within the plants, enabling optimization of phytoremediation strategies.

9.4. Wireless Nanosensor Network for Soil Moisture and Salinity Monitoring

- A collaborative project between a nanotechnology research center and an agricultural institute developed a wireless nanosensor network for real-time monitoring of soil moisture and salinity levels in arid and semi-arid regions. (Kansotia, Kavita, et al. 2024)
- The nanosensor nodes comprised of carbon nanotube-based strain sensors for moisture monitoring and ion-selective nanosensors for salinity detection.
- The wireless network was deployed in a large agricultural area, providing high-resolution data on soil moisture and salinity levels, enabling efficient irrigation scheduling and management of saline soils.

9.5. Multiplexed Nanosensor Array for Comprehensive Soil Analysis

Researchers at a multidisciplinary center developed a multiplexed nanosensor array capable of simultaneous monitoring of various soil parameters, including pH, nutrient levels, heavy metal contaminants, and microbial activity. The nanosensor array integrated electrochemical, optical, and nanobiosensors on a single platform, leveraging different nanomaterials and transduction mechanisms. The multiplexed nanosensor array was deployed in various soil environments, including agricultural fields, contaminated sites, and natural ecosystems, providing comprehensive soil analysis and informing sustainable soil management strategies (Ang et al., 2024).

Integrating IoT, ML and Cloud Computing for Sustainable Agriculture: Opportunities and Challenges:

In addition to the advancements in nanosensor technology for soil monitoring and analysis, there is growing interest in integrating Internet of Things (IoT), machine learning (ML), and cloud computing to further enhance agricultural sustainability (Eswaran & Khang, 2024). This integration offers opportunities for real-time data collection, analysis, and decision-making, optimizing resource usage and improving crop yields. However, it also presents challenges such as data security, interoperability, and scalability. Addressing these challenges requires interdisciplinary collaboration and innovative solutions to harness the full potential of IoT, ML, and cloud computing in sustainable agriculture (Eswaran & Eswaran, 2023).

These case studies demonstrate the diverse applications of nanosensors in soil monitoring and analysis, highlighting their potential to revolutionize precision agriculture, environmental monitoring, and soil remediation efforts. As nanosensor technologies continue to evolve and integrate with emerging technologies, their impact on soil science and sustainable land management practices will become increasingly significant.

10. ETHICAL AND REGULATORY CONSIDERATIONS IN NANOSENSOR DEPLOYMENT

While nanosensors offer numerous benefits for soil monitoring and analysis, their deployment and use in soil environments raise important ethical and regulatory considerations that must be addressed to ensure responsible innovation and minimize potential risks.

10.1. Environmental impacts and risk assessment:

- **Release of nanomaterials:** The potential release of nanomaterials from nanosensors into soil ecosystems during deployment or disposal raises concerns about their environmental fate, transport, and potential toxicity to soil organisms and plants.
- **Bioaccumulation and biomagnification:** The potential for nanomaterials to bioaccumulate and biomagnify along the food chain, potentially impacting human health and ecological systems, must be carefully evaluated.
- **Life cycle assessment**: Comprehensive life cycle assessments of nanosensor production, deployment, and disposal are necessary to identify and mitigate potential environmental impacts throughout their entire life cycle.

10.2. Human health and safety considerations:

- **Occupational exposure**: Workers involved in the production, deployment, and maintenance of nanosensors may be at risk of exposure to nanomaterials, necessitating the development of appropriate safety protocols and personal protective equipment.
- **Consumer exposure**: If nanosensors are integrated into consumer products or food production systems, potential exposure pathways and impacts on human health must be thoroughly investigated.

10.3. Ethical and societal implications:

- **Privacy and data ownership**: The deployment of nanosensor networks for soil monitoring and precision agriculture may raise concerns about privacy, data ownership, and the potential misuse of sensitive information.
- **Equitable access and digital divide**: The widespread adoption of nanosensor technologies in agriculture and soil management may exacerbate existing inequalities and the digital divide between resource-rich and resource-constrained communities or regions.
- **Public perception and acceptance**: The introduction of nanosensor technologies in agriculture and soil management may face public skepticism or resistance, necessitating effective communication, education, and stakeholder engagement strategies.

10.4. Regulatory frameworks and governance:

- **Nanosensor regulation**: Developing robust regulatory frameworks and guidelines for nanosensor production, deployment, and disposal is crucial to ensure their safe and responsible use in soil science applications.
- **Standardization and validation**: Establishing standardized protocols for nanosensor calibration, validation, and performance evaluation is necessary to ensure data quality and reliability across different applications and regions.
- **International cooperation and harmonization**: Given the global nature of soil management and agricultural practices, international cooperation and harmonization of nanosensor regulations and standards are essential for consistent implementation and responsible innovation.

10.5. Ethical guidelines and responsible innovation:

- **Ethical guidelines**: Developing ethical guidelines and codes of conduct for nanosensor research, development, and deployment in soil science applications can help ensure responsible innovation and address potential ethical concerns.
- **Stakeholder engagement**: Engaging diverse stakeholders, including researchers, policymakers, industry representatives, and community members, in the development and implementation of nanosensor technologies is crucial for fostering trust, transparency, and accountable decision-making.
- **Interdisciplinary collaboration**: Promoting interdisciplinary collaboration between nanotechnology experts, soil scientists, environmental scientists,

ethicists, and social scientists can help anticipate and address potential ethical and societal implications of nanosensor deployment.

Addressing these ethical and regulatory considerations through proactive measures, rigorous risk assessment, stakeholder engagement, and responsible innovation practices is essential to harness the full potential of nanosensors for sustainable soil management while minimizing potential risks and fostering public trust.

11. CONCLUSION: HARNESSING NANOSENSOR TECHNOLOGIES FOR SUSTAINABLE SOIL MANAGEMENT

The integration of nanosensors into soil science practices has the potential to revolutionize our understanding and management of soil ecosystems. By providing real-time, high-resolution data on soil characteristics, environmental conditions, and contaminant presence, nanosensors offer unprecedented opportunities for precision agriculture, environmental monitoring, and soil remediation efforts.

Nanosensors' unique capabilities, including high sensitivity, selectivity, and the ability to detect and quantify various analytes at the nanoscale, have opened new avenues for innovative soil monitoring and analysis techniques. From nutrient monitoring and precision fertilizer application to contaminant detection and phytoremediation tracking, nanosensors are poised to play a pivotal role in optimizing resource utilization, minimizing environmental impacts, and promoting sustainable land management practices.

Despite the significant potential of nanosensors, several challenges must be addressed, including calibration and standardization issues, long-term performance and reliability concerns, integration and scalability considerations, and data management and interpretation complexities. Overcoming these challenges will require collaborative efforts among researchers, engineers, policymakers, and stakeholders, leveraging interdisciplinary expertise and fostering responsible innovation practices.

As nanosensor technologies continue to evolve, future directions in nanomaterial engineering, advanced transduction mechanisms, sensor networking, and integration with emerging technologies like artificial intelligence, internet of things, and energy harvesting systems will further enhance their capabilities and broaden their applications in soil science.

However, the deployment and use of nanosensors in soil environments must be accompanied by rigorous risk assessment, ethical considerations, and robust regulatory frameworks. Addressing environmental impacts, human health and safety concerns, ethical implications, and developing appropriate governance mechanisms are crucial for ensuring the responsible and sustainable implementation of nanosensor technologies in soil science.

By harnessing the power of nanosensors and embracing responsible innovation practices, we can unlock new frontiers in soil monitoring, analysis, and management, paving the way for sustainable agricultural practices, effective environmental stewardship, and the preservation of our planet's precious soil resources for future generations.

REFERENCES

Aboagye, P. D., & Sharifi, A. (2024). Urban climate adaptation and mitigation action plans: A critical review. *Renewable & Sustainable Energy Reviews*, *189*, 113886. DOI: 10.1016/j.rser.2023.113886

Amani, S., Keshtdar, A., & Ghiassi, R. (2024). Investigation of the effectiveness of electrical resistivity tomography in monitoring and detecting hydrocarbon contamination in sandy soil at various moisture: Pilot scale. *Water, Air, and Soil Pollution*, *235*(3), 1–10. DOI: 10.1007/s11270-024-06984-y

Amelung, W., Tang, N., Siebers, N., Aehnelt, M., Eusterhues, K., Felde, V. J. M. N. L., Guggenberger, G., Kaiser, K., Kögel-Knabner, I., Klumpp, E., Knief, C., Kruse, J., Lehndorff, E., Mikutta, R., Peth, S., Ray, N., Prechtel, A., Ritschel, T., Schweizer, S. A., & Totsche, K. U. (2024). Architecture of soil microaggregates: Advanced methodologies to explore properties and functions. *Journal of Plant Nutrition and Soil Science*, *187*(1), 17–50. DOI: 10.1002/jpln.202300149

Anburuvel, A. (2024). The engineering behind soil stabilization with additives: A state-of-the-art review. *Geotechnical and Geological Engineering*, *42*(1), 1–42. DOI: 10.1007/s10706-023-02554-x

Ang, M. C.-Y., Saju, J. M., Porter, T. K., Mohaideen, S., Sarangapani, S., Khong, D. T., Wang, S., Cui, J., Loh, S. I., Singh, G. P., Chua, N.-H., Strano, M. S., & Sarojam, R. (2024). Decoding early stress signaling waves in living plants using nanosensor multiplexing. *Nature Communications*, *15*(1), 2943. DOI: 10.1038/s41467-024-47082-1 PMID: 38580637

Baig, N., Kammakakam, I., & Falath, W. (2021). Nanomaterials: A review of synthesis methods, properties, recent progress, and challenges. *Materials Advances*, *2*(6), 1821–1871. DOI: 10.1039/D0MA00807A

Dhanapal, A. R., Thiruvengadam, M., Vairavanathan, J., Venkidasamy, B., Easwaran, M., & Ghorbanpour, M. (2024). Nanotechnology approaches for the remediation of agricultural polluted soils. *ACS Omega*, *9*(12), 13522–13533. DOI: 10.1021/acsomega.3c09776 PMID: 38559935

Eswaran, U., & Eswaran, V. (2023). Real-time monitoring of environmental parameters in agriculture using WSN. i-manager's Journal on Embedded Systems, 11(2), 13–20.

Eswaran, U., Eswaran, V., Murali, K., & Eswaran, V. (2023). Precision agriculture revolution: A low-cost smart irrigation system empowered by IoT for small-scale farmers. *Journal of Mechatronics and Automation*, *10*(3), 21–29.

Eswaran, U., Eswaran, V., Murali, K., & Eswaran, V. (2024). Application of sensors for smart farming. In Khang, A. (Ed.), *Agriculture and aquaculture applications of biosensors and bioelectronics* (pp. 18–44). IGI Global. DOI: 10.4018/979-8-3693-2069-3.ch002

Eswaran, U., & Khang, A. (2024). Application of internet of things (IoT) and sensors technologies for agriculture. In *Advanced IoT technologies and applications in the Industry 4.0 digital economy* (1st ed.). CRC Press. DOI: 10.1201/9781003434269-8

Feng, X., Bi, S., Li, H., Qi, Y., Chen, S., & Shao, L. (2024). Soil moisture forecasting for precision irrigation management using real-time electricity consumption records. *Agricultural Water Management*, *291*, 108656. DOI: 10.1016/j.agwat.2023.108656

Higgs, V., Ahmed, H., & Flavier, J. A. (2024). Transforming asthma diagnosis: Point of care (POC) breath test with deep neural networks (DNNs) and nanosensors. In A34. Late breaking abstracts in asthma and allergic inflammation (pp. A1380–A1380). American Thoracic Society.

Nunes, M. R., Veum, K. S., Parker, P. A., Holan, S. H., Amsili, J. P., van Es, H. M., Wills, S. A., Seybold, C. A., & Karlen, D. L. (2024). SHAPEv1.0 scoring curves and peer group benchmarks for dynamic soil health indicators. *Soil Science Society of America Journal*, *88*(3), 858–875. DOI: 10.1002/saj2.20668

Parameswari, P., Belagalla, N., Singh, B. V., Abhishek, G. J., Rajesh, G. M., Katiyar, D., Hazarika, B., & Paul, S. (2024). Nanotechnology-based sensors for real-time monitoring and assessment of soil health and quality: A review. *Asian Journal of Soil Science and Plant Nutrition*, *10*(2), 157–173. DOI: 10.9734/ajsspn/2024/v10i2272

Shah, S. S., Niaz, F., Ehsan, M. A., Das, H. T., Younas, M., Khan, A. S., Rahman, H. U., Nayem, S. M. A., Oyama, M., & Aziz, M. A. (2024). Advanced strategies in electrode engineering and nanomaterial modifications for supercapacitor performance enhancement: A comprehensive review. *Journal of Energy Storage*, *79*, 110152. DOI: 10.1016/j.est.2023.110152

Tharini, C., Sachdev, A., & Manikandan, E. (2024). An efficient sensing system using ion-selective membrane on Ni_2O_3/rGO nanocomposite for electrochemical detection of nitrate ions. *Journal of Alloys and Compounds*, *980*, 173414. DOI: 10.1016/j.jallcom.2023.173414

Wen, S., Su, Y., Dai, C., Jia, J., Fan, G.-C., Jiang, L.-P., Song, R.-B., & Zhu, J.-J. (2019). Plasmon coupling-enhanced Raman sensing platform integrated with exonuclease-assisted target recycling amplification for ultrasensitive and selective detection of microRNA-21. *Analytical Chemistry*, *91*(19), 12298–12306. DOI: 10.1021/acs.analchem.9b02476 PMID: 31486639

Yang, W., Li, T., Liao, S., Zhou, J., & Huang, L. (2024). Organ-on-a-chip platforms integrated with biosensors for precise monitoring of the cells and cellular microenvironment. *Trends in Analytical Chemistry*, *172*, 117569. DOI: 10.1016/j.trac.2024.117569

Chapter 9
Global Web of Science-Based Bibliometric Analysis of Heavy Metals:
Current Research Gaps and Biochar-Nanotechnology-Based Solutions for Armenia

Karen Ghazaryan
http://orcid.org/0000-0002-8973-2595
Faculty of Biology, Yerevan State University, Armenia

Abhishek Singh
http://orcid.org/0000-0002-0049-6910
Yerevan State University, Armenia

ABSTRACT

Heavy metals are one of the major persistent environmental threats to both human health and environmental quality, so that the determination of baseline and background becomes an indispensable step before reliable risk evaluation. This paper provides a full bibliometric analysis of the Web of Science-indexed literature (2000–2025) to analyzed global and Armenia-contextualized research on heavy metals. Global findings show that the publication landscape is very concentrated, with a few countries contributing most of the output and focussing on baseline environmental characterisation. While, Armenia stands out as a topically specialized and highly productive contributor with most of its publication trends leaning toward contamination sources, ecological risk, and human health impacts.

DOI: 10.4018/979-8-3693-7006-3.ch009

1. INTRODUCTION

Soil heavy metals have been recognized as integral elements of the environment and food chain, and have significant effects on human health (Mielke et al., 2004; Santos-Francés et al., 2011). The original physical-chemical properties of the elements in nature have been changed by humans. It is very important to study the content differences of soil elements, ascertain the extent of this change and seek a scientific and reasonable reference standard of soil element content in order to investigate the degree of soil pollution and evolution trends (Guo et al., 2025). At present, the guidelines of soil geochemical background value not only include *"geochemical background value", "reference value",* and *"baseline value"* (Xue et al., 2022; Mai et al., 2025). They might look similar but they are quite different in the real world. It is easy to cause "excessive" or "inadequate" evaluation when evaluating soil pollution according to inappropriate assessment criteria (Valdés et al., 2021). The geochemical baseline—an official definition, first published in 1993 within the framework of the International Geological Correlation Program as Global Geochemical Baselines—is defined as the natural variation within concentration of an element in a surface environment at a specified location and time. Natural geographic concentrations (background level) and diffuse anthropogenic load in soils are included in this concept.

Background level is a value to distinguish between the concentration of the natural compound and concentrations influenced by anthropogenic in an environmental sample (García-Sánchez et al., 2010). The natural soil background levels of heavy metals are determined by the geological substrates and the pedogenetic processes responsible for soil formation (Kabata-Pendias and Pendias, 1992). Heavy metals content of soil is strongly affected by rocks and the critical value of its concentration could be exceeded (Albanese et al., 2007). However, it is extremely difficult to define the natural background levels which are due to a geological concentration of one system and therefore soil geochemical composition in the virgin soils could not be determined because aerial pollution inherently contains trace elements that can pollute deposit soil (Wei and Wen, 2012a). The determination of the geochemical baseline is thus more appropriate, because it reflects a situation where a certain human pressure to environment is already present (Salonen and Korkka-Niemi, 2007). Contaminants are being deposited in large quantities into the soil compartment from various sources on an annual basis. Accordingly, this compartment serves as a sink to innumerable emissions including many heavy metals that are toxic. The environment becomes polluted with heavy metals through aerial deposition of particles derived from a variety of human activities such as mining.

The determination of environmental geochemical baselines is fundamental for evaluating environmental status and legislation, especially in the case of contaminated soil, as well as for risk calculation (Alonso Rojo et al., 2004). There are two principal methods for the assessment of local background levels of heavy metals: one using direct (geochemical), and the other indirect (statistical) method (Baize and Sterckeman, 2001). It is therefore important to understand geochemical baselines for each heavy metal in a variety of geological areas to avoid the misclassification of contaminated soils (Guo et al., 2025).

Bibliometrics is an approach to surveying the past publication history and identifying current trends among research papers and has been applied toward finding trends in scientific researches of many fields, including engineering. Nevertheless, relatively little research has addressed risks of health associated with heavy metals by means of bibliometric analysis especially for heavy metals of food products (Han et al., 2020). This includes combining different framework, tools and techniques for analyzing and studying citations in publications to create different types of metrics. These measures are utilized to gain an insight into the intellectual landscape within a variety of scholarly disciplines, and as metrics for assessing the impact of scientific journals, research articles and scientists (Akhavan et al., 2016).

In this research, our objective is to examine the current vision for research trends and identify gaps in the context of background and baseline, with a primary focus on Armenia and the international arena (Figure 1). Our bibliometric analysis, utilizing the Web of Science (WoS) database, highlights a research gap in Armenia, indicating the necessity to establish standard reference values for geochemical background and baseline at both local and national levels, in alignment with international standards These analyses further indicate that although numerous studies have focused on identifying and evaluating risks related to heavy metals, there is a significant lack of research on geochemical background and baseline values.

Figure 1. The current illustration presents an analysis of the background and baseline levels of soil heavy metals, along with a bibliometric study

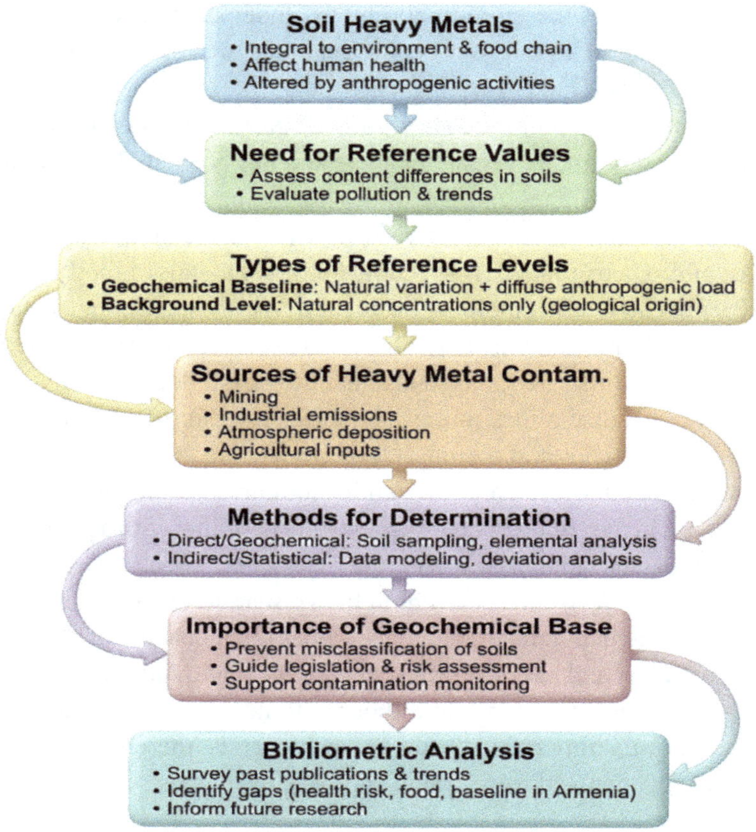

2. DATA COLLECTION AND SEARCH STRATEGY

In the current study, bibliometric analysis was used for data extraction and bibliometrics is a mathematical and statistical approach to analysing documents that can assist in revealing interaction among them (Gao et al., 2022). Information for articles (i.e., affiliations, countries, authors, journals and research field, keywords) was acquired from WoS. Search was performed on 28 January 2026 using the search terms: "Geochemical Background" (all fields), "Geochemical Baseline" (all fields), "Trace Elements" (all fields), "Heavy Metal" (all fields) and content filter" Heavy

Metal Armenia" (all field) and publication year range "2000–2025". Such words were acted as search terms to retrieve related papers with the scope including title of papers, abstract and keywords, affiliations, authors, sources and cited references. The analysis included 72 articles (Figure 2) (Background and Baseline) and 68 articles (Figure 3) (Main focus on Heavy Metal Armenia) selected according to the inclusion criteria. We employed the MetaInfoSci bibliometric platform (https://metainfosci.bmu.edu.in) as our primary resource for conducting bibliometric analysis (Figure 2). Our research methodology involves channeling all collected data through MetaInfoSci, a versatile platform that delivers the bibliographic information crucial for further analyses (Sharmaa et al., 2025).

Figure 2. Analysis and extraction using the MetaInfoSci bibliometric platform based on data collected for background and baseline

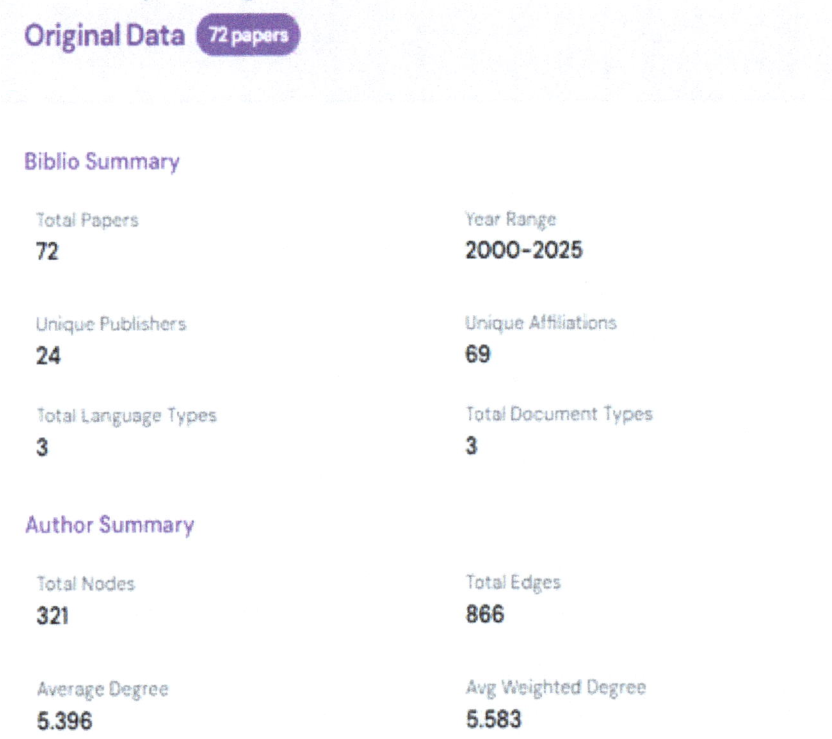

Figure 3. Analysis and extraction using the MetalnfoSci bibliometric platform based on data collected heavy metal Armenia from WoS

3. RESULTS AND DISCUSSION

3.1 The General Trends of Publications and Citations

The study related to background and baseline for heavy metal publication activity and impact for a 25-year time span (2000–2025) where in citation was taken as the primary measure of influence (Fig 3 A). Total productivity were 72 publications in this period, revealing a more or less constant output that averaged ~2.7 articles a year. Cumulative impact related background and baseline for heavy metal works have together been cited 2414 times indicating the considerably wide spread acceptance and acknowledgment of this work in the academic world. Outstanding citation strength is a one of the most impressive statistics is the mean number (33.53) of citations per manuscript related background and baseline for heavy metal. This high average means that the research that is output is highly regarded and becomes cited often and widely (Figure 4). Research profile would be marked by high quality, influential publications rather than a large body of background and baseline for heavy metal work. The high and stable average citation rate indicates the continued importance of the background and baseline for heavy metal research work after more than 20 years.

68 original articles were published in academic journals play a stable work output over the 25-year interval (2000-2025) (Figure 5). These publications as a whole had high academic impact, with total citations reaching an impressive 1912. One noteworthy result is the high mean of 28.12 citations per paper. This number indicates that the average work is influential and highly-cited within heavy metal research area in Armenia. This high average number of citations indicates that research output was consistently impactful in terms of both reach and acknowledgement throughout the complete timespan. Individual year-to-year productivity peaks (or dips) and increases or declines in citation impact are not reported in the aggregate scores shown, though overall they represent a solid, high-impact publication profile from 2000 to 2025. Baseline and background studies of heavy metals are developed to define the natural or current concentration of pollutants, while risk assessment studies concern about how those concentrations could affect human health or ecosystem (Santos-Francés et al., 2017).

Figure 4. Annual number of publications and citations

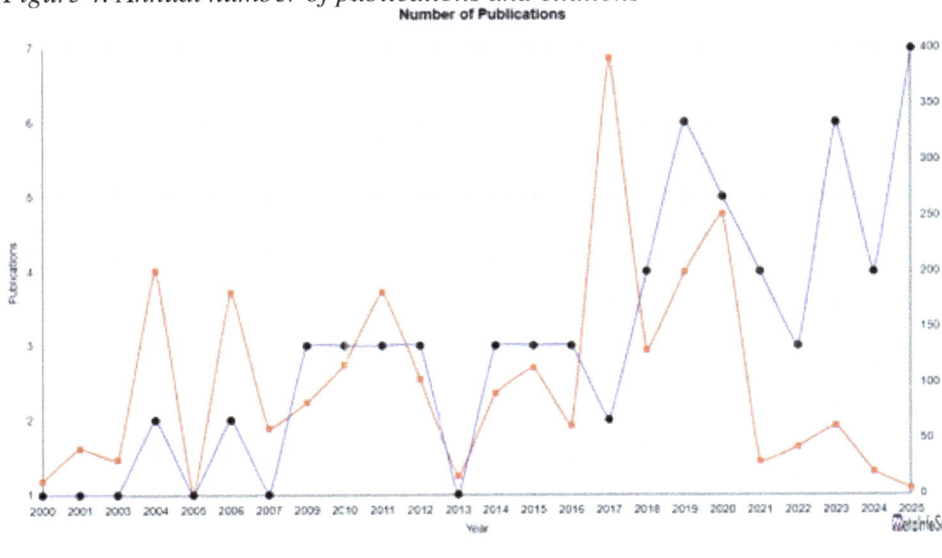

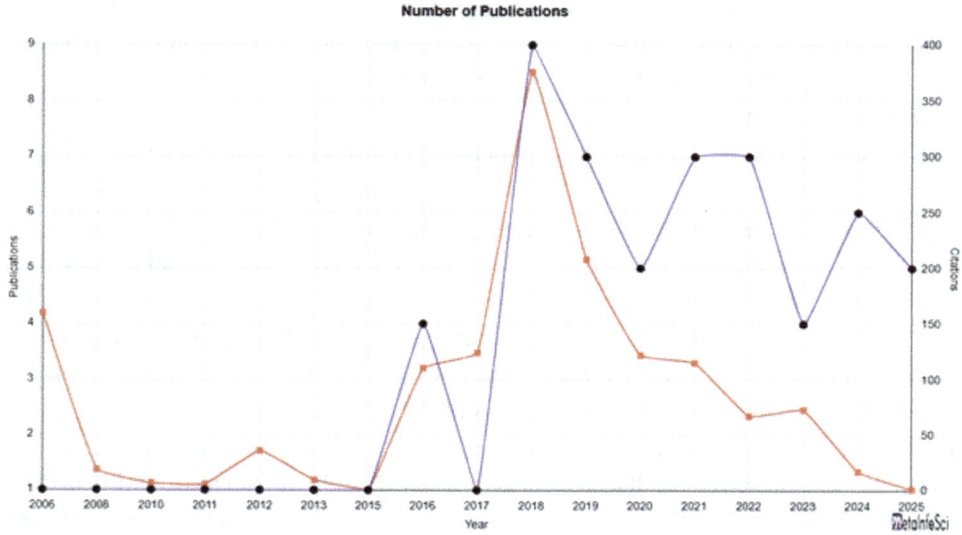

Geochemical baseline concentration is hence more practical, since it represents 95% of the expected range of background concentrations (Gough et al., 1994; Facchinelli et al., 2001). Several research works were performed worldwide to determine geochemical background and the baseline concentration of heavy metals (Wei and Wen, 2012b; Blonda and Valenzano, 2014). Despite reviewing 68 articles in the WoS database, none addressed the geochemical baseline and background values of metals in Armenia key mining areas and nearby agricultural soils; they mainly focused on ecological risk assessments. Therefore, this study's findings will offer crucial insights for authorities monitoring the impact of human activities on heavy metal contamination in these areas. Furthermore, the results will support the development of effective management plans to reduce environmental pollution and clean up heavy metal-contaminated soils in Armenia mining regions.

3.2 Influential and Productive Countries

This data, covering 40 scientific publications from 10 countries between 2000 and 2025, reveals a highly concentrated distribution of output (Figure 6). Spain is the clear leader, contributing 9 publications (22.5% of the total), significantly above the 4.00 average. The distribution shows a distinct tiered pattern. Following Spain, France (6 publications) forms a second tier, while Italy and India (5 each) constitute a third. These top four countries alone account for 62.5% of all publications, indicating a substantial concentration of research output within a select few nations. A "long tail" of countries contributes fewer publications, with Brazil matching the average

(4) and the remaining five countries (Poland, Turkey, Chile, Cameroon, Australia) each contributing only 2 or 3 publications. This bottom half of the listed countries collectively accounts for just 27.5% of the total. While geographic diversity is present, spanning Europe, Asia, South America, Africa, and Oceania, the publication output is heavily skewed towards a few dominant nations, particularly Spain, showcasing significant disparity in contribution (Figure 6).

This WoS database spanning 93 papers across 10 countries from 2000 to 2025, surfaces a highly skewed distribution. Armenia is by far the largest producer, with 62 papers or about two-thirds (67%) of this output (Figure 6). The strong leadership of Armenia in this list makes it approach six times the world average, which is 9.3 publications per country. It is followed by Russia with 11 and Italy with 7. This is followed by a middle tier consisting of Norway, Germany and Azerbaijan with 3 publications each. The other four countries, India, Iran, Egypt and Slovakia are negligible and contribute at most one publication irrespective of the threshold. The distribution itself has a clear "long tail" in which one country is extremely prolific, and all others very unproductive, suggesting a highly centric or specialized disposition of research that is dominantly influenced by Armenia.

Figure 6. Top most nations

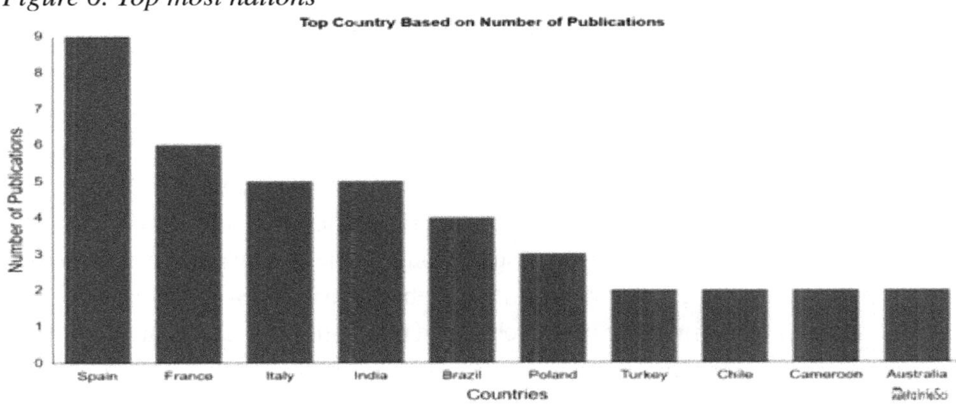

Figure 7.

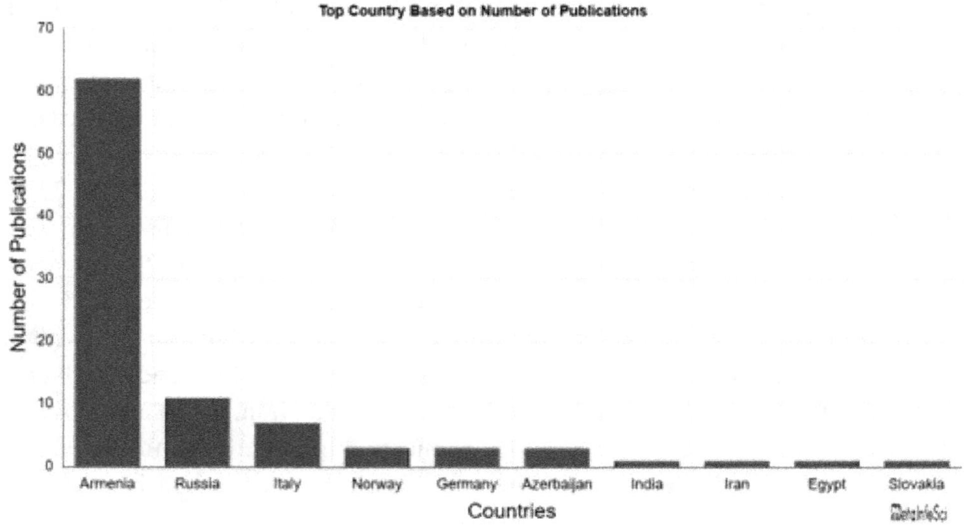

A global bibliometric analysis (Figure 6) includes 40 articles in all from 10 countries (2000–2025) and shows a highly focused research panorama. The volume of output is dominated by Spain (9 papers = 22.5%), France (6), Italy (5) and India (5) providing together for >62.5% of the total output. Despite the geographical diversity, exploring countries mainly focus on basic environmental characterization and background levels of heavy metals indicative for reliable risk assessments. In comparison, a thematic-limited WoS search with "Heavy metal AND Armenia" yielded 68 articles, and 62 of those are related to the contamination and risk assessments by heavy metal pollutants suggested strong theme dependency. Nevertheless, there is a lack of baseline and background reference data in WoS-indexed Armenian studies, indicating an important research gap. This bibliometric data indicates that, and although it seems that Armenia is leading the impact-induced heavy metals research, a systematic baseline measurement would be an important chance for subsequent national and comparative investigations (Wen et al., 2024).

3.3 What are the Current Hot Topics and How are Trends Developing According to Keyword Profiling?

Analysis of keywords can provide information about the main topics and research trends in the domain of background and baseline heavy metals (Chen et al., 2016; van Nunen et al., 2018). This wordcloud is mainly stressed on environmental geochemistry and pollution assessment, and risk in evaluation especially as regards metals in geological materials (Fig.5 A) The terms "geochemical baseline", heavy metals" and "trace elements" were the most dominant themes, while the use of "sediments" and "geochemical background," without any modifier adds a layer of certainty to what is at its roots an uncontrollable variable. Their elevated TF-IDF scores suggest that they are both frequent and unique to the analyzed text. We observe here a clear trend regarding the measurement, evaluation, and comparison of elemental concentrations. "Geochemical baseline," "geochemical background," "baseline," and other equivalent names are essential, emphasizing the necessity to characterize natural or less-impacted concentrations. The primary data can then be used to identify and quantify "heavy metals" and the contamination of certain materials, such as "sediments" and "soil," by the so-called "trace elements." The bottom-line goal seems to be environmental damage prevention and remediation indicated by "ecological risk" and "soil pollution." The information illustrates what would appear to be a sound assessment of contamination against natural reference conditions.

Figure 8. Word cloud maps of most-frequently appeared keywords (A) geochemical background-baseline (B) heavy metal

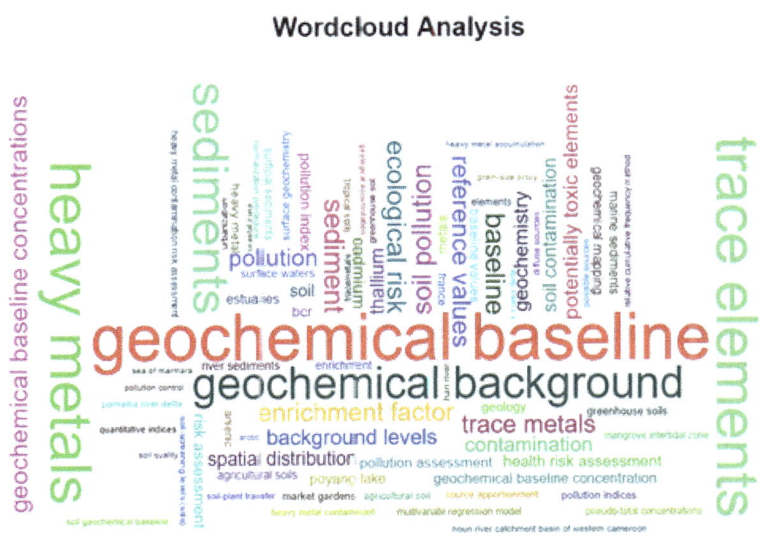

Figure 9.

Wordcloud Analysis

This informetric study through word clouds, shows a high level of specialization on environmental pollution and more specifically soil contamination and cause (Fig 5 B). The heavy metals appear as the undisputable main topic, as indicated by its considerably higher TF-IDF score (4.028) and frequency (16), when compared to any other term, meaning that it is the essential theme of the content analysed. This central tenet was repeatedly resounded by semantic co-occurrences such as “heavy metal,” “potentially toxic elements,” and “trace elements.” The background of this broader context is obviously soil pollution, which is shown by the high indices of soil-today or soil-pollution as well as 'pollution' and 'soil' (and rank 1 at the seed). “Risk assessment” and “dietary exposure” are more than facilitators of identifying the contaminants; they signal a second and pivotal theme: concern for human health implications, as well as methodological approaches concerning those risks. TF-IDF skeps these terms being frequent buts especially valuable to the original data.

The Armenia-focused word-cloud analysis (Fig. 5 B) shows clear thematic specialization in studies dealing with environmental geochemistry, heavy metal contamination and risk assessment of which “heavy metals”, “trace elements” and “potentially toxic elements” are dominated by a high TF-IDF score. Commonly known terminologies such as the “geochemical baseline,” “geochemical background,” or simply, “baseline” emphasize their accepted role in providing reference conditions to assess soil and sediment contamination. All these are showing a lack of empirical baseline-background datasets for Armenia. This is also evidenced in the prominence of "soil pollution," "ecological risk" and "dietary exposure", which indicate a focus on impact and human health effects over fundamental conduct. Overall, the word-cloud analysis demonstrates that Armenian research is heavy metal focused and risk orientated; and systematic baseline with background geochemical frameworks stands out as a major research gap required for contamination assessment studies of robustness with respect to long-term environment management (Han et al., 2020).

4. CONCLUSION AND FUTURE PERSPECTIVES

It is found from this bibliometric analysis that there are few countries which have the largest share in world's scientific research of heavy metals from 2000 to 2025. Baselines characterization and background reference conditions are prioritized by leading countries (Spain, France, Italy, India), while Armenia stands as a quantitatively major content contributor specialized in theme issues (most of WoS-indexed papers are on heavy metal contamination and associated ecological risk and human health counterpoint). However, despite its significant output, systematic quantitative baseline and background geochemical data sets for Armenia published in WoS literature are still relatively poor which hampers long-term comparability and risk interpretation.

Prospective research is needed to set up nationwide representative baseline and background datasets, employ multi-compartment (soil, sediment, water, plants) assessments and develop long-term monitoring programs. The project will fill in these gaps, increasing the scientific robustness of contamination assessments and providing evidence-based environmental policy formulation.

The combination of biochar, nano-biochar, and biogenic nanoparticle–based technologies are expected to play a pivotal role in the development of future research on heavy metals in Armenia when coupled with baseline and risk assessment approaches. Biochar and nano-biochar are the promising options in immobilizing heavy metals, enhancing soil physicochemical properties as well as decreasing metal bioavailability, which can also be used as a long-term monitoring matrix to assess efficiency of remediation. Likewise, biogenic and metallic nanoparticles could improve the efficiency of phytoremediation by modulating metal uptake and translocation and increasing plant tolerance against pollutants and stress conditions. Such advanced materials need to be tested in baseline establishment, field-scale trials and long-term monitoring programs as part of the overall waste-contaminated site scenario in Armenia so that contamination assessment can lead to solution-based sustainable remediation research. This will help ensure climate-resilient agriculture, food safety and evidence-based options for environmental management that establish

Armenia as the regional leader in green remediation and nanotechnology-induced heavy metal mitigation.

ACKNOWLEDGEMENT

AS is supported by the internal grant project of Yerevan State University.

REFERENCES

Akhavan, P., Ebrahim, N. A., Fetrati, M. A., and Pezeshkan, A. (2016). Major trends in knowledge management research: a bibliometric study. *Scientometrics 2016 107:3* 107, 1249–1264. DOI: 10.1007/s11192-016-1938-x

Albanese, S., De Vivo, B., Lima, A., & Cicchella, D. (2007). Geochemical background and baseline values of toxic elements in stream sediments of Campania region (Italy). *Journal of Geochemical Exploration*, *93*(1), 21–34. DOI: 10.1016/j.gexplo.2006.07.006

Alonso Rojo, P., Santos Frances, F., Garcia-Sanchez, A., & Alvarez-Ayuso, E. (2004). Baseline concentrations of heavy metals in native soils of the Salamanca and Valladolid provinces, Spain. *Arid Land Research and Management*, *18*(3), 241–250. DOI: 10.1080/15324980490451320

Baize, D., & Sterckeman, T. (2001). Of the necessity of knowledge of the natural pedo-geochemical background content in the evaluation of the contamination of soils by trace elements. *The Science of the Total Environment*, *264*(1-2), 127–139. DOI: 10.1016/S0048-9697(00)00615-X PMID: 11213175

Blonda, M., & Valenzano, B. (2014). Proposal of procedure to determine metals and metalloids background values in contaminated soils. Case study of a national interest site in South Italy. *Desalination and Water Treatment*, *52*(4-6), 1171–1176. DOI: 10.1080/19443994.2013.848649

Chen, C. F., Ju, Y. R., Chen, C. W., & Di Dong, C. (2016). Vertical profile, contamination assessment, and source apportionment of heavy metals in sediment cores of Kaohsiung Harbor, Taiwan. *Chemosphere*, *165*, 67–79. DOI: 10.1016/j.chemosphere.2016.09.019 PMID: 27639462

Facchinelli, A., Sacchi, E., & Mallen, L. (2001). Multivariate statistical and GIS-based approach to identify heavy metal sources in soils. *Environmental Pollution*, *114*(3), 313–324. DOI: 10.1016/S0269-7491(00)00243-8 PMID: 11584630

Gao, J. ;, Faheem, M. ;, Yu, X., Gao, J., Faheem, M., and Yu, X. (2022). Global Research on Contaminated Soil Remediation: A Bibliometric Network Analysis. *Land 2022, Vol. 11, Page 1581* 11, 1581. DOI: 10.3390/land11091581

García-Sánchez, A., Alonso-Rojo, P., & Santos-Francés, F. (2010). Distribution and mobility of arsenic in soils of a mining area (Western Spain). *The Science of the Total Environment*, *408*(19), 4194–4201. DOI: 10.1016/j.scitotenv.2010.05.032 PMID: 20538319

Gough, L. P., Severson, R. C., & Jackson, L. L. (1994). Baseline element concentrations in soils and plants, Bull Island, Cape Romain National Wildlife Refuge, South Carolina, U.S.A. *Water, Air, and Soil Pollution*, *74*(1-2), 1–17. DOI: 10.1007/BF01257145

Guo, Y., Zhang, J., Li, H., Liu, Y., Gui, H., Chen, S., et al. (2025). Determination and application of soil heavy metal geochemical baseline in the southern region of Wushan County in the Yangtze River Basin, China. *Scientific Reports 2025 15:1* 15, 10889-. DOI: 10.1038/s41598-025-86404-1

Han, R., Zhou, B., Huang, Y., Lu, X., Li, S., & Li, N. (2020). Bibliometric overview of research trends on heavy metal health risks and impacts in 1989–2018. *Journal of Cleaner Production*, *276*, 123249. DOI: 10.1016/j.jclepro.2020.123249

Kabata-Pendias. A.., & Pendias, H. (1992). *Trace elements in soils and plants. 2nd ed.* CRC Press.

Mai, X., Tang, J., Tang, J., Zhu, X., Yang, Z., Liu, X., Zhuang, X., Feng, G., & Tang, L. (2025). Research progress on the environmental risk assessment and remediation technologies of heavy metal pollution in agricultural soil. *Journal of Environmental Sciences (China)*, *149*, 1–20. DOI: 10.1016/j.jes.2024.01.045 PMID: 39181626

Mielke, H. W., Wang, G., Gonzales, C. R., Powell, E. T., Le, B., & Quach, V. N. (2004). PAHs and metals in the soils of inner-city and suburban New Orleans, Louisiana, USA. *Environmental Toxicology and Pharmacology*, *18*(3), 243–247. DOI: 10.1016/j.etap.2003.11.011 PMID: 21782755

Salonen, V. P., & Korkka-Niemi, K. (2007). Influence of parent sediments on the concentration of heavy metals in urban and suburban soils in Turku, Finland. *Applied Geochemistry*, *22*(5), 906–918. DOI: 10.1016/j.apgeochem.2007.02.003

Santos-Francés, F., García-Sánchez, A., Alonso-Rojo, P., Contreras, F., & Adams, M. (2011). Distribution and mobility of mercury in soils of a gold mining region, Cuyuni river basin, Venezuela. *Journal of Environmental Management*, *92*(4), 1268–1276. DOI: 10.1016/j.jenvman.2010.12.003 PMID: 21215510

Santos-Francés, F., Martínez-Graña, A., Rojo, P. A., and Sánchez, A. G. (2017). Geochemical Background and Baseline Values Determination and Spatial Distribution of Heavy Metal Pollution in Soils of the Andes Mountain Range (Cajamarca-Huancavelica, Peru). *International Journal of Environmental Research and Public Health 2017, Vol. 14, Page 859* 14, 859. DOI: 10.3390/ijerph14080859

Sharmaa, K., Khurana, P., & Uddina, Z. (2025). MetaInfoSci: An Integrated Web Tool for Scholarly Data Analysis. Available at: https://arxiv.org/pdf/2506.09056 (Accessed January 18, 2026).

Valdés, A., Guillermo, D., Katja, A., Cyrus, D., Dante, K., Nario, C. A., et al. (2021). The environmental geochemical baseline, background and sources of metal and metalloids present in urban, peri-urban and rural soils in the O´Higgins region, Chile. *Environmental Geochemistry and Health 2021 44:10* 44, 3173–3189. DOI: 10.1007/s10653-021-01098-4

van Nunen, K., Li, J., Reniers, G., & Ponnet, K. (2018). Bibliometric analysis of safety culture research. *Safety Science*, *108*, 248–258. DOI: 10.1016/j.ssci.2017.08.011

Wei, C., & Wen, H. (2012a). Geochemical baselines of heavy metals in the sediments of two large freshwater lakes in China: Implications for contamination character and history. *Environmental Geochemistry and Health*, *34*(6), 737–748. DOI: 10.1007/s10653-012-9492-9 PMID: 23001472

Wei, C., and Wen, H. (2012b). Geochemical baselines of heavy metals in the sediments of two large freshwater lakes in China: implications for contamination character and history. *Environmental Geochemistry and Health 2012 34:6* 34, 737–748. DOI: 10.1007/s10653-012-9492-9

Wen, W., Yang, F., Xie, S., Wang, C., Song, Y., Zhang, Y., et al. (2024). Determination of Geochemical Background and Baseline and Research on Geochemical Zoning in the Desert and Sandy Areas of China. *Applied Sciences 2024, Vol. 14, Page 10612* 14, 10612. DOI: 10.3390/app142210612

Xue, Z., Wang, Z., Tan, L., and Fan, M. (2022). Geochemical baseline assessment and ecological risk evaluation of potentially toxic elements in soils: a case in Han River, China. *Journal of Soils and Sediments 2022 23:2* 23, 745–764. DOI: 10.1007/s11368-022-03339-0

Compilation of References

Aazami, M. A., Mehrabani, L. V., Hashemi, T., Hassanpouraghdam, M. B., & Rasouli, F. (2022). Soil-based nano-graphene oxide and foliar selenium and nano-Fe influence physiological responses of' Sultana' grape under salinity. *Scientific Reports*, *12*(1), 4234. DOI: 10.1038/s41598-022-08251-8 PMID: 35273327

Abasi, F., Raja, N. I., Mashwani, Z. U. R., Amjad, M. S., Ehsan, M., Mustafa, N., Haroon, M., & Proćków, J. (2022). Biogenic silver nanoparticles as a stress alleviator in plants: A mechanistic overview. *Molecules (Basel, Switzerland)*, *27*(11), 3378. DOI: 10.3390/molecules27113378 PMID: 35684312

Abbas, Q., Liu, G., Ali, M. U., Han, J., El-Naggar, A., & Yousaf, B. (2024). Nanomaterials in environment: release, compartmental transformation, and nanotoxicity. In *Handbook of Nanomaterials* (Vol. 2, pp. 745–789). Elsevier. DOI: 10.1016/B978-0-323-95513-3.00010-1

Abd-Alla, M. H., Nafady, N. A., & Khalaf, D. M. (2016). Assessment of silver nanoparticles contamination on faba bean-Rhizobium leguminosarum bv. viciae-Glomus aggregatum symbiosis: Implications for induction of autophagy process in root nodule. *Agriculture, Ecosystems & Environment*, *218*, 163–177. DOI: 10.1016/j.agee.2015.11.022

Abdel-Aziz, H. M. M., Hasaneen, M. N. A., & Omer, A. M. (2016). Nano chitosan-NPK fertilizer enhances the growth and productivity of wheat plants grown in sandy soil. *Spanish Journal of Agricultural Research*, *14*(1), e0902. DOI: 10.5424/sjar/2016141-8205

Abdul Aziz, M., Brini, F., Rouached, H., & Masmoudi, K. (2022). Genetically engineered crops for sustainably enhanced food production systems. *Frontiers in Plant Science*, *13*, 1027828. DOI: 10.3389/fpls.2022.1027828 PMID: 36426158

Aboagye, P. D., & Sharifi, A. (2024). Urban climate adaptation and mitigation action plans: A critical review. *Renewable & Sustainable Energy Reviews*, *189*, 113886. DOI: 10.1016/j.rser.2023.113886

Acharya, A., & Pal, P. K. (2020). Agriculture nanotechnology: Translating research outcome to field applications by influencing environmental sustainability. *NanoImpact*, *19*, 100232. DOI: 10.1016/j.impact.2020.100232

Acharya, D., Singha, K. M., Pandey, P., Mohanta, B., Rajkumari, J., & Singha, L. P. (2018). Shape dependent physical mutilation and lethal effects of silver nanoparticles on bacteria. *Scientific Reports*, *8*(1), 201. DOI: 10.1038/s41598-017-18590-6 PMID: 29317760

Ahari, H., & Naeimabadi, M. (2021). Employing Nanoemulsions in Food Packaging: Shelf Life Enhancement. *Food Engineering Reviews*, *13*(4), 858–883. DOI: 10.1007/s12393-021-09282-z

Ahmad, A., Javad, S., Iqbal, S., Shahid, T., Naz, S., Shah, A. A., Shaffique, S., & Gatasheh, M. K. (2024). Efficacy of soil drench and foliar application of iron nanoparticles on tomato exposed to cadmium stress. *Scientific Reports*, *14*, 27920. DOI: 10.1038/s41598-024-79270-w PMID: 39538010

Ahmad, M., Rajapaksha, A. U., Lim, J. E., Zhang, M., Bolan, N., Mohan, D., Vithanage, M., Lee, S. S., & Ok, Y. S. (2014). Biochar as a sorbent for contaminant management in soil and water: A review. *Chemosphere*, *99*, 19–33. DOI: 10.1016/j.chemosphere.2013.10.071 PMID: 24289982

Ahmadvand, M., Soltani, J., Hashemi, S. A. H., & Nabipour, M. (2018). The relationship between the characteristics of biochar produced at different temperatures and its impact on the uptake of NO3−. *Environmental Science and Pollution Research International*, *25*(4), 3388–3400.

Ahmar, S., Mahmood, T., Fiaz, S., Mora-Poblete, F., Shafique, M. S., Chattha, M. S., & Jung, K.-H. (2021). Advantage of Nanotechnology-Based Genome Editing System and Its Application in Crop Improvement. *Frontiers in Plant Science*, *12*, 663849. Advance online publication. DOI: 10.3389/fpls.2021.663849 PMID: 34122485

Ahmed, A., He, P., He, P., Wu, Y., He, Y., & Munir, S. (2023). Environmental effect of agriculture-related manufactured nano-objects on soil microbial communities. *Environment International*, *173*, 107819. DOI: 10.1016/j.envint.2023.107819 PMID: 36842382

Ahmed, B., Ameen, F., Rizvi, A., Ali, K., Sonbol, H., Zaidi, A., Khan, M. S., & Musarrat, J. (2020). Destruction of cell topography, morphology, membrane, inhibition of respiration, biofilm formation, and bioactive molecule production by nanoparticles of Ag, ZnO, CuO, TiO2, and Al2O3 toward beneficial soil bacteria. *ACS Omega*, *5*(14), 7861–7876. DOI: 10.1021/acsomega.9b04084 PMID: 32309695

Ahmed, B., Rizvi, A., Ali, K., Lee, J., Zaidi, A., Khan, M. S., & Musarrat, J. (2021). Nanoparticles in the soil–plant system: A review. *Environmental Chemistry Letters*, *19*(2), 1545–1609. DOI: 10.1007/s10311-020-01138-y

Ahmed, T., Noman, M., Gardea-Torresdey, J. L., White, J. C., & Li, B. (2023). Dynamic interplay between nano-enabled agrochemicals and the plant-associated microbiome. *Trends in Plant Science*, *28*(11), 1310–1325. DOI: 10.1016/j.tplants.2023.06.001 PMID: 37453924

Akanmu, A. O., Akol, A. M., Ndolo, D. O., Kutu, F. R., & Babalola, O. O. (2023). Agroecological techniques: Adoption of safe and sustainable agricultural practices among smallholder farmers in Africa. *Frontiers in Sustainable Food Systems*, *7*, 1143061. DOI: 10.3389/fsufs.2023.1143061

Akhavan, P., Ebrahim, N. A., Fetrati, M. A., and Pezeshkan, A. (2016). Major trends in knowledge management research: a bibliometric study. *Scientometrics 2016 107:3* 107, 1249–1264. DOI: 10.1007/s11192-016-1938-x

Akhtar, N., Ilyas, N., Mashwani, Z., Hayat, R., Yasmin, H., Noureldeen, A., & Ahmad, P. (2021). Synergistic effects of plant growth-promoting rhizobacteria and silicon dioxide nanoparticles to ameliorate drought stress in wheat. *Plant Physiology and Biochemistry*, *166*, 160–176. DOI: 10.1016/j.plaphy.2021.05.039 PMID: 34116336

Akhtar, S. S., Andersen, M. N., & Liu, F. (2015). Residual effects of biochar on improving growth, physiology and yield of wheat under salt stress. *Agricultural Water Management*, *158*, 61–68. DOI: 10.1016/j.agwat.2015.04.010

Akhtar, S. S., Li, G., Andersen, M. N., & Liu, F. (2015). Biochar enhances yield and quality of tomato under reduced irrigation. *Agricultural Water Management*, *138*, 37–44. DOI: 10.1016/j.agwat.2014.02.016

Akyildiz, I. F., & Jornet, J. M. (2010). Electromagnetic wireless nanosensor networks. *Nano Communication Networks*, *1*(1), 3–19. DOI: 10.1016/j.nancom.2010.04.001

Albanese, S., De Vivo, B., Lima, A., & Cicchella, D. (2007). Geochemical background and baseline values of toxic elements in stream sediments of Campania region (Italy). *Journal of Geochemical Exploration*, *93*(1), 21–34. DOI: 10.1016/j.gexplo.2006.07.006

Alfei, S., Marengo, B., & Zuccari, G. (2020). Nanotechnology application in food packaging: A plethora of opportunities versus pending risks assessment and public concerns. *Food Research International*, *137*, 109664. DOI: 10.1016/j.foodres.2020.109664 PMID: 33233243

Alharbi, K., Hafez, E. M., Omara, A. E. D., Rashwan, E., & Alshaal, T. (2023). Zinc oxide nanoparticles and PGPR strengthen salinity tolerance and productivity of wheat irrigated with saline water in sodic–saline soil. *Plant and Soil*, *493*(1-2), 475–495. DOI: 10.1007/s11104-023-06245-7

Ali, S., Mehmood, A., & Khan, N. (2021). Uptake, translocation, and consequences of nanomaterials on plant growth and stress adaptation. *Journal of Nanomaterials*, *6677616*, 1–17. Advance online publication. DOI: 10.1155/2021/6677616

Alkilany, A. M., & Murphy, C. J. (2010). Toxicity and cellular uptake of gold nanoparticles: What we have learned so far? *Journal of Nanoparticle Research*, *12*(7), 2313–2333. DOI: 10.1007/s11051-010-9911-8 PMID: 21170131

Alkubaisi, N. A., & Aref, N. M. A. (2017). Dispersed gold nanoparticles potentially ruin gold barley yellow dwarf virus and eliminate virus infectivity hazards. *Applied Nanoscience*, *7*(1-2), 31–40. DOI: 10.1007/s13204-016-0540-0

Alonso Rojo, P., Santos Frances, F., Garcia-Sanchez, A., & Alvarez-Ayuso, E. (2004). Baseline concentrations of heavy metals in native soils of the Salamanca and Valladolid provinces, Spain. *Arid Land Research and Management*, *18*(3), 241–250. DOI: 10.1080/15324980490451320

Al-Wabel, M. I., Al-Omran, A., El-Naggar, A. H., Nadeem, M., & Usman, A. R. (2013). Pyrolysis temperature induced changes in characteristics and chemical composition of biochar produced from conocarpus wastes. *Bioresource Technology*, *131*, 374–379. DOI: 10.1016/j.biortech.2012.12.165 PMID: 23376202

Amani, S., Keshtdar, A., & Ghiassi, R. (2024). Investigation of the effectiveness of electrical resistivity tomography in monitoring and detecting hydrocarbon contamination in sandy soil at various moisture: Pilot scale. *Water, Air, and Soil Pollution*, *235*(3), 1–10. DOI: 10.1007/s11270-024-06984-y

Ameen, F., Alsamhary, K., Alabdullatif, J. A., & ALNadhari, S. (2021). A review on metal-based nanoparticles and their toxicity to beneficial soil bacteria and fungi. *Ecotoxicology and Environmental Safety*, *213*, 112027. DOI: 10.1016/j.ecoenv.2021.112027 PMID: 33578100

Amelung, W., Tang, N., Siebers, N., Aehnelt, M., Eusterhues, K., Felde, V. J. M. N. L., Guggenberger, G., Kaiser, K., Kögel-Knabner, I., Klumpp, E., Knief, C., Kruse, J., Lehndorff, E., Mikutta, R., Peth, S., Ray, N., Prechtel, A., Ritschel, T., Schweizer, S. A., & Totsche, K. U. (2024). Architecture of soil microaggregates: Advanced methodologies to explore properties and functions. *Journal of Plant Nutrition and Soil Science*, *187*(1), 17–50. DOI: 10.1002/jpln.202300149

Anawar, H. M., Akter, F., Solaiman, Z. M., & Strezov, V. (2015). Biochar: An emerging panacea for remediation of soil contaminants from mining, industry and sewage wastes. *Pedosphere*, *25*(5), 654–665. DOI: 10.1016/S1002-0160(15)30046-1

Anburuvel, A. (2024). The engineering behind soil stabilization with additives: A state-of-the-art review. *Geotechnical and Geological Engineering*, *42*(1), 1–42. DOI: 10.1007/s10706-023-02554-x

An, C., Sun, C., Li, N., Huang, B., Jiang, J., Shen, Y., Wang, C., Zhao, X., Cui, B., Wang, C., Li, X., Zhan, S., Gao, F., Zeng, Z., Cui, H., & Wang, Y. (2022). Nanomaterials and nanotechnology for the delivery of agrochemicals: Strategies towards sustainable agriculture. *Journal of Nanobiotechnology*, *20*(1), 11. DOI: 10.1186/s12951-021-01214-7 PMID: 34983545

Andrade Díaz, C., Albers, A., Zamora-Ledezma, E., & Hamelin, L. (2024). The interplay between bioeconomy and the maintenance of long-term soil organic carbon stock in agricultural soils: A systematic review. *Renewable & Sustainable Energy Reviews*, *189*, 113890. DOI: 10.1016/j.rser.2023.113890

Andries, M., Pricop, D., Oprica, L., Creanga, D. E., & Iacomi, F. (2016). The effect of visible light on gold nanoparticles and some bioeffects on environmental fungi. *International Journal of Pharmaceutics*, *505*(1-2), 255–261. DOI: 10.1016/j.ijpharm.2016.04.004 PMID: 27063667

Ang, M. C.-Y., & Lew, T. T. S. (2022). Non-destructive Technologies for Plant Health Diagnosis. *Frontiers in Plant Science*, *13*, 884454. Advance online publication. DOI: 10.3389/fpls.2022.884454 PMID: 35712566

Ang, M. C.-Y., Saju, J. M., Porter, T. K., Mohaideen, S., Sarangapani, S., Khong, D. T., Wang, S., Cui, J., Loh, S. I., Singh, G. P., Chua, N.-H., Strano, M. S., & Sarojam, R. (2024). Decoding early stress signaling waves in living plants using nanosensor multiplexing. *Nature Communications*, *15*(1), 2943. DOI: 10.1038/s41467-024-47082-1 PMID: 38580637

Anikwe, M. A., & Ife, K. (2023). The role of soil ecosystem services in the circular bioeconomy. *Frontiers in Soil Science*, *3*, 1209100. DOI: 10.3389/fsoil.2023.1209100

Ansari, M. A. (2023). Nanotechnology in Food and Plant Science: Challenges and Future Prospects. *Plants*, *12*(13), 2565. DOI: 10.3390/plants12132565 PMID: 37447126

Aria, M., & Cuccurullo, C. (2017). bibliometrix: An R-tool for comprehensive science mapping analysis. *Journal of Informetrics*, *11*(4), 959–975. DOI: 10.1016/j.joi.2017.08.007

Arnold, M. C., Badireddy, A. R., Wiesner, M. R., Di Giulio, R. T., & Meyer, J. N. (2013). Cerium oxide nanoparticles are more toxic than equimo-lar bulk cerium oxide in Caenorhabditis elegans. *Archives of Environmental Contamination and Toxicology*, *65*(2), 224–233. DOI: 10.1007/s00244-013-9905-5 PMID: 23619766

Asadishad, B., Chahal, S., Akbari, A., Cianciarelli, V., Azodi, M., Ghoshal, S., & Tufenkji, N. (2018). Amendment of agricultural soil with metal nanoparticles: Effects on soil enzyme activity and microbial community composition. *Environmental Science & Technology*, *52*(4), 1908–1918. DOI: 10.1021/acs.est.7b05389 PMID: 29356510

Ashraf, S. A., Siddiqui, A. J., Elkhalifa, A. E. O., Khan, M. I., Patel, M., Alreshidi, M., Moin, A., Singh, R., Snoussi, M., & Adnan, M. (2021). Innovations in nanoscience for the sustainable development of food and agriculture with implications on health and environment. *The Science of the Total Environment*, *768*, 144990. DOI: 10.1016/j.scitotenv.2021.144990 PMID: 33736303

Aslani, F., Bagheri, S., Muhd Julkapli, N., Juraimi, A. S., Hashemi, F. S. G., & Baghdadi, A. (2014). Effects of engineered nanomaterials on plant growth: An overview. *TheScientificWorldJournal*, *2014*, 641759. Advance online publication. DOI: 10.1155/2014/641759 PMID: 25202734

Aswathi, V. P., Meera, S., Maria, C. A., & Nidhin, M. (2023). Green synthesis of nanoparticles from biodegradable waste extracts and their applications: A critical review. *Nanotechnology for Environmental Engineering*, *8*(2), 377–397. DOI: 10.1007/s41204-022-00276-8

Atanda, S. A., Shaibu, R. O., & Agunbiade, F. O. (2025). Nanoparticles in agriculture: balancing food security and environmental sustainability. *Discover Agriculture 2025 3:1*, *3*(1), 26-. DOI: 10.1007/s44279-025-00159-x

Atapattu, A. J., Ranasinghe, C. S., Nuwarapaksha, T. D., Udumann, S. S., & Dissanayaka, N. S. (2024a). Sustainable Agriculture and Sustainable Development Goals (SDGs). In *Emerging Technologies and Marketing Strategies for Sustainable Agriculture* (pp. 1-27). IGI Global. https://doi.org/DOI: 10.4018/979-8-3693-4864-2.ch001

Atapattu, A. J., Udumann, S. S., Nuwarapaksha, T. D., & Dissanayaka, N. S. (2024b). Upcycling Coconut Husk By-Products: Transitioning from Traditional Applications to Emerging High-Value Usages. In Neelancherry, R., Gao, B., & Wisniewski, A.Jr., (Eds.), *Agricultural Waste to Value-Added Products*. Springer., DOI: 10.1007/978-981-97-2535-9_12

Atkinson, C. J., Fitzgerald, J. D., & Hipps, N. A. (2010). Potential mechanisms for achieving agricultural benefits from biochar application to temperate soils: A review. *Plant and Soil*, *337*(1), 1–18. DOI: 10.1007/s11104-010-0464-5

Avila-Quezada, G. D., Ingle, A. P., Golińska, P., & Rai, M. (2022). Strategic applications of nano-fertilizers for sustainable agriculture: Benefits and bottlenecks. *Nanotechnology Reviews*, *11*(1), 2123–2140. DOI: 10.1515/ntrev-2022-0126

Babu, S., & Joseph, S. (2016). Effect of nanomaterials on properties of soft soil. *International Journal of Scientific Research*, *5*, 634–637.

Badawy, A. A., Abdelfattah, N. A., Salem, S. S., Awad, M. F., & Fouda, A. (2021). Efficacy assessment of biosynthesized copper oxide nanoparticles on stored grain insects and their impacts on wheat. *Biology (Basel)*, *10*(3), 233. DOI: 10.3390/biology10030233 PMID: 33802973

Baig, N., Kammakakam, I., & Falath, W. (2021). Nanomaterials: A review of synthesis methods, properties, recent progress, and challenges. *Materials Advances*, *2*(6), 1821–1871. DOI: 10.1039/D0MA00807A

Baize, D., & Sterckeman, T. (2001). Of the necessity of knowledge of the natural pedo-geochemical background content in the evaluation of the contamination of soils by trace elements. *The Science of the Total Environment*, *264*(1-2), 127–139. DOI: 10.1016/S0048-9697(00)00615-X PMID: 11213175

Bandala, E. R., & Berli, M. (2018). Nanomaterials: New Agrotechnology Tools to Improve Soil Quality? *Agricultural Nanobiotechnology*, 127–140. DOI: 10.1007/978-3-319-96719-6_7

Bandyopadhyay, S., Plascencia-Villa, G., Mukherjee, A., Rico, C. M., José-Yacamán, M., Peralta-Videa, J. R., & Gardea-Torresdey, J. L. (2015). Comparative phytotoxicity of ZnO NPs, bulk ZnO, and ionic zinc onto the alfalfa plants symbiotically associated with Sinorhizobium meliloti in soil. *The Science of the Total Environment*, *515*, 60–69. DOI: 10.1016/j.scitotenv.2015.02.014 PMID: 25698520

Banerjee, K., Pramanik, P., Maity, A., Joshi, D. C., Wani, S. H., & Krishnan, P. (2019). Methods of Using Nanomaterials to Plant Systems and Their Delivery to Plants (Mode of Entry, Uptake, Translocation, Accumulation, Biotransformation and Barriers). In *Advances in Phytonanotechnology* (pp. 123–152). Elsevier. DOI: 10.1016/B978-0-12-815322-2.00005-5

Baqar, M., Saleem, R., Zhao, M., Zhao, L., Cheng, Z., Chen, H., Yao, Y., & Sun, H. (2024). Combustion of high-calorific industrial waste in conventional brick kilns: An emerging source of PFAS emissions to agricultural soils. *The Science of the Total Environment*, *906*, 167612. DOI: 10.1016/j.scitotenv.2023.167612 PMID: 37804982

Barbu, E., Molnàr, É., Tsibouklis, J., & Górecki, D. C. (2009). The potential for nanoparticle-based drug delivery to the brain: Overcoming the blood–brain barrier. *Expert Opinion on Drug Delivery*, *6*(6), 553–565. DOI: 10.1517/17425240902939143 PMID: 19435406

Bashir, I., Lone, F. A., Bhat, R. A., Mir, S. A., Dar, Z. A., & Dar, S. A. (2020). Concerns and threats of contamination on aquatic ecosystems. In Hakeem, K., Bhat, R., & Qadri, H. (Eds.), *Bioremediation and Biotechnology* (Vol. 27, pp. 1–26). Springer. DOI: 10.1007/978-3-030-35691-0_1

Basso, A. S., Miguez, F. E., Laird, D. A., Horton, R., & Westgate, M. (2013). Assessing potential of biochar for increasing water-holding capacity of sandy soils. *Global Change Biology. Bioenergy*, *5*(2), 132–143. DOI: 10.1111/gcbb.12026

Basu, P. (2010). *Biomass gasification and pyrolysis: practical design and theory*. Academic press.

Basu, P., Sadhukhan, A. K., Gupta, P., Rao, S., Dhungana, A., & Acharya, B. (2013). An experimental and theoretical investigation on torrefaction of a large wet wood particle. *Bioresource Technology*, *127*, 449–455. PMID: 24657751

Beegum, S., & Das, S. (2022). Nanosensors in agriculture. In *Agricultural Nanobiotechnology* (pp. 465–478). Woodhead Publishing. DOI: 10.1016/B978-0-323-91908-1.00012-2

Beesley, L., Moreno-Jiménez, E., Gomez-Eyles, J. L., Harris, E., Robinson, B., & Sizmur, T. (2011). A review of biochars' potential role in the remediation, revegetation and restoration of contaminated soils. *Environmental Pollution*, *159*(12), 3269–3282. DOI: 10.1016/j.envpol.2011.07.023 PMID: 21855187

Bergström, L., Bowman, B. T., & Sims, J. T. (2005). Definition of sustainable and unsustainable issues in nutrient management of modern agriculture. *Soil Use and Management*, *21*(1), 76–81. DOI: 10.1111/j.1475-2743.2005.tb00111.x

Bertola, M., Ferrarini, A., & Visioli, G. (2021). Improvement of Soil Microbial Diversity through Sustainable Agricultural Practices and Its Evaluation by -Omics Approaches: A Perspective for the Environment, Food Quality and Human Safety. *Microorganisms*, *9*(7), 1400. DOI: 10.3390/microorganisms9071400 PMID: 34203506

Bhagat, R., Ingle, A. P., & Chen, H. (2023). Nanosensors and nanobiosensors for sustainable agriculture. In *Nanotechnology in Agriculture and Agroecosystems* (pp. 93–112). Elsevier., DOI: 10.1016/B978-0-323-99446-0.00014-3

Bian, R., Joseph, S., Cui, L., Pan, G., Li, L., Liu, X., Zhang, A., Rutlidge, H., Wong, S., Chia, C., Marjo, C., Gong, B., Munroe, P., & Donne, S. (2014). A three-year experiment confirms continuous immobilization of cadmium and lead in contaminated paddy field with biochar amendment. *Journal of Hazardous Materials*, *272*, 121–128. DOI: 10.1016/j.jhazmat.2014.03.017 PMID: 24685528

Biederman, L. A., & Harpole, W. S. (2013). Biochar and its effects on plant productivity and nutrient cycling: A meta-analysis. *Global Change Biology. Bioenergy*, *5*(2), 202–214. DOI: 10.1111/gcbb.12037

Bindraban, P. S., Dimkpa, C., Nagarajan, L., Roy, A., & Rab-binge, R. (2015). Revisiting fertilisers and fertilisation strategies for improved nutrient uptake byplants. *Biology and Fertility of Soils*, *51*(8), 897–911. DOI: 10.1007/s00374-015-1039-7

Blanco-Canqui, H. (2017). Biochar and soil physical properties. *Soil Science Society of America Journal*, *81*(4), 687–711. DOI: 10.2136/sssaj2017.01.0017

Blonda, M., & Valenzano, B. (2014). Proposal of procedure to determine metals and metalloids background values in contaminated soils. Case study of a national interest site in South Italy. *Desalination and Water Treatment*, *52*(4-6), 1171–1176. DOI: 10.1080/19443994.2013.848649

Bongiovanni, R., & Lowenberg-Deboer, J. (2004). Precision Agriculture and Sustainability. *Precision Agriculture*, *5*(4), 359–387. DOI: 10.1023/B:PRAG.0000040806.39604.aa

Borchard, N., Schirrmann, M., Cayuela, M. L., Kammann, C., Wrage-Mönnig, N., Estavillo, J. M., Fuertes-Mendizábal, T., Sigua, G., Spokas, K., Ippolito, J. A., & Novak, J. (2019). Biochar, soil and land-use interactions that reduce nitrate leaching and N2O emissions: A meta-analysis. *The Science of the Total Environment*, *651*, 2354–2364. DOI: 10.1016/j.scitotenv.2018.10.060 PMID: 30336425

Boretti, A., & Rosa, L. (2019). Reassessing the projections of the World Water Development Report. *npj. NPJ Clean Water*, *2*(1), 1–6. DOI: 10.1038/s41545-019-0039-9

Bosona, T., & Gebresenbet, G. (2023). The Role of Blockchain Technology in Promoting Traceability Systems in Agri-Food Production and Supply Chains. *Sensors (Basel)*, *23*(11), 5342. DOI: 10.3390/s23115342 PMID: 37300069

Brevik, E. C., Pereg, L., Steffan, J. J., & Burgess, L. C. (2018). Soil ecosystem services and human health. *Current Opinion in Environmental Science & Health*, *5*, 87–92. DOI: 10.1016/j.coesh.2018.07.003

Brewer, C. E., Chuang, V. J., Masiello, C. A., Gonnermann, H., Gao, X., Dugan, B., Driver, L. E., Panzacchi, P., Zygourakis, K., & Davies, C. A. (2014). New approaches to measuring biochar density and porosity. *Biomass and Bioenergy*, *66*, 176–185. DOI: 10.1016/j.biombioe.2014.03.059

Brewer, C. E., Schmidt-Rohr, K., Satrio, J. A., & Brown, R. C. (2009). Characterization of biochar from fast pyrolysis and gasification systems. *Environmental Progress & Sustainable Energy*, *28*(3), 386–396. DOI: 10.1002/ep.10378

Bridgwater, A. V. (2012). Review of fast pyrolysis of biomass and product upgrading. *Biomass and Bioenergy*, *38*, 68–94. DOI: 10.1016/j.biombioe.2011.01.048

Brown, T. R., Wright, M. M., & Brown, R. C. (2011). Estimating profitability of two biochar production scenarios: Slow pyrolysis vs fast pyrolysis. *Biofuels, Bioproducts & Biorefining*, *5*(1), 54–68. DOI: 10.1002/bbb.254

Brunetti, G., Donner, E., Laera, G., Sekine, R., Scheckel, K. G., Khaksar, M., Vasilev, K., De Mastro, G., & Lombi, E. (2015). Fate of zinc and silver engineered nanoparticles in sewerage networks. *Water Research*, *77*, 72–84. DOI: 10.1016/j.watres.2015.03.003 PMID: 25841090

Cambra Baseca, C., Sendra, S., Lloret, J., & Tomas, J. (2019). A Smart Decision System for Digital Farming. *Agronomy (Basel)*, *9*(5), 216. DOI: 10.3390/agronomy9050216

Camps-Arbestain, M., Amonette, J. E., Singh, B., Wang, T., & Schmidt, H. P. (2015). A biochar classification system and associated test methods. Biochar for environmental management: science, technology and implementation, 165-193.

Canesi, L., Auguste, M., & Bebianno, M. J. (2019). Sublethal effects of nanoparticles on aquatic invertebrates, from molecular to organism level. In *Ecotoxicology of Nanoparticles in Aquatic Systems* (pp. 38–61). CRC press. DOI: 10.1201/9781315158761-2

Cantrell, K. B., Hunt, P. G., Uchimiya, M., Novak, J. M., & Ro, K. S. (2012). Impact of pyrolysis temperature and manure source on physicochemical characteristics of biochar. *Bioresource Technology*, *107*, 419–428. DOI: 10.1016/j.biortech.2011.11.084 PMID: 22237173

Cao, C., Huang, J., Yan, C., Liu, J., Hu, Q., & Guan, W. (2018). Shifts of system performance and microbial community structure in a constructed wetland after exposing silver nanoparticles. *Chemosphere*, *199*, 661–669. DOI: 10.1016/j.chemosphere.2018.02.031 PMID: 29471236

Cao, M., Lv, W., Wang, F., Ma, S., Geng, H., Li, J., Gao, Z., Xu, Q., Guo, J., Leng, W., Chen, K., Tan, Z., Zhang, P., Sun, K., & Xing, B. (2024). Foliar application of zinc oxide nanoparticles alleviates phenanthrene and cadmium-induced phytotoxicity in lettuce. *Environmental Science & Technology*, *59*(1), 730–743. DOI: 10.1021/acs.est.4c07881 PMID: 39704184

Cao, X., & Harris, W. (2010). Properties of dairy-manure-derived biochar pertinent to its potential use in remediation. *Bioresource Technology*, *101*(14), 5222–5228. DOI: 10.1016/j.biortech.2010.02.052 PMID: 20206509

Cao, X., Ma, L., Gao, B., & Harris, W. (2009). Dairy-manure derived biochar effectively sorbs lead and atrazine. *Environmental Science & Technology*, *43*(9), 3285–3291. DOI: 10.1021/es803092k PMID: 19534148

Case, S. D. C., McNamara, N. P., Reay, D. S., & Whitaker, J. (2015). The effect of biochar on nitrous oxide and carbon dioxide emissions from temperate agricultural soils: A meta-analysis. *Soil Biology & Biochemistry*, *89*, 1–11.

Cayuela, M. L., Van Zwieten, L., Singh, B. P., Jeffery, S., Roig, A., & Sánchez-Monedero, M. A. (2014). Biochar's role in mitigating soil nitrous oxide emissions: A review and meta-analysis. *Agriculture, Ecosystems & Environment*, *191*, 5–16. DOI: 10.1016/j.agee.2013.10.009

Cela-Dablanca, R., Míguez-González, A., Barreiro, A., Rodríguez-López, L., Arias-Estévez, M., Núñez-Delgado, A., Fernández-Sanjurjo, M. J., Castillo-Ramos, V., & Álvarez-Rodríguez, E. (2024). Simultaneous adsorption of amoxicillin and ciprofloxacin on agricultural soils and by-products used as bio-adsorbents: Unraveling the interactions in complex systems. *Environmental Research*, *240*, 117535. DOI: 10.1016/j.envres.2023.117535 PMID: 39492492

Cetin, E., Moghtaderi, B., Gupta, R., & Wall, T. F. (2004). Influence of pyrolysis conditions on the structure and gasification reactivity of biomass chars. *Fuel*, *83*(16), 2139–2150. DOI: 10.1016/j.fuel.2004.05.008

Chaganti, V. N., Crohn, D. M., & Šimůnek, J. (2015). Leaching and reclamation of a biochar and compost amended saline-sodic soil with moderate SAR reclaimed water. *Agricultural Water Management*, *158*, 255–265. DOI: 10.1016/j.agwat.2015.05.016

Chandran, R., & Chandran MC, S. (2024). Green finance and sustainability: mapping research development through bibliometric analysis. *Discover Sustainability 2024 5:1, 5*(1), 339-. DOI: 10.1007/s43621-024-00549-z

Chang, M., Liu, Y., Xu, M., Li, H., & Li, S. W. (2024). Particle morphology and soil properties affect the retention of copper oxide nanoparticles in agricultural soils. *Environmental Geochemistry and Health, 46*(8), 281. DOI: 10.1007/s10653-024-02057-5 PMID: 38963650

Chan, K. Y., Van Zwieten, L., Meszaros, I., Downie, A., & Joseph, S. (2008). Using poultry litter biochars as soil amendments. *Soil Research (Collingwood, Vic.), 46*(5), 437–444. DOI: 10.1071/SR08036

Chanu, N. B., Alice, A. K., & Thokchom, A.. (2021). Engineered nanomaterials and their interactions with the plant–soil system. *Nanotechnology for Environmental Engineering, 6*, 36. Advance online publication. DOI: 10.1007/s41204-021-00130-3

Chao, G. A. O., Sun, B., & Zhang, T. L. (2006). Sustainable Nutrient Management in Chinese Agriculture: Challenges and Perspective. *Pedosphere, 16*(2), 253–263. DOI: 10.1016/S1002-0160(06)60051-9

Chauhan, P., Sharma, N., Tapwal, A., Kumar, A., Verma, G. S., Meena, M., Seth, C. S., & Swapnil, P. (2023). Soil microbiome: Diversity, benefits, and interactions with plants. *Sustainability (Basel), 15*(19), 14643. DOI: 10.3390/su151914643

Chen, C. F., Ju, Y. R., Chen, C. W., & Di Dong, C. (2016). Vertical profile, contamination assessment, and source apportionment of heavy metals in sediment cores of Kaohsiung Harbor, Taiwan. *Chemosphere, 165*, 67–79. DOI: 10.1016/j.chemosphere.2016.09.019 PMID: 27639462

Cheng, C. H., & Lehmann, J. (2009). Ageing of black carbon along a temperature gradient. *Chemosphere, 75*(8), 1021–1027. DOI: 10.1016/j.chemosphere.2009.01.045 PMID: 19223059

Cheng, C. H., Lehmann, J., & Engelhard, M. H. (2008). Natural oxidation of black carbon in soils: Changes in molecular form and surface charge along a climosequence. *Geochimica et Cosmochimica Acta, 72*(6), 1598–1610. DOI: 10.1016/j.gca.2008.01.010

Cheng, C. H., Lehmann, J., Thies, J. E., Burton, S. D., & Engelhard, M. H. (2006). Oxidation of black carbon by biotic and abiotic processes. *Organic Geochemistry, 37*(11), 1477–1488. DOI: 10.1016/j.orggeochem.2006.06.022

Chen, T., Zhang, Y., Wang, H., Lu, W., Zhou, Z., Zhang, Y., & Ren, L. (2014). Influence of pyrolysis temperature on characteristics and heavy metal adsorptive performance of biochar derived from municipal sewage sludge. *Bioresource Technology*, *164*, 47–54. DOI: 10.1016/j.biortech.2014.04.048 PMID: 24835918

Chen, Y. T., Wang, Y., & Yeh, K. C. (2017). Role of root exudates in metal acquisition and tolerance. *Current Opinion in Plant Biology*, *39*, 66–72. DOI: 10.1016/j.pbi.2017.06.004 PMID: 28654805

Chen, Z., Chen, B., Zhou, D., & Chen, W. (2011). Bisolute sorption and thermodynamic behavior of organic pollutants to biomass-derived biochars at two pyrolytic temperatures. *Environmental Science & Technology*, *46*(24), 12476–12483. PMID: 23121559

Chhipa, H. (2017). Nanofertilizers and Nanopesticides for Agriculture. *Environmental Chemistry Letters*, *15*(1), 15–22. DOI: 10.1007/s10311-016-0600-4

Chia, C. H., Gong, B., Joseph, S. D., Marjo, C. E., Munroe, P., & Rich, A. M. (2012). Imaging of mineral-enriched biochar by FTIR, Raman and SEM–EDX. *Vibrational Spectroscopy*, *62*, 248–257. DOI: 10.1016/j.vibspec.2012.06.006

Choi, O., Deng, K. K., Kim, N. J., Ross, L.Jr, Surampalli, R. Y., & Hu, Z. (2008). The inhibitory effects of silver nanoparticles, silver ions, and silver chloride colloids on microbial growth. *Water Research*, *42*(12), 3066–3074. DOI: 10.1016/j.watres.2008.02.021 PMID: 18359055

Chouhan, R. S., Qureshi, A., Yagci, B., Gulgun, M. A., Ozguz, V., & Niazi, J. H. (2016). Biotransformation of multi-walled carbon nanotubes mediated by nanomaterial resistant soil bacteria. *Chemical Engineering Journal*, *298*, 1–9. DOI 10.1016/j.cej.2016.04.019

Christel, W., Bruun, S., Magid, J., Kwapinski, W., & Jensen, L. S. (2016). Pig slurry acidification, separation technology and thermal conversion affect phosphorus availability in soil amended with the derived solid fractions, chars or ashes. *Plant and Soil*, *401*(1), 93–107. DOI: 10.1007/s11104-015-2519-0

Clare, A., Shackley, S., Joseph, S., Hammond, J., Pan, G., & Bloom, A. (2015). Competing uses for China's straw: The economic and carbon abatement potential of biochar. *Global Change Biology. Bioenergy*, *7*(6), 1272–1282. DOI: 10.1111/gcbb.12220

Clough, T. J., & Condron, L. M. (2010). Biochar and the nitrogen cycle: Introduction. *Journal of Environmental Quality*, *39*(4), 1218–1223. DOI: 10.2134/jeq2010.0204 PMID: 20830909

Clough, T. J., Condron, L. M., Kammann, C., & Müller, C. (2013). A review of biochar and soil nitrogen dynamics. *Agronomy (Basel)*, *3*(2), 275–293. DOI: 10.3390/agronomy3020275

Cohen, A. W., Hnasko, R., Schubert, W., & Lisanti, M. P. (2004). Role of Caveolae and Caveolins in Health and Disease. *Physiological Reviews*, *84*(4), 1341–1379. DOI: 10.1152/physrev.00046.2003 PMID: 15383654

Coles, D., & Frewer, L. J. (2013). Nanotechnology applied to European food production – A review of ethical and regulatory issues. *Trends in Food Science & Technology*, *34*(1), 32–43. DOI: 10.1016/j.tifs.2013.08.006

Concha-Guerrero, S. I., Brito, E. M. S., Piñón-Castillo, H. A., Tarango-Rivero, S. H., Caretta, C. A., Luna-Velasco, A., Duran, R., & Orrantia-Borunda, E. (2014). Effect of CuO nanoparticles over isolated bacterial strains from agricultural soil. *Journal of Nanomaterials*, *2014*(1), 148743. DOI: 10.1155/2014/148743

Cornelis, G., Hund-Rinke, K., Kuhlbusch, T., Van den Brink, N., & Nickel, C. (2014). Fate and bioavailability of engineered nanoparticles in soils: A review. *Critical Reviews in Environmental Science and Technology*, *44*(24), 2720–2764. DOI: 10.1080/10643389.2013.829767

Costa, J., Oliveira, R. S., Tiago, I., Ma, Y., Galhano, C., Freitas, H., & Castro, P. (2018). Soil Microorganisms. *Advances in Plant Ecophysiology Techniques*, 457–482. DOI: 10.1007/978-3-319-93233-0_27

Crombie, K., Mašek, O., Sohi, S. P., Brownsort, P., & Cross, A. (2013). The effect of pyrolysis conditions on biochar stability as determined by three methods. *Global Change Biology. Bioenergy*, *5*(2), 122–131. DOI: 10.1111/gcbb.12030

Cullen, L. G., Tilston, E. L., Mitchell, G. R., Collins, C. D., & Shaw, L. J. (2011). Assessing the impact of nano-and micro-scale zerovalent iron particles on soil microbial activities: Particle reactivity interferes with assay conditions and interpretation of genuine microbial effects. *Chemosphere*, *82*(11), 1675–1682. DOI: 10.1016/j.chemosphere.2010.11.009 PMID: 21122887

D'Acunto, L., Iglesias, M. A., Poggio, S. L., & Semmartin, M. (2024). Land cover, plant residue and soil microbes as drivers of soil functioning in temperate agricultural lands. A microcosm study. *Applied Soil Ecology*, *193*, 105133. DOI: 10.1016/j.apsoil.2023.105133

Dai, W., Liu, Y., Yao, D., Wang, N., Shao, J., Ye, X., Cui, Z., Zong, H., Tian, L., Chen, X. Z., & Wang, H. (2024). Biogeographic distribution, assembly processes and potential nutrient cycling functions of myxobacteria communities in typical agricultural soils in China. *The Science of the Total Environment*, *906*, 167255. DOI: 10.1016/j.scitotenv.2023.167255 PMID: 37741390

Dang, K., Wang, Y., Tian, H., Bai, J., Cheng, X., Guo, L., & Shao, X. (2024). Impact of ZnO nanoparticles on rice photosynthesis in saline–sodic soil. *Scientific Reports*, *14*, 16233. DOI: 10.1038/s41598-024-66935-9 PMID: 39004658

Dasauni, K., & Nailwal, T. K. (2022). Nano–delivery system: In the agriculture sector. In *Nano-enabled Agrochemicals in Agriculture* (pp. 467–484). Elsevier., DOI: 10.1016/B978-0-323-91009-5.00014-8

de Oliveira, C., Mulinari, J., Reichert, F., & Júnior, A. (2020, September). Nano-delivery systems of pesticides active agents for agriculture applications-An overview. In Ciência, Tecnologia e Inovação: Do Campo à Mesa, Proceeding of International Agribusiness Congress, Recife, Brazil (pp. 25-27). DOI: 10.31692/ICIAGRO.2020.0051

De Sousa, A., Saleh, A. M., Habeeb, T. H., Hassan, Y. M., Zrieq, R., & Wadaan, M. A. AbdElgawad, H. (. (2019). Silicon dioxide nanoparticles ameliorate aluminum phytotoxicity in maize. *The Science of the Total Environment*, *693*, 133636. DOI: 10.1016/j.scitotenv.2019.133636 PMID: 31377375

Del Buono, D., Luzi, F., & Puglia, D. (2021). Lignin nanoparticles: A promising tool to improve maize physiological, biochemical, and chemical traits. *Nanomaterials (Basel, Switzerland)*, *11*(4), 846. DOI: 10.3390/nano11040846 PMID: 33810279

DeLuca, T. H., MacKenzie, M. D., & Gundale, M. J. (2009). Biochar effects on soil nutrient transformations. In Lehmann, J., & Joseph, S. (Eds.), *Biochar for environmental management: Science and technology* (pp. 137–182). Earthscan.

Deng, H., Dutta, P., & Liu, J. (2018). Stochastic simulations of nanoparticle internalization through transferrin receptor dependent clathrin-mediated endocytosis. *Biochimica et Biophysica Acta (BBA) -. Biochimica et Biophysica Acta. G, General Subjects*, *1862*(9), 2104–2111. DOI: 10.1016/j.bbagen.2018.06.018 PMID: 29959983

Dhanapal, A. R., Thiruvengadam, M., Vairavanathan, J., Venkidasamy, B., Easwaran, M., & Ghorbanpour, M. (2024). Nanotechnology approaches for the remediation of agricultural polluted soils. *ACS Omega*, *9*(12), 13522–13533. DOI: 10.1021/acsomega.3c09776 PMID: 38559935

Dhanaraju, M., Chenniappan, P., Ramalingam, K., Pazhanivelan, S., & Kaliaperumal, R. (2022). Smart Farming: Internet of Things (IoT)-Based Sustainable Agriculture. *Agriculture, 12*(10), 1745. DOI: 10.3390/agriculture12101745

Dickinson, D., Balduccio, L., Buysse, J., Ronsse, F., van Huylenbroeck, G., & Prins, W. (2015). Cost-benefit analysis of using biochar to improve cereals agriculture. *Global Change Biology. Bioenergy*, *7*(4), 850–864. DOI: 10.1111/gcbb.12180

Dinesh, R., Anandaraj, M., Srinivasan, V., & Hamza, S. (2012). Engineered nanoparticles in the soil and their potential implications to microbial activity. *Geoderma, 173–174*, 19–27. DOI: 10.1016/j.geoderma.2011.12.018

Dispenza, V., De Pasquale, C., Fascella, G., Mammano, M. M., & Alonzo, G. (2016). Use of biochar as peat substitute for growing substrates of Euphorbia × lomi potted plants. *Spanish Journal of Agricultural Research, 14*(4), e0908. DOI: 10.5424/sjar/2016144-9082

Dissanayaka, D. M. N. S., Dissanayake, D. K. R. P. L., Udumann, S. S., Nuwarapaksha, T. D., & Atapattu, A. J. (2023a). Agroforestry a key tool in the climate-smart agriculture context: A review on coconut cultivation in Sri Lanka. *Frontiers in Agronomy*, *5*, 1162750. DOI: 10.3389/fagro.2023.1162750

Dissanayaka, N. S., Udumann, S. S., Nuwarapaksha, T. D., & Atapattu, A. J. (2023b). Agroforestry: An Avenue for Resilient and Productive Farming through Integrated Crops and Livestock Production. In Kiba, D. I. (Ed.), *Transitioning to zero hunger* (pp. 115–136). MDPI Books., DOI: 10.3390/books978-3-03897-863-3-5

Dong, X., Ma, L. Q., & Li, Y. (2011). Characteristics and mechanisms of hexavalent chromium removal by biochar from sugar beet tailing. *Journal of Hazardous Materials, 190*(1-3), 909–915. DOI: 10.1016/j.jhazmat.2011.04.008 PMID: 21550718

Drobne, D., Ciornii, D., Hodoroaba, V. D., Bohmer, N., Novak, S., Kranjc, E., Kononenko, V., & Reuther, R. (2023). Knowledge, Information, and Data Readiness Levels (KaRLs) for Risk Assessment, Communication, and Governance of Nano-, New, and Other Advanced Materials. *Global Challenges (Hoboken, NJ), 7*(7), 2200211. DOI: 10.1002/gch2.202200211 PMID: 37483420

Dutta, D., & Singh, O. Shivangi (2023). Carbon Footprint of Different Energy-Intensive Systems. In: Rakshit, A., Biswas, A., Sarkar, D., Meena, V.S., Datta, R. (eds) Handbook of Energy Management in Agriculture. Springer, Singapore. DOI: 10.1007/978-981-19-7736-7_5-1

Dutta, D., Meena, A. L., Bhanu, C., Ghasal, P. C., Choudhary, J., Kumar, S., Mishra, R. P., Ansari, M. A., Raghavendra, K. J., Prusty, A. K., Jat, P. C., Kashyap, P., Punia, P., Dixit, M., Singh, O., Rai, A. K., Meena, A. K., Rathi, S., & Yadav, P. (2024). Sustainable soil management for climate resilience: Long- term management effects on soil carbon sequestration and nitrogen dynamics in a semi-arid tropical Inceptisol of India. *Journal of Soil Science and Plant Nutrition*, *24*(3), 4407–4426. DOI: 10.1007/s42729-024-01844-4

Dutta, H., & Dutta, A. (2016). The microbial aspect of climate change. *Energy, Ecology & Environment*, *1*(4), 209–232. DOI: 10.1007/s40974-016-0034-7

Du, W., Sun, Y., Ji, R., Zhu, J., Wu, J., & Guo, H. (2011). TiO 2 and ZnO nanoparticles negatively affect wheat growth and soil enzyme activities in agricultural soil. *Journal of Environmental Monitoring*, *13*(4), 822–828. DOI: 10.1039/c0em00611d PMID: 21267473

El Akrami, N., Hanine, M., Flores, E. S., Aray, D. G., & Ashraf, I. (2023). Unleashing the Potential of Blockchain and Machine Learning: Insights and Emerging Trends From Bibliometric Analysis. *IEEE Access : Practical Innovations, Open Solutions*, *11*, 78879–78903. DOI: 10.1109/ACCESS.2023.3298371

El Wakeil, N., Alkahtani, S., & Gaafar, N. (2017). Is nanotechnology a promising field for insect pest control in IPM programs? In *New Pesticides and Soil Sensors* (pp. 273–309). Elsevier., DOI: 10.1016/B978-0-12-804299-1.00008-4

Elad, Y., David, D. R., Harel, Y. M., Borenshtein, M., Kalifa, H. B., Silber, A., & Graber, E. R. (2010). Induction of systemic resistance in plants by biochar, a soil-applied carbon sequestering agent. *Phytopathology*, *100*(9), 913–921. DOI: 10.1094/PHYTO-100-9-0913 PMID: 20701489

El-Badri, A. M., Batool, M., Mohamed, I. A., Khatab, A., Sherif, A., Wang, Z., & Zhou, G. (2021). Nano-priming with ZnO mitigates salinity stress in rapeseed. *Plant Physiology and Biochemistry*, *166*, 376–392. DOI: 10.1016/j.plaphy.2021.05.040 PMID: 34153882

Elemike, E., Uzoh, I., Onwudiwe, D., & Babalola, O. (2019). The role of nanotechnology in the fortifica-tion of plant nutrients and improvement of crop production. *Applied Sciences (Basel, Switzerland)*, *9*(3), 499. DOI: 10.3390/app9030499

El-Ramady, H., Prokisch, J., Sári, D., Singh, A., Ghazaryan, K., Rajput, V. D., & Brevik, E. C. (2024). Nanotechnology in the soil system: An ecological approach towards sustainable management. *Applied Soil Ecology*, *204*, 105669. Advance online publication. DOI: 10.1016/j.apsoil.2024.105669

Elsayed, M. E., Ayoub, H. A., Helal, M. I., Sang, W., Shen, Z., & Abdelhafeez, I. A. (2025). Nanotechnology-enabled soil management for sustainable agriculture: Interactions, challenges, and prospects. *Environmental Science. Nano*, *12*(4), 2128–2153. DOI: 10.1039/D4EN00943F

Enders, A., Hanley, K., Whitman, T., Joseph, S., & Lehmann, J. (2012). Characterization of biochars to evaluate recalcitrance and agronomic performance. *Bioresource Technology*, *114*, 644–653. DOI: 10.1016/j.biortech.2012.03.022 PMID: 22483559

Eswaran, U., & Eswaran, V. (2023). Real-time monitoring of environmental parameters in agriculture using WSN. i-manager's Journal on Embedded Systems, 11(2), 13–20.

Eswaran, U., Eswaran, V., Murali, K., & Eswaran, V. (2023). Precision agriculture revolution: A low-cost smart irrigation system empowered by IoT for small-scale farmers. *Journal of Mechatronics and Automation*, *10*(3), 21–29.

Eswaran, U., Eswaran, V., Murali, K., & Eswaran, V. (2024). Application of sensors for smart farming. In Khang, A. (Ed.), *Agriculture and aquaculture applications of biosensors and bioelectronics* (pp. 18–44). IGI Global. DOI: 10.4018/979-8-3693-2069-3.ch002

Eswaran, U., & Khang, A. (2024). Application of internet of things (IoT) and sensors technologies for agriculture. In *Advanced IoT technologies and applications in the Industry 4.0 digital economy* (1st ed.). CRC Press. DOI: 10.1201/9781003434269-8

Etesami, H., Jeong, B. R., & Glick, B. R. (2021). Contribution of Arbuscular Mycorrhizal Fungi, Phosphate–Solubilizing Bacteria, and Silicon to P Uptake by Plant. *Frontiers in Plant Science*, *12*, 699618. Advance online publication. DOI: 10.3389/fpls.2021.699618 PMID: 34276750

European Biochar Certificate (EBC). (2012). *Guidelines for a Sustainable Production of Biochar. European Biochar Foundation*. EBC.

Facchinelli, A., Sacchi, E., & Mallen, L. (2001). Multivariate statistical and GIS-based approach to identify heavy metal sources in soils. *Environmental Pollution*, *114*(3), 313–324. DOI: 10.1016/S0269-7491(00)00243-8 PMID: 11584630

Fajardo, C., Saccà, M. L., Costa, G., Nande, M., & Martin, M. (2014). Impact of Ag and Al2O3 nanoparticles on soil organisms: In vitro and soil experiments. *The Science of the Total Environment*, *473*, 254–261. DOI: 10.1016/j.scitotenv.2013.12.043 PMID: 24374587

Fan, R., Huang, Y. C., Grusak, M. A., Huang, C. P., & Sherrier, D. J. (2014). Effects of nano-TiO2 on the agronomically-relevant Rhizobium–legume symbiosis. *The Science of the Total Environment*, *466*, 503–512. DOI: 10.1016/j.scitotenv.2013.07.032 PMID: 23933452

Fanzo, J., McLaren, R., Davis, C., & Choufani, J. (2017). *Climate change and variability: What are the risks for nutrition, diets, and food systems?* International Food Policy Research Institute.

FAO. (2003). Prospects for aggregate agriculture and major commodity groups. *World Agriculture: Towards 2015/2030 An FAO Perspective*, 57–123.

FAO. W. I. U. W. (2017). The State of Food Security and Nutrition in the World 2017. *United Nations Human Settlements Programme (UN-Habitat): Addis Ababa, Ethiopia. (2021), 1*, 1–132. Retrieved from https://openknowledge.fao.org/handle/20.500.14283/i7695en

Farhangi-Abriz, S., & Ghassemi-Golezani, K. (2023). Nutrient Availability Regulates Root System Behavior. In *Plant Ionomics* (pp. 96–119). Wiley., DOI: 10.1002/9781119803041.ch6

Fatima, A., Singh, S., & Prasad, S. M. (2020). Interaction between copper oxide nanoparticles and plants: Uptake, accumulation and phytotoxicity. In *Sustainable Agriculture Reviews 41: Nanotechnology for Plant Growth and Development* (pp. 143–161). Springer International Publishing. DOI: 10.1007/978-3-030-33996-8_7

Fedorenko, A. G., Minkina, T. M., Chernikova, N. P., Fedorenko, G. M., Mandzhieva, S. S., Rajput, V. D., & Soldatov, A. V. (2021). Toxic effects of CuO nanoparticles on barley cells. *Environmental Geochemistry and Health*, *43*(4), 1673–1687. DOI: 10.1007/s10653-020-00530-5 PMID: 32026274

Feng, X., Bi, S., Li, H., Qi, Y., Chen, S., & Shao, L. (2024). Soil moisture forecasting for precision irrigation management using real-time electricity consumption records. *Agricultural Water Management*, *291*, 108656. DOI: 10.1016/j.agwat.2023.108656

Feng, Y., Cui, X., He, S., Dong, G., Chen, M., Wang, J., & Lin, X. (2013). The role of metal nanoparticles in influencing arbuscular mycorrhizal fungi effects on plant growth. *Environmental Science & Technology*, *47*(16), 9496–9504. DOI: 10.1021/es402109n PMID: 23869579

Ferdous, Z., & Nemmar, A. (2020). Health impact of silver nanoparticles: A review of the biodistribution and toxicity following various routes of exposure. *International Journal of Molecular Sciences*, *21*(7), 2375. DOI: 10.3390/ijms21072375 PMID: 32235542

Ferreira, G. M. D., Ferreira, G. M. D., Franca, J. R., & Soares, J. R. (2022). Carbonaceous Materials for Nanoremediation of Polluted and Nutrient-Depleted Soils. In Nanotechnology for Environmental Pollution Decontamination (pp. 345-408). Apple Academic Press. DOI: 10.1201/9781003279563-15

Francia, V., Montizaan, D., & Salvati, A. (2020). Interactions at the cell membrane and pathways of internalization of nano-sized materials for nanomedicine. *Beilstein Journal of Nanotechnology*, *11*, 338–353. DOI: 10.3762/bjnano.11.25 PMID: 32117671

Francis, D. V., Abdalla, A. K., Mahakham, W., Sarmah, A. K., & Ahmed, Z. F. (2024). Interaction of plants and metal nanoparticles. *Environment International*, *190*, 108859. DOI: 10.1016/j.envint.2024.108859 PMID: 38970982

Frenk, S., Ben-Moshe, T., Dror, I., Berkowitz, B., & Minz, D. (2013). Effect of metal oxide nanoparticles on microbial community structure and function in two different soil types. *PLoS One*, *8*(12), e84441. DOI: 10.1371/journal.pone.0084441 PMID: 24349575

Fu, P. P., Xia, Q., Hwang, H. M., Ray, P. C., & Yu, H. (2014). Mechanisms of Nanotoxicity: Generation of reactive oxygen species. *Yao Wu Shi Pin Fen Xi*, *22*(1), 64–75. DOI: 10.1016/j.jfda.2014.01.005 PMID: 24673904

Futa, B., Gmitrowicz-Iwan, J., Skersienė, A., Šlepetienė, A., & Parašotas, I. (2024). Innovative Soil Management Strategies for Sustainable Agriculture. *Sustainability 2024, Vol. 16, Page 9481, 16*(21), 9481. DOI: 10.3390/su16219481

Galinato, S. P., Yoder, J. K., & Granatstein, D. (2011). The economic value of biochar in crop production and carbon sequestration. *Energy Policy*, *39*(10), 6344–6350. DOI: 10.1016/j.enpol.2011.07.035

Gambino, M., Marzano, V., Villa, F., Vitali, A., Vannini, C., Landini, P., & Cappitelli, F. (2015). Effects of sublethal doses of silver nanoparticles on Bacillus subtilis planktonic and sessile cells. *Journal of Applied Microbiology*, *118*(5), 1103–1115. DOI: 10.1111/jam.12779 PMID: 25702880

Gan, N., Yang, X., Xie, D., Wu, Y., & Wen, W. A. (2010). Disposable organophosphorus pesticides enzyme biosensor based on magnetic composite nanoparticles modified screen printed carbon electrode. *Sensors (Basel)*, *10*(1), 625–638. DOI: 10.3390/s100100625 PMID: 22315558

Gao, J. ;, Faheem, M. ;, Yu, X., Gao, J., Faheem, M., and Yu, X. (2022). Global Research on Contaminated Soil Remediation: A Bibliometric Network Analysis. *Land 2022, Vol. 11, Page 1581* 11, 1581. DOI: 10.3390/land11091581

Gao, Y., Zhou, H., He, C., Huang, L., Gao, H., & Ruan, R. (2019). Biochar-supported nanoscale iron sulfide composite for enhanced Cr (VI) removal from aqueous solution. *Bioresource Technology*, *281*, 417–424.

García-Sánchez, A., Alonso-Rojo, P., & Santos-Francés, F. (2010). Distribution and mobility of arsenic in soils of a mining area (Western Spain). *The Science of the Total Environment*, *408*(19), 4194–4201. DOI: 10.1016/j.scitotenv.2010.05.032 PMID: 20538319

Garvey, T. (2016). *Identifying the Potential environmental impacts of engineered nanomaterials*. Rochester Institute of Technology.

Gaskin, J. W., Steiner, C., Harris, K., Das, K. C., & Bibens, B. (2008). Effect of low-temperature pyrolysis conditions on biochar for agricultural use. *Transactions of the ASABE*, *51*(6), 2061–2069. DOI: 10.13031/2013.25409

Ghafari, N., Narchin, P., & Niazi, A. (2022). Slow-release fertilizer based on nano-biochar: A review on synthesis, characterization, and applications. *Journal of Cleaner Production*, *353*, 131643.

Ghasemian, S., Masoudian, N., Nematpour, F. S., & Afshar, A. S. (2021). Selenium nanoparticles alleviate salt stress in *Melissa officinalis*. *Biologia*, *76*(10), 2879–2888. DOI: 10.1007/s11756-021-00854-2

Ghazaryan, K., Pandey, D., Singh, S., Varagyan, V., Alexiou, A., Petropoulos, D., Kriemadis, A., Rajput, V., Minkina, T., Singh, R., Sousa, J., Kumar, S., El-Ramady, H., Singh, O., & Singh, A. (2024). Enhancing Crop Production: Unveiling the Role of Nanofertilizers in Sustainable Agriculture and Precision Nutrient Management. *Egyptian Journal of Soil Science*, *64*(3), 981–1007. DOI: 10.21608/ejss.2024.282778.1748

Ghouri, M. Z., Khan, Z., Khan, S. H., Ismail, M., Aftab, S. O., Sultan, Q., & Ahmad, A. (2020). Nanotechnology: Transformation of agriculture and food security. *Bioscience*, *3*, 19.

Gladkova, M. M., & Terekhova, V. A. (2013). Engineered nanomaterials in soil: Sources of entry and migration pathways. *Moscow University Soil Science Bulletin*, *68*(3), 129–134. DOI: 10.3103/S0147687413030046

Glaser, B., & Birk, J. J. (2012). State of the scientific knowledge on properties and genesis of Anthropogenic Dark Earths in Central Amazonia (terra preta de Índio). *Geochimica et Cosmochimica Acta*, *82*, 39–51. DOI: 10.1016/j.gca.2010.11.029

Glaser, B., Haumaier, L., Guggenberger, G., & Zech, W. (2001). The 'Terra Preta' phenomenon: A model for sustainable agriculture in the humid tropics. *Naturwissenschaften*, *88*(1), 37–41. DOI: 10.1007/s001140000193 PMID: 11302125

Gogos, A., Moll, J., Klingenfuss, F., van der Heijden, M., Irin, F., Green, M. J., Zenobi, R., & Bucheli, T. D. (2016). Vertical transport and plant uptake of nanoparticles in a soil mesocosm experiment. *Journal of Nanobiotechnology*, *14*(1), 40. DOI: 10.1186/s12951-016-0191-z PMID: 27278090

Gomez-Eyles, J. L., Sizmur, T., Collins, C. D., & Hodson, M. E. (2011). Effects of biochar and the earthworm Eisenia fetida on the bioavailability of polycyclic aromatic hydrocarbons and potentially toxic elements. *Environmental Pollution*, *159*(2), 616–622. DOI: 10.1016/j.envpol.2010.09.037 PMID: 21035930

Gonçalves, S. P. C., Delite, F. D., Côa, F., Neto, L. L. R., da Silva, G. H., Bortolozzo, L. D., Ferreira, A. G., de Medeiros, A. M. Z., Strauss, M., & Martinez, D. S. T. (2019). Chapter 7 - Biotransformation of Nanomaterials in the Soil Environment: Nanoecotoxicology and Nanosafety Implications, Editor(s): do Nascimento RF, Ferreira OP, De Paula AJ, Neto VDOS, In Advanced Nanomaterials, Nanomaterials Applications for Environmental Matrices, Elsevier, pp 265-304, DOI: 10.1016/B978-0-12-814829-7.00007-0

González, J. F., Román, S., Encinar, J. M., & Martínez, G. (2009). Pyrolysis of various biomass residues and char utilization for the production of activated carbons. *Journal of Analytical and Applied Pyrolysis*, *85*(1-2), 134–141. DOI: 10.1016/j.jaap.2008.11.035

Gough, L. P., Severson, R. C., & Jackson, L. L. (1994). Baseline element concentrations in soils and plants, Bull Island, Cape Romain National Wildlife Refuge, South Carolina, U.S.A. *Water, Air, and Soil Pollution*, *74*(1-2), 1–17. DOI: 10.1007/BF01257145

Grieger, K. D., Hansen, S. F., & Baun, A. (2009). The known unknowns of nanomaterials: Describing and characterizing uncertainty within environmental, health and safety risks. *Nanotoxicology*, *3*(3), 222–233. DOI: 10.1080/17435390902944069

Grillo, R., Fraceto, L. F., & de Melo, N. F. S. (2021). Biochar as a carrier for pesticides and herbicides: A sustainable and promising approach for agriculture. *Journal of Agricultural and Food Chemistry*, *69*(26), 7241–7243.

Grün, A. L., Manz, W., Kohl, Y. L., Meier, F., Straskraba, S., Jost, C., Drexel, L., & Emmerling, C. (2019). Impact of silver nanoparticles (AgNP) on soil microbial community depending on functionalization, concentration, exposure time, and soil texture. *Environmental Sciences Europe*, *31*(1), 1–22. DOI: 10.1186/s12302-019-0196-y

Guan, H., Chi, D., Yu, J., & Li, X. (2008). A novel photodegradable insecticide: Preparation, characterization and properties evaluation of nano-Imidacloprid. *Pesticide Biochemistry and Physiology*, *92*(2), 83–91. DOI: 10.1016/j.pestbp.2008.06.008

Guo, Y., Zhang, J., Li, H., Liu, Y., Gui, H., Chen, S., et al. (2025). Determination and application of soil heavy metal geochemical baseline in the southern region of Wushan County in the Yangtze River Basin, China. *Scientific Reports 2025 15:1* 15, 10889-. DOI: 10.1038/s41598-025-86404-1

Guo, H., White, J. C., Wang, Z., & Xing, B. (2018). Nano-enabled fertilizers to control the release and use efficiency of nutrients. *Current Opinion in Environmental Science & Health*, *6*, 77–83. DOI: 10.1016/j.coesh.2018.07.009

Gu, X., Wang, Y., Lai, C., Qiu, J., Li, S., Hou, Y., Martens, W., Mahmood, N., & Zhang, S. (2015). Microporous bamboo biochar for lithium-sulfur batteries. *Nano Research*, *8*(1), 129–139. DOI: 10.1007/s12274-014-0601-1

Hachicho, N., Hoffmann, P., Ahlert, K., & Heipieper, H. J. (2014). Effect of silver nanoparticles and silver ions on growth and adaptive response mechanisms of Pseudomonas putida mt-2. *FEMS Microbiology Letters*, *355*(1), 71–77. DOI: 10.1111/1574-6968.12460 PMID: 24801753

Hajji-Hedfi, L., & Chhipa, H. (2021). Nano-based pesticides: Challenges for pest and disease management. *Euro-Mediterranean Journal for Environmental Integration*, *6*(3), 69. DOI: 10.1007/s41207-021-00279-y

Hale, S. E., Alling, V., Martinsen, V., Mulder, J., Breedveld, G. D., & Cornelissen, G. (2013). The sorption of PAHs to seven different biochars as a function of their physicochemical properties. *Chemosphere*, *91*(10), 1412–1419.

Hammond, J., Shackley, S., Sohi, S., & Brownsort, P. (2011). Prospective life cycle carbon abatement for pyrolysis biochar systems in the UK. *Energy Policy*, *39*(5), 2646–2655. DOI: 10.1016/j.enpol.2011.02.033

Han, R., Zhou, B., Huang, Y., Lu, X., Li, S., & Li, N. (2020). Bibliometric overview of research trends on heavy metal health risks and impacts in 1989–2018. *Journal of Cleaner Production*, *276*, 123249. DOI: 10.1016/j.jclepro.2020.123249

Hansen, V., Müller-Stöver, D., Ahrenfeldt, J., Holm, J. K., Henriksen, U. B., & Hauggaard-Nielsen, H. (2015). Gasification biochar as a valuable by-product for carbon sequestration and soil amendment. *Biomass and Bioenergy*, *72*, 300–308. DOI: 10.1016/j.biombioe.2014.10.013

Hao, L., Chen, L., Hao, J., & Zhong, N. (2013). Bioaccumulation and sub-acutetoxicity of zinc oxide nanoparticles in juvenile carp (Cyprinuscarpio): A comparative study with its bulk counterparts. *Ecotoxicology and Environmental Safety*, *91*, 52–60. DOI: 10.1016/j.ecoenv.2013.01.007 PMID: 23375439

Hardy, A., Benford, D., Halldorsson, T., Jeger, M. J., Knutsen, H. K., More, S., Naegeli, H., Noteborn, H., Ockleford, C., Ricci, A., Rychen, G., Schlatter, J. R., Silano, V., Solecki, R., Turck, D., Younes, M., Chaudhry, Q., Cubadda, F., Gott, D., & Mortensen, A. (2018). Guidance on risk assessment of the application of nanoscience and nanotechnologies in the food and feed chain: Part 1, human and animal health. *EFSA Journal*, *16*(7). Advance online publication. DOI: 10.2903/j.efsa.2018.5327 PMID: 32625968

Harish, V., Tewari, D., Gaur, M., Yadav, A. B., Swaroop, S., Bechelany, M., & Barhoum, A. (2022). Review on Nanoparticles and Nanostructured Materials: Bioimaging, Biosensing, Drug Delivery, Tissue Engineering, Antimicrobial, and Agro-Food Applications. *Nanomaterials (Basel, Switzerland)*, *12*(3), 457. DOI: 10.3390/nano12030457 PMID: 35159802

Hartmann, N. I. B., Skjolding, L. M., Hansen, S. F., Baun, A., Kjølholt, J., & Gottschalk, F. (2014). *Environmental fate and behaviour of nanomaterials: New knowledge on important transfomation processes*. Danish Environmental Protection Agency.

Hartmann, T., Mühling, M., Wolf, A., Mariana, F., Maskow, T., Mertens, F., Neu, T. R., & Lerchner, J. (2013). A chip-calorimetric approach to the analysis of Ag nanoparticle caused inhibition and inactivation of beads-grown bacterial biofilms. *Journal of Microbiological Methods*, *95*(2), 129–137. DOI: 10.1016/j.mimet.2013.08.003 PMID: 23968644

Haydar, M. S., Ghosh, D., & Roy, S. (2024). Slow and controlled release nanofertilizers as an efficient tool for sustainable agriculture: Recent understanding and concerns. *Plant Nano Biology*, *7*, 100058. DOI: 10.1016/j.plana.2024.100058

Haydar, M. S., Kaur, S., Singh, P., & Talukdar, D. (2025). From synthesis to soil: Transformations and impacts of nanoparticles on soil ecosystems. *The Science of the Total Environment*, *944*, 173926. DOI: 10.1016/j.scitotenv.2025.180374 PMID: 40914116

Hayles, J., Johnson, L., Worthley, C., & Losic, D. (2017). Nanopesticides: a review of current research and perspectives. In *New Pesticides and Soil Sensors* (pp. 193–225). Elsevier. DOI: 10.1016/B978-0-12-804299-1.00006-0

Hedley, C. (2015). The role of precision agriculture for improved nutrient management on farms. *Journal of the Science of Food and Agriculture*, *95*(1), 12–19. DOI: 10.1002/jsfa.6734 PMID: 24816925

Hee Joo, S., & Cheng, I. F. (2006). *Nanotechnology for environmental remediation* (Hee Joo, S., & Cheng, I. F., Eds.). Springer. DOI: 10.1007/b137366

He, S., Feng, Y., Ni, J., Sun, Y., Xue, L., Feng, Y., Yu, Y., Lin, X., & Yang, L. (2016). Different responses of soil microbial metabolic activity to silver and iron oxide nanoparticles. *Chemosphere*, *147*, 195–202. DOI: 10.1016/j.chemosphere.2015.12.055 PMID: 26766356

Higgs, V., Ahmed, H., & Flavier, J. A. (2024). Transforming asthma diagnosis: Point of care (POC) breath test with deep neural networks (DNNs) and nanosensors. In A34. Late breaking abstracts in asthma and allergic inflammation (pp. A1380–A1380). American Thoracic Society.

Hobman, J. L., & Crossman, L. C. (2015). Bacterial antimicrobial metal ion resistance. *Journal of Medical Microbiology*, *64*(5), 471–497. DOI: 10.1099/jmm.0.023036-0 PMID: 25418738

Hofmann, T., Lowry, G. V., Ghoshal, S., Tufenkji, N., Brambilla, D., Dutcher, J. R., Gilbertson, L. M., Giraldo, J. P., Kinsella, J. M., Landry, M. P., Lovell, W., Naccache, R., Paret, M., Pedersen, J. A., Unrine, J. M., White, J. C., & Wilkinson, K. J. (2020). Technology readiness and overcoming barriers to sustainably implement nanotechnology-enabled plant agriculture. *Nature Food*, *1*(7), 416–425. DOI: 10.1038/s43016-020-0110-1

Hong, J., Rico, C. M., Zhao, L., Adeleye, A. S., Keller, A. A., Peralta-Videa, J. R., & Gardea-Torresdey, J. L. (2015). Toxic effects of copper-based nanoparticles or compounds to lettuce (*Lactuca sativa*) and alfalfa (*Medicago sativa*). *Environmental Science. Processes & Impacts*, *17*(1), 177–185. DOI: 10.1039/C4EM00551A PMID: 25474419

Hoppe, M., Mikutta, R., Utermann, J., Duijnisveld, W., Kaufhold, S., Stange, C. F., & Guggenberger, G. (2015). Remobilization of sterically stabilized silver nanoparticles from farmland soils determined by column leaching. *European Journal of Soil Science*, *66*(5), 898–909. DOI: 10.1111/ejss.12270

Hossain, M. K., Strezov, V., Chan, K. Y., Ziolkowski, A., & Nelson, P. F. (2011). Influence of pyrolysis temperature on production and nutrient properties of wastewater sludge biochar. *Journal of Environmental Management*, *92*(1), 223–228. DOI: 10.1016/j.jenvman.2010.09.008 PMID: 20870338

Hossain, M. S. (2019). Present scenario of global salt-affected soils, its management and importance of salinity research. *International Research Journal of Biological Sciences Perspectives*, *1*(1), 1–3.

Huang, D., Liu, H., Yin, X., Yuan, F., Song, B., Teng, Y., Ding, Y., Chen, Q., & Liu, T. (2024). A nanoformulation that enhance the efficiency of camptothecin against lepidopteran pests by controlled release and 'poison bait' strategies. *Chemical Engineering Journal*, *486*, 150358. DOI: 10.1016/j.cej.2024.150358

Huang, Q., Yu, H., & Ru, Q. (2010). Bioavailability and Delivery of Nutraceuticals Using Nanotechnology. *Journal of Food Science*, *75*(1). Advance online publication. DOI: 10.1111/j.1750-3841.2009.01457.x PMID: 20492195

Hu, D., Bai, Y., Li, L., Ai, M., Jin, J., Song, K., & Liu, D. (2022). Phytotoxicity assessment of gold nanoclusters in broad bean (*Vicia faba* L.). *Environmental Pollutants and Bioavailability*, *34*(1), 284–296. DOI: 10.1080/26395940.2022.2101543

Huggins, T., Wang, H., Kearns, J., Jenkins, P., & Ren, Z. J. (2014). Biochar as a sustainable electrode material for electricity production in microbial fuel cells. *Bioresource Technology*, *157*, 114–119. DOI: 10.1016/j.biortech.2014.01.058 PMID: 24534792

Hussain, I., Singh, A., Singh, N. B., Singh, P., & Singh, P. (2019). Plant–nanoceria interaction. *South African Journal of Botany*, *121*, 239–247. DOI: 10.1016/j.sajb.2018.11.013

Ibrahim, A. A., Abd-Ellatif, S., Abdel Razik, E. S. S., Hamedo, H. A., & Salem, K. F. (2024). Nanosensors for enhancing plant growth and productivity. In *Innovative Methods in Horticultural Crop Improvement: Biosensors and Nanosensors* (pp. 25–50). Springer Nature Switzerland. https://doi.org/DOI: 10.1007/978-3-031-61095-0_2

Igiehon, N. O., & Babalola, O. O. (2018). Below-ground–above-ground plant–microbial interactions in soybean. *The Open Microbiology Journal*, *12*, 261–279. DOI: 10.2174/1874285801812010261 PMID: 30197700

Inoue, Y. (2020). Satellite- and drone-based remote sensing of crops and soils for smart farming – a review. *Soil Science and Plant Nutrition*, *66*(6), 798–810. DOI: 10.1080/00380768.2020.1738899

International Biochar Initiative (IBI). (2015). Standardized Product Definition and Product Testing Guidelines for Biochar That Is Used in Soil. Accessed from: https://biochar-international.org/

Inyang, M. I., Gao, B., Yao, Y., Xue, Y., Zimmerman, A., Mosa, A., Pullammanappallil, P., Ok, Y. S., & Cao, X. (2016). A review of biochar as a low-cost adsorbent for aqueous heavy metal removal. *Critical Reviews in Environmental Science and Technology*, *46*(4), 406–433. DOI: 10.1080/10643389.2015.1096880

Inyang, M., Gao, B., Pullammanappallil, P., Ding, W., & Zimmerman, A. R. (2010). Biochar from anaerobically digested sugarcane bagasse. *Bioresource Technology*, *101*(22), 8868–8872. DOI: 10.1016/j.biortech.2010.06.088 PMID: 20634061

Ippolito, J. A., Laird, D. A., & Busscher, W. J. (2012). Environmental benefits of biochar. *Journal of Environmental Quality*, *41*(4), 967–972. DOI: 10.2134/jeq2012.0151 PMID: 22751039

Ippolito, J. A., Spokas, K. A., Novak, J. M., Lentz, R. D., & Cantrell, K. B. (2020). Biochar elemental composition and factors influencing nutrient retention. In Lehmann, J., & Joseph, S. (Eds.), *Biochar for environmental management: Science, technology and implementation* (3rd ed., pp. 137–162). Routledge.

Isibor, P. O. (2024). Regulations and policy considerations for nanoparticle safety. In *Environmental Nanotoxicology: Combatting the Minute Contaminants* (pp. 295–316). Springer Nature Switzerland. DOI: 10.1007/978-3-031-54154-4_14

Ismail, A., Munsi, H., Yusuf, A. M., & Hijjang, P. (2025). Mapping One Decade of Identity Studies: A Comprehensive Bibliometric Analysis of Global Trends and Scholarly Impact. *Social Sciences 2025, Vol. 14, Page 92, 14*(2), 92. DOI: 10.3390/socsci14020092

Jacoby, R., Peukert, M., Succurro, A., Koprivova, A., & Kopriva, S. (2017). The role of soil microorganisms in plant mineral nutrition—Current knowledge and future directions. *Frontiers in Plant Science*, *8*, 1617. DOI: 10.3389/fpls.2017.01617 PMID: 28974956

Jaiswal, A. K., Elad, Y., Graber, E. R., & Frenkel, O. (2015). Contribution of biochar to suppression of Fusarium oxysporum f. sp. lycopersici in tomato plants. *Israel Journal of Plant Sciences*, *62*(1-2), 76–85.

Jakhar, A. M., Aziz, I., Kaleri, A. R., Hasnain, M., Haider, G., Ma, J., & Abideen, Z. (2022). Nano-fertilizers: A sustainable technology for improving crop nutrition and food security. *NanoImpact*, *27*, 100411. DOI: 10.1016/j.impact.2022.100411 PMID: 35803478

Jansson, J. K. (2023). Soil viruses: Understudied agents of soil ecology. *Environmental Microbiology*, *25*(1), 143–146. DOI: 10.1111/1462-2920.16258 PMID: 36271323

Jatav, H. S., Rajput, V. D., Raza, T., & Maurya, P. (2025). Mitigation strategies to reduce the climate change impact in agriculture. *Journal of Innovative Solutions for Eco-Environmental Sustainability*, *1*(1), 13493. DOI: 10.46991/JISEES.2025.01.1.13493

Jeffery, S., Abalos, D., Prodana, M., Bastos, A. C., van Groenigen, J. W., Hungate, B. A., & Verheijen, F. (2017). Biochar boosts tropical but not temperate crop yields. *Environmental Research Letters*, *12*(5), 053001. DOI: 10.1088/1748-9326/aa67bd

Jeffery, S., Verheijen, F. G., Kammann, C., & Abalos, D. (2016). Biochar effects on methane emissions from soils: A meta-analysis. *Soil Biology & Biochemistry*, *101*, 251–258. DOI: 10.1016/j.soilbio.2016.07.021

Jeffery, S., Verheijen, F. G., Van Der Velde, M., & Bastos, A. C. (2011). A quantitative review of the effects of biochar application to soils on crop productivity using meta-analysis. *Agriculture, Ecosystems & Environment*, *144*(1), 175–187. DOI: 10.1016/j.agee.2011.08.015

Jiang, J., Xu, R. K., Jiang, T. Y., & Li, Z. (2012). Immobilization of Cu(II), Pb(II) and Cd(II) by the addition of rice straw derived biochar to a simulated polluted Ultisol. *Journal of Hazardous Materials*, *229*, 145–150. DOI: 10.1016/j.jhazmat.2012.05.086 PMID: 22704774

Joseph, S. D., Camps-Arbestain, M., Lin, Y., Munroe, P., Chia, C. H., Hook, J., van Zwieten, L., Kimber, S., Cowie, A., Singh, B. P., Lehmann, J., Foidl, N., Smernik, R. J., & Amonette, J. E. (2010). An investigation into the reactions of biochar in soil. *Soil Research (Collingwood, Vic.)*, *48*(7), 501–515. DOI: 10.1071/SR10009

Joudeh, N., & Linke, D. (2022). Nanoparticle classification, physicochemical properties, characterization, and applications: A comprehensive review for biologists. *Journal of Nanobiotechnology*, *20*(1), 262. Advance online publication. DOI: 10.1186/s12951-022-01477-8 PMID: 35672712

Judy, J. D., Kirby, J. K., Creamer, C., McLaughlin, M. J., Fiebiger, C., Wright, C., Cavagnaro, T., & Bertsch, P. M. (2015). Effects of silver sulfide nanomaterials on mycorrhizal colonization of tomato plants and soil microbial communities in biosolid-amended soil. *Environmental Pollution*, *206*, 256–263. DOI: 10.1016/j.envpol.2015.07.002 PMID: 26196315

Kabata-Pendias. A.., & Pendias, H. (1992). *Trace elements in soils and plants. 2nd ed.* CRC Press.

Kahlon, S. K., Sharma, G., Julka, J. M., Kumar, A., Sharma, S., & Stadler, F. J. (2018). Impact of heavy metals and nanoparticles on aquatic biota. *Environmental Chemistry Letters*, *16*(3), 919–946. DOI: 10.1007/s10311-018-0737-4

Kah, M., Kookana, R. S., Gogos, A., & Bucheli, T. D. (2018). A critical evaluation of nanopesticides and nanofertilizers against their conventional analogues. *Nature Nanotechnology*, *13*(8), 677–684. DOI: 10.1038/s41565-018-0131-1 PMID: 29736032

Kakaei, H., Nourmoradi, H., Bakhtiyari, S., Jalilian, M., & Mirzaei, A. (2022). Effect of COVID-19 on food security, hunger, and food crisis. In *COVID-19 and the Sustainable Development Goals* (pp. 3–29). Elsevier., DOI: 10.1016/B978-0-323-91307-2.00005-5

Kalia, A., Sharma, S. P., Kaur, H., & Kaur, H. (2020). Novel nanocomposite-based controlled-release fertilizer and pesticide formulations: Prospects and challenges. In *Multifunctional Hybrid Nanomaterials for Sustainable Agri-Food and Ecosystems* (pp. 99–134). Elsevier., DOI: 10.1016/B978-0-12-821354-4.00005-4

Kalwani, M., Chakdar, H., Srivastava, A., Pabbi, S., & Shukla, P. (2022). Effects of nanofertilizers on soil and plant-associated microbial communities: Emerging trends and perspectives. *Chemosphere*, *287*, 132107. DOI: 10.1016/j.chemosphere.2021.132107 PMID: 34492409

Kaman, P. K., & Dutta, P. (2019). Synthesis, characterization and antifungal activity of biosynthesized silver nanoparticle. *Indian Phytopathology*, *72*(1), 79–88. DOI: 10.1007/s42360-018-0081-4

Kambo, H. S., & Dutta, A. (2015). A comparative review of biochar and hydrochar in terms of production, physico-chemical properties and applications. *Renewable & Sustainable Energy Reviews*, *45*, 359–378. DOI: 10.1016/j.rser.2015.01.050

Kang, M., Bai, X., Liu, Y., Weng, Y., Wang, H., & Ye, Z. (2024). Zinc oxide nanoparticle size and hydrophobicity regulate eco-corona formation. *Environmental Science & Technology*, *58*(22), 9875–9886. DOI: 10.1021/acs.est.4c01731 PMID: 38722770

Kannan, S. (2010). *Foliar Fertilization for Sustainable Crop Production.*, DOI: 10.1007/978-90-481-8741-6_13

Karimi, M., Eslami, M., Sahandi-Zangabad, P., Mirab, F., Farajisafiloo, N., Shafaei, Z., Ghosh, D., Bozorgomid, M., Dashkhaneh, F., & Hamblin, M. R. (2016). <scp>pH</scp> -Sensitive stimulus-responsive nanocarriers for targeted delivery of therapeutic agents. *Wiley Interdisciplinary Reviews. Nanomedicine and Nanobiotechnology*, *8*(5), 696–716. DOI: 10.1002/wnan.1389 PMID: 26762467

Kasel, D., Bradford, S. A., Simunek, J., Putz, T., Vereecken, H., & Klumpp, E. (2013). Limited transport of functionalized multi-walled carbon nanotubes in two natural soils. *Environmental Pollution*, *180*, 152–158. DOI: 10.1016/j.envpol.2013.05.031 PMID: 23770315

Kausar, A., Ahmad, I., Maaza, M., & Eisa, M. H. (2022). State-of-the-Art Nanoclay Reinforcement in Green Polymeric Nanocomposite: From Design to New Opportunities. *Minerals (Basel)*, *12*(12), 1495. DOI: 10.3390/min12121495

Kausar, A., Rafique, I., & Muhammad, B. (2017). Aerospace Application of Polymer Nanocomposite with Carbon Nanotube, Graphite, Graphene Oxide, and Nanoclay. *Polymer-Plastics Technology and Engineering*, *56*(13), 1438–1456. DOI: 10.1080/03602559.2016.1276594

Kaźmierczak, Z., Szostak-Paluch, K., Przybyło, M., Langner, M., Witkiewicz, W., Jędruchniewicz, N., & Dąbrowska, K. (2020). Endocytosis in cellular uptake of drug delivery vectors: Molecular aspects in drug development. *Bioorganic & Medicinal Chemistry*, *28*(18), 115556. DOI: 10.1016/j.bmc.2020.115556 PMID: 32828419

Keller, A. A., Wang, H., Zhou, D., Lenihan, H. S., Cherr, G., Cardinale, B. J., Miller, R., & Ji, Z. (2010). Stability and aggregation of metal oxide nanoparticles in natural aqueous matrices. *Environmental Science & Technology*, *44*(6), 1962–1967. DOI: 10.1021/es902987d PMID: 20151631

Keller, A. A., Zheng, Y., Praetorius, A., Quik, J. T., & Nowack, B. (2024). Predicting environmental concentrations of nanomaterials for exposure assessment-a review. *NanoImpact*, *33*, 100496. DOI: 10.1016/j.impact.2024.100496 PMID: 38266914

Khanna, K., Kohli, S. K., Handa, N., Kaur, H., Ohri, P., Bhardwaj, R., Yousaf, B., Rinklebe, J., & Ahmad, P. (2021). Enthralling the impact of engineered nanoparticles on soil microbiome: A concentric approach towards environmental risks and cogitation. *Ecotoxicology and Environmental Safety*, *222*, 112459. DOI: 10.1016/j.ecoenv.2021.112459 PMID: 34217114

Khan, S. B., Faisal, M., Rahman, M. M., & Jamal, A. (2011). Exploration of CeO2nanoparticles as a chemi-sensor and photo-catalyst for environ-mental applications. *The Science of the Total Environment*, *409*(15), 2987–2992. DOI: 10.1016/j.scitotenv.2011.04.019 PMID: 21570707

Khan, S., Naushad, M., Al-Gheethi, A., & Iqbal, J. (2021). Engineered nanoparticles for removal of pollutants from wastewater: Current status and future prospects of nanotechnology for remediation strategies. *Journal of Environmental Chemical Engineering*, *9*(5), 106160. DOI: 10.1016/j.jece.2021.106160

Kirkegaard, M. L., Kines, P., Jeschke, K. C., & Jensen, K. A. (2020). Risk perceptions and safety cultures in the handling of nanomaterials in academia and industry. *Annals of Work Exposures and Health*, *64*(5), 479–489. DOI: 10.1093/annweh/wxaa022 PMID: 32155241

Klaine, S. J., Alvarez, P. J., Batley, G. E., Fernandes, T. F., Handy, R. D., Lyon, D. Y., Mahendra, S., McLaughlin, L. J., & Lead, J. R. (2008). Nanomaterials in the environment: Behavior, fate, bioavailability, and effects. *Environmental Toxicology and Chemistry*, *27*(9), 1825–1851. DOI: 10.1897/08-090.1 PMID: 19086204

Klarin, A. (2024). How to conduct a bibliometric content analysis: Guidelines and contributions of content co-occurrence or co-word literature reviews. *International Journal of Consumer Studies*, *48*(2), e13031. DOI: 10.1111/ijcs.13031

Klitzke, S., Metreveli, G., Peters, A., Schaumann, G. E., & Lang, F. (2015). The fate of silver nanoparticles in soil solution—Sorption of solutes and aggregation. *The Science of the Total Environment*, *535*, 54–60. DOI: 10.1016/j.scitotenv.2014.10.108 PMID: 25434472

Kloss, S., Zehetner, F., Dellantonio, A., Hamid, R., Ottner, F., Liedtke, V., Schwanninger, M., Gerzabek, M. H., & Soja, G. (2012). Characterization of slow pyrolysis biochars: Effects of feedstocks and pyrolysis temperature on biochar properties. *Journal of Environmental Quality*, *41*(4), 990–1000. DOI: 10.2134/jeq2011.0070 PMID: 22751041

Kou, L., Sun, J., Zhai, Y., & He, Z. (2013). The endocytosis and intracellular fate of nanomedicines: Implication for rational design. *Asian Journal of Pharmaceutical Sciences*, *8*(1), 1–10. DOI: 10.1016/j.ajps.2013.07.001

Kuang, L., Burgess, B., Cuite, C. L., Tepper, B. J., & Hallman, W. K. (2020). Sensory acceptability and willingness to buy foods presented as having benefits achieved through the use of nanotechnology. *Food Quality and Preference*, *83*, 103922. DOI: 10.1016/j.foodqual.2020.103922

Kudureti, A., Zhao, S., Zhakyp, D., & Tian, C. (2023). Responses of soil fauna community under changing environmental conditions. *Journal of Arid Land*, *15*(5), 620–636. DOI: 10.1007/s40333-023-0009-4

Kuhn, R., Bryant, I. M., Jensch, R., & Böllmann, J. (2022). Applications of Environmental Nanotechnologies in Remediation, Wastewater Treatment, Drinking Water Treatment, and Agriculture. *Applied Nanoscience*, *3*(1), 54–90. DOI: 10.3390/applnano3010005

Kumar, A., Gupta, K., Dixit, S., Mishra, K., & Srivastava, S. (2018). A review on positive and negative impacts of nanotechnology in agriculture. *International Journal of Environmental Science and Technology*, *16*(4), 2175–2184. DOI: 10.1007/s13762-018-2119-7

Kumar, C. V., Karthick, V., Kumar, V. G., Inbakandan, D., Rene, E. R., Suganya, K. U., & Sowmiya, P. (2022). The impact of engineered nanomaterials on the environment: Release mechanism, toxicity, transformation, and remediation. *Environmental Research*, *212*, 113202. DOI: 10.1016/j.envres.2022.113202 PMID: 35398077

Kumar, D., Shandilya, A. K., & Srivastava, S. (2023). The journey of F1000Research since inception: through bibliometric analysis. *F1000Research 2023 12:516, 12*, 516. DOI: 10.12688/f1000research.134244.1

Kumari, A., Lakshmi, G. A., Krishna, G. K., Patni, B., Prakash, S., Bhattacharyya, M., Singh, S. K., & Verma, K. K. (2022). Climate change and its impact on crops: A comprehensive investigation for sustainable agriculture. *Agronomy (Basel)*, *12*(12), 3008. DOI: 10.3390/agronomy12123008

Kumari, A., Rana, V., Yadav, S. K., & Kumar, V. (2023). Nanotechnology as a powerful tool in plant sciences: Recent developments, challenges and perspectives. *Plant Nano Biology*, *5*, 100046. DOI: 10.1016/j.plana.2023.100046

Kumari, A., Singh, S. K., Mathpal, B., Verma, K. K., Garg, V. K., Bhattacharyya, M., & Bhatt, R. (2023). The biosynthesis, mechanism of action, and physiological functions of melatonin in horticultural plants: A review. *Horticulturae*, *9*(8), 913. DOI: 10.3390/horticulturae9080913

Kumari, K. G. I. D., Moldrup, P., Paradelo, M., Elsgaard, L., Hauggaard-Nielsen, H., & de Jonge, L. W. (2014). Effects of biochar on air and water permeability and colloid and phosphorus leaching in soils from a natural calcium carbonate gradient. *Journal of Environmental Quality*, *43*(2), 647–657. DOI: 10.2134/jeq2013.08.0334 PMID: 25602666

Kumari, K., Rani, N., & Hooda, V. (2024). Effects of nano-SiO_2, nano-TiO_2 and nanocomposites on maize growth and soil health. *Scientific Reports*, *14*, 13996. DOI: 10.1038/s41598-024-61456-x PMID: 38886471

Kumari, R., Suman, K., Karmakar, S., Mishra, V., Lakra, S. G., Saurav, G. K., & Mahto, B. K. (2023). Regulation and safety measures for nanotechnology-based agri-products. *Frontiers in Genome Editing*, *5*, 1200987. Advance online publication. DOI: 10.3389/fgeed.2023.1200987 PMID: 37415849

Kumar, L., Chhogyel, N., Gopalakrishnan, T., Hasan, M. K., Jayasinghe, S. L., Kariyawasam, C. S., Kogo, B. K., & Ratnayake, S. (2022). Climate change and future of agri-food production. In *Future Foods* (pp. 49–79). Elsevier., DOI: 10.1016/B978-0-323-91001-9.00009-8

Kuzyakov, Y., Subbotina, I., Chen, H., Bogomolova, I., & Xu, X. (2014). Black carbon decomposition and incorporation into soil microbial biomass estimated by ^{14}C labeling. *Soil Biology & Biochemistry*, *73*, 44–51.

Laird, D. A., Fleming, P., Davis, D. D., Horton, R., Wang, B., & Karlen, D. L. (2010). Impact of biochar amendments on the quality of a typical Midwestern agricultural soil. *Geoderma*, *158*(3-4), 443–449. DOI: 10.1016/j.geoderma.2010.05.013

Lal, R. (2015). Restoring soil quality to mitigate soil degradation. *Sustainability (Basel)*, *7*(5), 5875–5895. DOI: 10.3390/su7055875

Lasagna-Reeves, C., Gonzalez-Romero, D., Barria, M. A., Olmedo, I., Clos, A., Ramanujam, V. S., Urayama, A., Vergara, L., Kogan, M. J., & Soto, C. (2010). Bioaccumulation and toxicity of gold nano-particles after repeated administration in mice. *Biochemical and Biophysical Research Communications*, *393*(4), 649–655. DOI: 10.1016/j.bbrc.2010.02.046 PMID: 20153731

Lead, J. R., Batley, G. E., Alvarez, P. J. J., Croteau, M.-N., Handy, R. D., McLaughlin, M. J., Judy, J. D., & Schirmer, K. (2018). Nanomaterials in the environment: Behavior, fate, bioavailability and effects—An updated review. *Environmental Toxicology and Chemistry*, *37*(8), 2029–2063. DOI: 10.1002/etc.4147 PMID: 29633323

Lecoanet, H. F., & Wiesner, M. R. (2004). Velocity effects on fullerene and oxide nanoparticle deposition in porous media. *Environmental Science & Technology*, *38*(16), 4377–4382. DOI: 10.1021/es035354f PMID: 15382867

Lehmann, J., Cowie, A., Masiello, C. A., Kammann, C., Woolf, D., Amonette, J. E., Cayuela, M. L., Camps-Arbestain, M., & Whitman, T. (2021). Biochar in climate change mitigation. *Nature Geoscience*, *14*(12), 883–892. DOI: 10.1038/s41561-021-00852-8

Lehmann, J., da Silva, J. P.Jr, Steiner, C., Nehls, T., Zech, W., & Glaser, B. (2003). Nutrient availability and leaching in an archaeological Anthrosol and a Ferralsol of the Central Amazon basin: Fertilizer, manure and charcoal amendments. *Plant and Soil*, *249*(2), 343–357. DOI: 10.1023/A:1022833116184

Lehmann, J., Gaunt, J., & Rondon, M. (2006). Bio-char sequestration in terrestrial ecosystems–a review. *Mitigation and Adaptation Strategies for Global Change*, *11*(2), 403–427. DOI: 10.1007/s11027-005-9006-5

Lehmann, J., & Joseph, S. (Eds.). (2015). *Biochar for environmental management: science, technology and implementation*. Routledge. DOI: 10.4324/9780203762264

Lehmann, J., Kern, D. C., Glaser, B., & Woods, W. I. (2003). *Amazonian dark earths: origin properties management*. Springer Science & Business Media. DOI: 10.1007/1-4020-2597-1

Lehmann, J., & Kleber, M. (2015). The contentious nature of soil organic matter. *Nature*, *528*(7580), 60–68. DOI: 10.1038/nature16069 PMID: 26595271

Lehmann, J., Rillig, M. C., Thies, J., Masiello, C. A., Hockaday, W. C., & Crowley, D. (2011). Biochar effects on soil biota–a review. *Soil Biology & Biochemistry*, *43*(9), 1812–1836. DOI: 10.1016/j.soilbio.2011.04.022

Lew, T. T. S., Koman, V. B., Gordiichuk, P., Park, M., & Strano, M. S. (2020). The Emergence of Plant Nanobionics and Living Plants as Technology. *Advanced Materials Technologies*, *5*(3), 1900657. Advance online publication. DOI: 10.1002/admt.201900657

Liang, B., Lehmann, J., Solomon, D., Kinyangi, J., Grossman, J., O'Neill, B., Skjemstad, J. O., Thies, J., Luizao, F. J., Petersen, J., & Neves, E. G. (2006). Black carbon increases cation exchange capacity in soils. *Soil Science Society of America Journal*, *70*(5), 1719–1730. DOI: 10.2136/sssaj2005.0383

Libra, J. A., Ro, K. S., Kammann, C., Funke, A., Berge, N. D., Neubauer, Y., Titirici, M.-M., Führer, C., Bens, O., Kern, J., & Emmerich, K. H. (2011). Hydrothermal carbonization of biomass residuals: A comparative review of the chemistry, processes and applications of wet and dry pyrolysis. *Biofuels*, *2*(1), 71–106. DOI: 10.4155/bfs.10.81

Li, L., Luo, Y., Li, R., Zhou, Q., Peijnenburg, W. J. G. M., Yin, N., Yang, J., Tu, C., & Zhang, Y. (2020). Uptake of submicrometre plastics by crop plants. *Nature Sustainability*, *3*(11), 929–937. DOI: 10.1038/s41893-020-0567-9

Li, M.. (2024). Dynamic transformation of nano-MoS_2 in a soil–plant system and its implications for biological nitrogen fixation and crop yield. *Environmental Science & Technology*, *58*(2), 1234–1246. DOI: 10.1021/acs.est.3c09004

Lim, S. L., Lee, L. H., & Wu, T. Y. (2016). Sustainability of using composting and vermicomposting technologies for organic solid waste biotransformation: Recent overview, greenhouse gases emissions and economic analysis. *Journal of Cleaner Production*, *111*, 262–278. DOI: 10.1016/j.jclepro.2015.08.083

Lin, Y., Chen, Z., Dai, M., Fang, S., Liao, Y., Yu, Z., & Ma, X. (2012). Rapid microwave-assisted pyrolysis of biomass using different heating modes. *Bioresource Technology*, *125*, 308–314.

Li, S., Liu, X. Q., Wang, F. Y., & Miao, Y. F. (2015). Effects of ZnO Nanoparticles, $ZnSO_4$ and Arbuscular Mycorrhizal Fungus on the Growth of Maize. *Huan Jing ke Xue=. Huan Jing Ke Xue*, *36*(12), 4615–4622. PMID: 27012001

Liu, L., Tsyusko, O. V., Unrine, J. M., Liu, S., Liu, Y., Guo, L., & Chen, C. (2023). Pristine and sulfidized zinc oxide nanoparticles promote organic carbon release and decomposition in the legume rhizosphere. *Environmental Science & Technology*, *57*(24), 8943–8953. DOI: 10.1021/acs.est.3c02071 PMID: 37285309

Liu, S., Meng, J., Jiang, L., Yang, X., Lan, Y., Cheng, X., & Chen, W. (2021). Rice husk biochar impacts on rice production and nitrogen and phosphorus uptake in a Chinese paddy soil. *Experimental Results*, •••, 2.

Liu, T., Qian, K., Fan, J., Qin, Z., Yi, L., Wang, J., & Dou, W. (2024). Nano-based smart pesticide formulations utilizing starch nanocrystals: Achieving high stability and extended effective duration. *Industrial Crops and Products*, *214*, 118470. DOI: 10.1016/j.indcrop.2024.118470

Liu, W., Zhang, M., & Bhandari, B. (2020). Nanotechnology – A shelf life extension strategy for fruits and vegetables. *Critical Reviews in Food Science and Nutrition*, *60*(10), 1706–1721. DOI: 10.1080/10408398.2019.1589415 PMID: 30896257

Liu, X., Zheng, J., Zhang, D., Cheng, K., Zhou, H., Zhang, A., Li, L., Joseph, S., Smith, P., Crowley, D., Kuzyakov, Y., & Pan, G. (2016). Biochar has no effect on soil respiration across Chinese agricultural soils. *The Science of the Total Environment*, *554*, 259–265. DOI: 10.1016/j.scitotenv.2016.02.179 PMID: 26950640

Lombardo, D., Kiselev, M. A., & Caccamo, M. T. (2019). Smart Nanoparticles for Drug Delivery Application: Development of Versatile Nanocarrier Platforms in Biotechnology and Nanomedicine. *Journal of Nanomaterials*, *2019*, 1–26. DOI: 10.1155/2019/3702518

Lombi, E., Donner, E., Tavakkoli, E., Turney, T. W., Naidu, R., Miller, B. M., & Scheckel, K. G. (2012). Fate of zinc oxide nanoparticles during anaerobic digestion of wastewater and post-treatment processing of sewage sludge. *Environmental Science & Technology*, *46*(16), 9089–9096. DOI: 10.1021/es301487s PMID: 22816872

Lorenz, K., & Lal, R. (2014). Biochar application to soil for climate change mitigation by soil organic carbon sequestration. *Journal of Plant Nutrition and Soil Science*, *177*(5), 651–670. DOI: 10.1002/jpln.201400058

Lowry, G. V., Gregory, K. B., Apte, S. C., & Lead, J. R. (2012). Transformations of nanomaterials in the environment. *Environmental Science & Technology*, *46*(13), 6893–6899. DOI: 10.1021/es300839e PMID: 22582927

Lu, H., Zhang, W., Yang, Y., Huang, X., Wang, S., & Qiu, R. (2012). Relative distribution of Pb2+ sorption mechanisms by sludge-derived biochar. *Water Research*, *46*(3), 854–862. DOI: 10.1016/j.watres.2011.11.058 PMID: 22189294

Luna Juncal, M. J., Masino, P., Bertone, E., & Stewart, R. A. (2023). Towards nutrient neutrality: A review of agricultural runoff mitigation strategies and the development of a decision-making framework. *The Science of the Total Environment*, *874*, 162408. DOI: 10.1016/j.scitotenv.2023.162408 PMID: 36828071

Luo, P. G., Yang, F., Yang, S. T., Sonkar, S. K., Yang, L., Broglie, J. J., Liu, Y., & Sun, Y. P. (2014). Carbon-based quantum dots for fluorescence imag-ing of cells and tissues. *RSC Advances*, *4*(21), 10791–10807. DOI: 10.1039/c3ra47683a

Luo, X., Li, J., Guo, S., Yu, H., Zeng, X., Zhou, Z., Shangguan, Y., He, M., Ouyang, Y., Chen, K., Chen, Z., & Qin, Y. (2025). Analysis of research status and trends on nano-agricultural application: A bibliometric study. *Frontiers in Plant Science*, *16*, 1530629. DOI: 10.3389/fpls.2025.1530629 PMID: 40034150

Lv, J., Christie, P., & Zhang, S. (2019). Uptake, translocation, and transformation of metal-based nanoparticles in plants. *Environmental Science. Nano*, *6*(1), 41–59. DOI: 10.1039/C8EN00645H

Lyu, H., Gao, B., He, F., Zimmerman, A. R., Ding, C., Huang, H., & Tang, J. (2018). Effects of ball milling on the physicochemical and sorptive properties of biochar: Experimental observations and governing mechanisms. *Environmental Pollution*, *233*, 54–63. DOI: 10.1016/j.envpol.2017.10.037 PMID: 29053998

Ma, C., Han, L., Shang, H., Hao, Y., Xu, X., White, J. C., Wang, Z., & Xing, B. (2023). Nanomaterials in agricultural soils: Ecotoxicity and application. *Current Opinion in Environmental Science & Health*, *31*, 100432. DOI: 10.1016/j.coesh.2022.100432

Machanuru, R., Shrivastava, M., Singh, R., Singh, B., Chakraborty, D., Ramalingappa, P. L., & Narayan, M. (2024). Plant enzymatic activity as an indicator of nano-TiO2 exposure in rice ecosystems. *Plant Nano Biology*, *10*, 100117. DOI: 10.1016/j.plana.2024.100117

Maestrini, B., Nannipieri, P., & Abiven, S. (2015). A meta-analysis on pyrogenic organic matter induced priming effect. *Global Change Biology. Bioenergy*, *7*(4), 577–590. DOI: 10.1111/gcbb.12194

Mahapatra, D. M., Satapathy, K. C., & Panda, B. (2022). Biofertilizers and nanofertilizers for sustainable agriculture: Phycoprospects and challenges. *The Science of the Total Environment*, *803*, 149990. DOI: 10.1016/j.scitotenv.2021.149990 PMID: 34492488

Mahor, A., Singh, P., Bharadwaj, P., Sharma, N., Yadav, S., Rosenholm, J. M., & Bansal, K. K. (2021). Carbon-based nanomaterials for delivery of biologicals and therapeutics: A cutting-edge technology. *C*, *7*(1), 19. DOI: 10.3390/c7010019

Mai, X., Tang, J., Tang, J., Zhu, X., Yang, Z., Liu, X., Zhuang, X., Feng, G., & Tang, L. (2025). Research progress on the environmental risk assessment and remediation technologies of heavy metal pollution in agricultural soil. *Journal of Environmental Sciences (China)*, *149*, 1–20. DOI: 10.1016/j.jes.2024.01.045 PMID: 39181626

Major, J., Rondon, M., Molina, D., Riha, S. J., & Lehmann, J. (2010). Maize yield and nutrition during 4 years after biochar application to a Colombian savanna oxisol. *Plant and Soil*, *333*(1), 117–128. DOI: 10.1007/s11104-010-0327-0

Malézieux, E. (2012). Designing cropping systems from nature. *Agronomy for Sustainable Development*, *32*(1), 15–29. DOI: 10.1007/s13593-011-0027-z

Mandal, S., Thangarajan, R., Bolan, N. S., Sarkar, B., Khan, N., Ok, Y. S., & Naidu, R. (2017). Biochar-induced concomitant decrease in ammonia volatilization and increase in nitrogen use efficiency by wheat. *Chemosphere*, *142*, 120–127. DOI: 10.1016/j.chemosphere.2015.04.086 PMID: 25959224

Manyà, J. J. (2012). Pyrolysis for biochar purposes: A review to establish current knowledge gaps and research needs. *Environmental Science & Technology*, *46*(15), 7939–7954. DOI: 10.1021/es301029g PMID: 22775244

Ma, Q., Zhang, Q., Yang, S., Yilihamu, A., Shi, M., Ouyang, B., Guan, X., & Yang, S. T. (2020). Toxicity of nanodiamonds to white rot fungi Phanerochaete chrysosporium through oxidative stress. *Colloids and Surfaces. B, Biointerfaces*, *187*, 110658. DOI: 10.1016/j.colsurfb.2019.110658 PMID: 31810567

Ma, R., Levard, C., Judy, J. D., Unrine, J. M., Durenkamp, M., Martin, B., Jefferson, B., & Lowry, G. V. (2013). Fate of zinc oxide and silver nanoparticles in a pilot wastewater treatment plant and in processed biosolids. *Environmental Science & Technology*, *48*(1), 104–112. DOI: 10.1021/es403646x PMID: 24266610

Marchiol, L., Iafisco, M., Fellet, G., & Adamiano, A. (2020). *Nanotechnology support the next agricultural revolution: Perspectives to enhancement of nutrient use efficiency* (pp. 27–116). DOI: 10 1016/bs.agron.2019.12.001

Martínez-Ballesta, M. C., Zapata, L., Chalbi, N., & Carvajal, M. (2016). Multi-walled carbon nanotubes enter broccoli cells enhancing growth and water uptake of plants exposed to salinity. *Journal of Nanobiotechnology*, *14*(1), 42. DOI: 10.1186/s12951-016-0199-4 PMID: 27278384

Masulili, A., Utomo, W. H., & Syechfani, M. S. (2010). Rice husk biochar for rice based cropping system in acid soil 1. The characteristics of rice husk biochar and its influence on the properties of acid sulfate soils and rice growth in West Kalimantan, Indonesia. *Journal of Agricultural Science (Toronto)*, *2*(1), 39–47. DOI: 10.5539/jas.v2n1p39

Mathews, D., Franzen-Castle, L., Colby, S., Kattelmann, K., Olfert, M., & White, A. (2015). Use of Word Clouds as a Novel Approach for Analysis and Presentation of Qualitative Data for Program Evaluation. *Journal of Nutrition Education and Behavior*, *47*(4), S26. DOI: 10.1016/j.jneb.2015.04.071

Matuštík, J., Hnátková, T., & Kočí, V. (2020). Life cycle assessment of biochar-to-soil systems: A review. *Journal of Cleaner Production*, *259*, 120998. DOI: 10.1016/j.jclepro.2020.120998

Ma, X., Geiser-Lee, J., Deng, Y., & Kolmakov, A. (2010). Interactions between engineered nanoparticles (ENPs) and plants: Phytotoxicity, uptake and accumulation. *The Science of the Total Environment*, *408*(16), 3053–3061. DOI: 10.1016/j.scitotenv.2010.03.031 PMID: 20435342

Ma, X., & Yan, J. (2018). Plant uptake of engineered metallic nanoparticles. *Current Opinion in Environmental Science & Health*, *6*, 16–20. DOI: 10.1016/j.coesh.2018.07.008

Mayor, S., & Pagano, R. E. (2007). Pathways of clathrin-independent endocytosis. *Nature Reviews. Molecular Cell Biology*, *8*(8), 603–612. DOI: 10.1038/nrm2216 PMID: 17609668

Mazari, S. A., Ali, E., Abro, R., Khan, F. S. A., Ahmed, I., Ahmed, M., Nizamuddin, S., Siddiqui, T. H., Hossain, N., Mubarak, N. M., & Shah, A. (2021). Nanomaterials: Applications, waste-handling, environmental toxicities, and future challenges–A review. *Journal of Environmental Chemical Engineering*, *9*(2), 105028. DOI: 10.1016/j.jece.2021.105028

Mazhar, M. W., Ishtiaq, M., Maqbool, M., Mazher, M., Amai, S., Siddiqui, M. H., & Bhatt, R. (2024). Management of yield losses in Vigna radiata (L.) R. Wilczek crop caused by charcoal-rot disease through synergistic application of biochar and zinc oxide nanoparticles as boosting fertilizers and nanofungicides. *BMC Plant Biology*, *24*(1), 1099. DOI: 10.1186/s12870-024-05813-y PMID: 39563266

McBeath, T. M., & McLaughlin, M. J. (2014). Efficacy of zinc oxides as fertilisers. *Plant and Soil*, *374*(1-2), 843–855. DOI: 10.1007/s11104-013-1919-2

McCormack, S. A., Ostle, N., Bardgett, R. D., Hopkins, D. W., & Vanbergen, A. J. (2013). Biochar in bioenergy cropping systems: Impacts on soil faunal communities and linked ecosystem processes. *Global Change Biology. Bioenergy*, *5*(2), 81–95. DOI: 10.1111/gcbb.12046

McGee, C. F., Clipson, N., & Doyle, E. (2020). Exploring the influence of raising soil pH on the ecotoxicological effects of silver nanoparticles and micron particles on soil microbial communities. *Water, Air, and Soil Pollution*, *231*(4), 1–8. DOI: 10.1007/s11270-020-04540-y

McKee, M. S., & Filser, J. (2016). Impacts of metal-based engineered nanomaterials on soil communities. *Environmental Science. Nano*, *3*(3), 506–533. DOI: 10.1039/C6EN00007J

Medici, S., Peana, M., Pelucelli, A., & Zoroddu, M. A. (2021, November). An updated overview on metal nanoparticles toxicity. [). Academic Press.]. *Seminars in Cancer Biology*, *76*, 17–26. DOI: 10.1016/j.semcancer.2021.06.020 PMID: 34182143

Medina-Pérez, G., Fernández-Luqueño, F., Vazquez-Nuñez, E., López-Valdez, F., Prieto-Mendez, J., Madariaga-Navarrete, A., & Miranda-Arámbula, M. (2019). Remediating polluted soils using nanotechnologies: Environmental benefits and risks. *Polish Journal of Environmental Studies*, *28*(3), 1013–1030. DOI: 10.15244/pjoes/87099

Meier, M. J., Dodge, A. E., Samarajeewa, A. D., & Beaudette, L. A. (2020). Soil exposed to silver nanoparticles reveals significant changes in community structure and altered microbial transcriptional profiles. *Environmental Pollution*, *258*, 113816. DOI: 10.1016/j.envpol.2019.113816 PMID: 31864930

Méndez, A., Gómez, A., Paz-Ferreiro, J., & Gascó, G. (2012). Effects of sewage sludge biochar on plant metal availability after application to a Mediterranean soil. *Chemosphere*, *89*(11), 1354–1359. DOI: 10.1016/j.chemosphere.2012.05.092 PMID: 22732302

Meyer, S., Bright, R. M., Fischer, D., Schulz, H., & Glaser, B. (2011). Albedo impact on the suitability of biochar systems to mitigate global warming. *Environmental Science & Technology*, *46*(24), 12726–12734. PMID: 23146092

Meyer, S., Genesio, L., Vogel, I., Schmidt, H. P., Soja, G., Someus, E., Shackley, S., Verheijen, F. G., & Glaser, B. (2017). Biochar standardization and legislation harmonization. *Journal of Environmental Engineering and Landscape Management, 25*(2), 175–191. DOI: 10.3846/16486897.2016.1254640

Mielke, H. W., Wang, G., Gonzales, C. R., Powell, E. T., Le, B., & Quach, V. N. (2004). PAHs and metals in the soils of inner-city and suburban New Orleans, Louisiana, USA. *Environmental Toxicology and Pharmacology, 18*(3), 243–247. DOI: 10.1016/j.etap.2003.11.011 PMID: 21782755

Mishra, S., Keswani, C., Abhilash, P. C., Fraceto, L. F., & Singh, H. B. (2017). Integrated approach of agri-nanotechnology: Challenges and futuretrends. *Frontiers in Plant Science, 8*. Advance online publication. DOI: 10.3389/fpls.2017.00471 PMID: 28421100

Mittal, D., Kaur, G., Singh, P., Yadav, K., & Ali, S. A. (2020). Nanoparticle-Based Sustainable Agriculture and Food Science: Recent Advances and Future Outlook. *Frontiers in Nanotechnology, 2*, 579954. Advance online publication. DOI: 10.3389/fnano.2020.579954

Mohan, D., Pittman, C. U.Jr, Bricka, M., Smith, F., Yancey, B., Mohammad, J., Steele, P. H., Alexandre-Franco, M. F., Gómez-Serrano, V., & Gong, H. (2007). Sorption of arsenic, cadmium, and lead by chars produced from fast pyrolysis of wood and bark during bio-oil production. *Journal of Colloid and Interface Science, 310*(1), 57–73. DOI: 10.1016/j.jcis.2007.01.020 PMID: 17331527

Mohan, D., Pittman, C. U.Jr, & Steele, P. H. (2006). Pyrolysis of wood/biomass for bio-oil: A critical review. *Energy & Fuels, 20*(3), 848–889. DOI: 10.1021/ef0502397

Moll, J., Klingenfuss, F., Widmer, F., Gogos, A., Bucheli, T. D., Hartmann, M., & van der Heijden, M. G. (2017). Effects of titanium dioxide nanoparticles on soil microbial communities and wheat biomass. *Soil Biology & Biochemistry, 111*, 85–93. DOI: 10.1016/j.soilbio.2017.03.019

Mondal, K. K., & Mani, C. (2012). Investigation of the antibacterial properties of nanocopper against *Xanthomonas axonopodis* pv. punicae, the incitant of pomegranate bacterial blight. *Annals of Microbiology, 62*(2), 889–893. DOI: 10.1007/s13213-011-0382-7

Monreal, C. M., DeRosa, M., Mallubhotla, S. C., Bindraban, P. S., & Dimkpa, C. (2016). Nanotechnologies for increasing the crop use efficiency of fertilizer-micronutrients. *Biology and Fertility of Soils, 52*(3), 423–437. DOI: 10.1007/s00374-015-1073-5

Mueller, N. C., & Nowack, B. (2010). Nanoparticles for remediation: Solving big problems with little particles. *Elements (Quebec)*, *6*(6), 395–400. DOI: 10.2113/gselements.6.6.395

Mugabo, J. P., Bhople, B. S., Kumar, A. K., Singh, V. K., Mahmoud, E. A., Ullah, F., & Singh, O. (2024). Insights into the tripartite interaction: Effects of Arbuscular mycorrhizae and Rhizobium on root morphology, soil enzymes, and biochemical properties in pea cultivation in alluvial soils of Punjab, India. *Cogent Food & Agriculture*, *10*(1), 2366384. DOI: 10.1080/23311932.2024.2366384

Mukherjee, A., Zimmerman, A. R., & Harris, W. (2011). Surface chemistry variations among a series of laboratory-produced biochars. *Geoderma*, *163*(3-4), 247–255. DOI: 10.1016/j.geoderma.2011.04.021

Mukhopadhyay, R., Sarkar, B., Khan, E., Alessi, D. S., Biswas, J. K., Manjaiah, K. M., Eguchi, M., Wu, K. C. W., Yamauchi, Y., & Ok, Y. S. (2022). Nanomaterials for sustainable remediation of chemical contaminants in water and soil. *Critical Reviews in Environmental Science and Technology*, *52*(15), 2611–2660. DOI: 10.1080/10643389.2021.1886891

Muneer, N. (2023). Emerging Materials and Technologies for the Next Generation of Engineering Solutions. *Liberal Journal of Language and Literature Review*, *1*(01), 39–46.

Mura, S., Seddaiu, G., Bacchini, F., Roggero, P. P., & Greppi, G. F. (2013). Advances of nanotechnology in agro-environmental studies. *Italian Journal of Agronomy*, *8*(3), e18–e18. DOI: 10.4081/ija.2013.e18

Murugadoss, S., Lison, D., Godderis, L., Van Den Brule, S., Mast, J., Brassinne, F., Sebaihi, N., & Hoet, P. H. (2017). Toxicology of silica nanoparticles: An update. *Archives of Toxicology*, *91*(9), 2967–3010. DOI: 10.1007/s00204-017-1993-y PMID: 28573455

Murugan, K., & Abd-Elsalam, K. A. (2022). Sustainable nanoemulsions for agri-food applications: Today and future trends. In *Bio-Based Nanoemulsions for Agri-Food Applications* (pp. 1–11). Elsevier., DOI: 10.1016/B978-0-323-89846-1.00012-7

Nadporozhskaya, M., Kovsh, N., Paolesse, R., & Lvova, L. (2022). Recent Advances in Chemical Sensors for Soil Analysis: A Review. *Chemosensors (Basel, Switzerland)*, *10*(1), 35. DOI: 10.3390/chemosensors10010035

Narong, D. K., & Hallinger, P. (2023). A Keyword Co-Occurrence Analysis of Research on Service Learning: Conceptual Foci and Emerging Research Trends. *Education Sciences*, *13*(4), 339. DOI: 10.3390/educsci13040339

Navarro, D. A., Kookana, R. S., McLaughlin, M. J., & Kirby, J. K. (2016). Fullerol as a potential pathway for mineralization of Fullerene nanoparticles in biosolid-amended soils. *Environmental Science & Technology Letters*, *3*(1), 7–12. DOI: 10.1021/acs.estlett.5b00292

Nayana, A. R., Joseph, B. J., Jose, A., & Radhakrishnan, E. K. (2020). Nanotechnological advances with PGPR applications. In Hayat, S., Pichtel, J., Faizan, M., & Fariduddin, Q. (Eds.), *Sustainable Agriculture Reviews* (Vol. 41). Springer., DOI: 10.1007/978-3-030-33996-8_9

Naz, S., Gul, A., & Zia, M. (2019). Toxicity of copper oxide nanoparticles: A review study. *IET Nanobiotechnology / IET*, *14*(1), 1–13. DOI: 10.1049/iet-nbt.2019.0176 PMID: 31935671

Negacz, K., Malek, Z., De Vos, A., & Vellinga, P. (2022). Saline soils worldwide: Identifying the most promising areas for saline agriculture. *Journal of Arid Environments*, *203*, 104775. DOI: 10.1016/j.jaridenv.2022.104775

Nielsen, M. N., & Winding, A. (2002). *Microorganisms as indicators of soil health* (Technical Report No. 388). National Environmental Research Institute, Denmark.

Ninkov, A. B., Frank, J. R., Costello, J. A., Artino, A. R.Jr, & Maggio, L. A. (2024). Bibliometric Networks for Researchers in Health Professions Education. *Academic Medicine*, *99*(9), 1048. DOI: 10.1097/ACM.0000000000005700 PMID: 38478805

Niño-Martínez, N., Salas Orozco, M. F., Martínez-Castañón, G. A., Torres Méndez, F., & Ruiz, F. (2019). Molecular mechanisms of bacterial resistance to metal and metal oxide nanoparticles. *International Journal of Molecular Sciences*, *20*(11), 2808. DOI: 10.3390/ijms20112808 PMID: 31181755

Nongbet, A., Mishra, A. K., Mohanta, Y. K., Mahanta, S., Ray, M. K., Khan, M., Baek, K.-H., & Chakrabartty, I. (2022). Nanofertilizers: A Smart and Sustainable Attribute to Modern Agriculture. *Plants*, *11*(19), 2587. DOI: 10.3390/plants11192587 PMID: 36235454

Novak, J. M., Busscher, W. J., Laird, D. L., Ahmedna, M., Watts, D. W., & Niandou, M. A. (2009). Impact of biochar amendment on fertility of a southeastern coastal plain soil. *Soil Science*, *174*(2), 105–112. DOI: 10.1097/SS.0b013e3181981d9a

Novak, J. M., Lima, I., Xing, B., Gaskin, J. W., Steiner, C., Das, K. C., & Schomberg, H. (2009). Characterization of designer biochar produced at different temperatures and their effects on a loamy sand. *Annals of Environmental Science (Boston, Mass.)*, *3*(1), 2.

Novaković, L., Guo, T., Bacic, A., Sampathkumar, A., & Johnson, K. L. (2018). Hitting the Wall—Sensing and Signaling Pathways Involved in Plant Cell Wall Remodeling in Response to Abiotic Stress. *Plants*, *7*(4), 89. DOI: 10.3390/plants7040089 PMID: 30360552

Nunes, M. R., Veum, K. S., Parker, P. A., Holan, S. H., Amsili, J. P., van Es, H. M., Wills, S. A., Seybold, C. A., & Karlen, D. L. (2024). SHAPEv1.0 scoring curves and peer group benchmarks for dynamic soil health indicators. *Soil Science Society of America Journal*, *88*(3), 858–875. DOI: 10.1002/saj2.20668

Nuwarapaksha, T. D., Dissanayake, W. K., Gunathilaka, W. S., Udumann, S. S., Dissanayaka, N. S., & Atapattu, A. J. (2024). Assessing the Optimum Harvesting Stage of *Tithonia diversifolia* as Climate Smart Soil Amendment for Coconut Plantations. *Biol. Life Sci. Forum, 30*, 1. DOI: 10.3390/IOCAG2023-15967

Nuwarapaksha, T. D., Udumann, S. S., Dissanayaka, D. M. N. S., Dissanayake, D. K. R. P. L., & Atapattu, A. J. (2022). Coconut based multiple cropping systems: An analytical review in Sri Lankan coconut cultivations. *Circular Agricultural Systems*, *2*(1), 8. DOI: 10.48130/CAS-2022-0008

Nuwarapaksha, T. D., Udumann, S. S., Dissanayaka, N. S., & Atapattu, A. J. (2023). Coconut-Based Livestock Farming: A Sustainable Approach to Enhancing Food Security in Sri Lanka. In Kiba, D. I. (Ed.), *Transitioning to Zero Hunger* (pp. 197–213). MDPI Books., DOI: 10.3390/books978-3-03897-863-3-8

Ogawa, M., & Okimori, Y. (2010). Pioneering works in biochar research, Japan. *Soil Research (Collingwood, Vic.)*, *48*(7), 489–500. DOI: 10.1071/SR10006

Ok, Y. S., Chang, S. X., Gao, B., & Wang, H. (2021). Nanobiochar: A new emerging material for environmental remediation and agricultural applications. *Environmental Science. Nano*, *8*(6), 1436–1437.

Omanovic-miklicanin, E., & Science, F. (2017). Green Inter-net of Things and Green Nanotechnology Rolein agriculture. VIII international scientific agri-culture symposium "AGROSYM 2017

Onay, O. (2007). Influence of pyrolysis temperature and heating rate on the production of bio-oil and char from safflower seed by pyrolysis, using a well-swept fixed-bed reactor. *Fuel Processing Technology*, *88*(5), 523–531. DOI: 10.1016/j fuproc.2007.01.001

Operti, M. C., Bernhardt, A., Grimm, S., Engel, A., Figdor, C. G., & Tagit, O. (2021). PLGA-based nanomedicines manufacturing: Technologies overview and challenges in industrial scale-up. *International Journal of Pharmaceutics*, *605*, 120807. DOI: 10.1016/j.ijpharm.2021.120807 PMID: 34144133

Osmond, M. J., & Mccall, M. J. (2010). Zinc oxide nanoparticles in modernsun-screens: An analysis of potential exposure and hazard. *Nanotoxicology*, *4*(1), 15–41. DOI: 10.3109/17435390903502028 PMID: 20795900

Ouyang, K., Yu, X. Y., Zhu, Y., Gao, C., Huang, Q., & Cai, P. (2017). Effects of humic acid on the interactions between zinc oxide nanoparticles and bacterial bio-films. *Environmental Pollution*, *231*, 1104–1111. DOI: 10.1016/j.envpol.2017.07.003 PMID: 28851497

Pan, B., & Xing, B. (2012). Applications and implications of manufactured nanopar-ticles in soils: A review. *European Journal of Soil Science*, *63*(4), 437–456. DOI: 10.1111/j.1365-2389.2012.01475.x

Pandey, S., Sharma, K., & Gundabala, V. (2022). Antimicrobial bio-inspired active packaging materials for shelf life and safety development: A review. *Food Bioscience*, *48*, 101730. DOI: 10.1016/j.fbio.2022.101730

Parameswari, P., Belagalla, N., Singh, B. V., Abhishek, G. J., Rajesh, G. M., Katiyar, D., Hazarika, B., & Paul, S. (2024). Nanotechnology-based sensors for real-time monitoring and assessment of soil health and quality: A review. *Asian Journal of Soil Science and Plant Nutrition*, *10*(2), 157–173. DOI: 10.9734/ajsspn/2024/v10i2272

Patil, R. R., Kumar, S., Rani, R., Agrawal, P., & Pippal, S. K. (2023). A Bibliomet-ric and Word Cloud Analysis on the Role of the Internet of Things in Agricultural Plant Disease Detection. *Applied System Innovation 2023, Vol. 6, Page 27, 6*(1), 27. DOI: 10.3390/asi6010027

Patni, B., Bhattacharyya, M., Kumari, A., & Purohit, V. K. (2022). Alarming influence of climate change and compromising quality of medicinal plants. *Plant Physiology Reports*, *27*(1), 1–10. DOI: 10.1007/s40502-021-00616-x

Paul, S. K., Mazumder, S., & Naidu, R. (2024). Herbicidal weed management practices: History and future prospects of nanotechnology in an eco-friendly crop production system. *Heliyon*, *10*(5), e26527. DOI: 10.1016/j.heliyon.2024.e26527 PMID: 38444464

Peng, C., Duan, D., Xu, C., Chen, Y., Sun, L., Zhang, H., Yuan, X., Zheng, L., Yang, Y., Yang, J., Zhen, X., Chen, Y., & Shi, J. (2015). Translocation and biotransformation of CuO nanoparticles in rice. *Environmental Pollution*, *197*, 99–107. DOI: 10.1016/j.envpol.2014.12.008 PMID: 25521412

Peralta-Videa, J. R., Hernandez-Viezcas, J. A., Zhao, L., Diaz, B. C., Ge, Y., Priester, J. H., & Gardea-Torresdey, J. L. (2014). CeO_2 and ZnO nanoparticles alter soybean nutrition. *Plant Physiology and Biochemistry*, *80*, 128–135. DOI: 10.1016/j.plaphy.2014.03.028 PMID: 24751400

Peyrot, C., Wilkinson, K. J., Desrosiers, M., & Sauvé, S. (2014). Effects of silver nanoparticles on soil enzyme activities with and without added organic matter. *Environmental Toxicology and Chemistry*, *33*(1), 115–125. DOI: 10.1002/etc.2398 PMID: 24115203

Pillay, V., Murugan, K., Choonara, Y. E., Kumar, P., Bijukumar, D., & du Toit, L. C. (2015). Parameters and characteristics governing cellular internalization and trans-barrier trafficking of nanostructures. *International Journal of Nanomedicine*, *2191*, 2191. Advance online publication. DOI: 10.2147/IJN.S75615

Pradhan, A., Seena, S., Pascoal, C., & Cássio, F. (2011). Can metal nanoparticles be a threat to microbial decomposers of plant litter in streams? *Microbial Ecology*, *62*(1), 58–68. DOI: 10.1007/s00248-011-9861-4 PMID: 21553058

Pradhan, N., Singh, S., Ojha, N., Shrivastava, A., Barla, A., Rai, V., & Bose, S. (2015). Facets of Nanotechnology as Seen in Food Processing, Packaging, and Preservation Industry. *BioMed Research International*, *2015*, 1–17. DOI: 10.1155/2015/365672 PMID: 26613082

Pradhan, S., & Mailapalli, D. R. (2020). *Nanopesticides for Pest Control.*, DOI: 10.1007/978-3-030-33281-5_2

Praetorius, A., Labille, J., Scheringer, M., Thill, A., Hungerbuhler, K., & Bottero, J. Y. (2014). Heteroaggregation of titanium dioxide nanoparticles with model natural colloids under environmentally relevant conditions. *Environmental Science & Technology*, *48*(18), 10690–10698. DOI: 10.1021/es501655v PMID: 25127331

Prakash, P., & S, S. C. (2023). Nano-phytoremediation of heavy metals from soil: A critical review. *Pollutants*, *3*(3), 360–380. DOI: 10.3390/pollutants3030025

Prasad, R., Bhattacharyya, A., & Nguyen, Q. D. (2017). Nanotechnology in sustainable agriculture: Recent developments, challenges, and perspectives. *Frontiers in Microbiology*, *8*, 1014. DOI: 10.3389/fmicb.2017.01014 PMID: 28676790

Prasad, R., Kumar, M., & Kumar, V. (Eds.). (2017). *Nanotechnology*. Springer Singapore., DOI: 10.1007/978-981-10-4573-8

Pulido-Reyes, G., Rodea-Palomares, I., Das, S., Sakthivel, T. S., Leganes, F., Rosal, R., Seal, S., & Fernández-Piñas, F. (2015). Untangling the bio-logical effects of cerium oxide nanoparticles: The role of surfacevalence states. *Scientific Reports*, *5*(1), 15613. DOI: 10.1038/srep15613 PMID: 26489858

Qian, K., Kumar, A., Zhang, H., Bellmer, D., & Huhnke, R. (2015). Recent advances in utilization of biochar. *Renewable & Sustainable Energy Reviews*, *42*, 1055–1064. DOI: 10.1016/j.rser.2014.10.074

Quilliam, R. S., Glanville, H. C., Wade, S. C., & Jones, D. L. (2013). The Trojan-horse effect: Biochar systems for nutrient co-delivery in soil. *Global Change Biology. Bioenergy*, *5*(2), 202–205.

Quintarelli, V., Ben Hassine, M., Radicetti, E., Stazi, S. R., Bratti, A., Allevato, E., Mancinelli, R., Jamal, A., Ahsan, M., Mirzaei, M., & Borgatti, D. (2024). Advances in Nanotechnology for Sustainable Agriculture: A Review of Climate Change Mitigation. *Sustainability 2024, Vol. 16, Page 9280, 16*(21), 9280. DOI: 10.3390/su16219280

Raffi, M. M., & Husen, A. (2019). Impact of Fabricated Nanoparticles on the Rhizospheric Microorganisms and Soil Environment. In *Nanomaterials and Plant Potential* (pp. 529–552). Springer International Publishing., DOI: 10.1007/978-3-030-05569-1_21

Rahman, K., & Zhang, D. (2018). Effects of Fertilizer Broadcasting on the Excessive Use of Inorganic Fertilizers and Environmental Sustainability. *Sustainability (Basel)*, *10*(3), 759. DOI: 10.3390/su10030759

Rahman, M., Haque, K. M. S., & Khan, M. Z. H. (2021). A review on application of controlled released fertilizers influencing the sustainable agricultural production: A Cleaner production process. *Environmental Technology & Innovation*, *23*, 101697. DOI: 10.1016/j.eti.2021.101697

Rai-Kalal, P., & Jajoo, A. (2021). ZnO nanoparticle priming improves wheat photosynthesis. *Plant Physiology and Biochemistry*, *160*, 341–351. DOI: 10.1016/j.plaphy.2021.01.032 PMID: 33548801

Rai, M., Ingle, A. P., Pandit, R., Paralikar, P., Shende, S., Gupta, I., Biswas, J. K., & da Silva, S. S. (2018). Copper and copper nanoparticles: Role in management of insect-pests and pathogenic microbes. *Nanotechnology Reviews*, *7*(4), 303–315. DOI: 10.1515/ntrev-2018-0031

Raja, M. A., & Husen, A. (2020). Role of nanomaterials in soil and water quality management. In *Nanomaterials for agriculture and forestry applications* (pp. 491–503). Elsevier. DOI: 10.1016/B978-0-12-817852-2.00020-2

Rajapaksha, A. U., Chen, S. S., Tsang, D. C., Zhang, M., Vithanage, M., Mandal, S., Gao, B., Bolan, N. S., & Ok, Y. S. (2016). Engineered/designer biochar for contaminant removal/immobilization from soil and water: Potential and implication of biochar modification. *Chemosphere*, *148*, 276–291. DOI: 10.1016/j.chemosphere.2016.01.043 PMID: 26820777

Rajput, V. D., Singh, A., Ghazaryan, K., Alexiou, A., & Minkina, T. M. (2025). Biogenic nanoparticles. In *Elsevier eBooks*. https://doi.org/DOI: 10.1016/C2024-0-00214-7

Rajput, V. D., Minkina, T., Mazarji, M., Shende, S., Sushkova, S., Mandzhieva, S., Burachevskaya, M., Chaplygin, V., Singh, A., & Jatav, H. (2020). Accumulation of nanoparticles in soil–plant systems and their effects on human health. *Annals of Agricultural Science*, *65*(2), 137–143. DOI: 10.1016/j.aoas.2020.08.001

Rajput, V. D., Minkina, T., Sushkova, S., Tsitsuashvili, V., Mandzhieva, S., Gorovtsov, A., Nevidomskyaya, D., & Gromakova, N. (2018). Effects of nanoparticles on crops and soil microbial communities. *Journal of Soils and Sediments*, *18*(6), 2179–2187. DOI: 10.1007/s11368-017-1793-2

Rajput, V. D., Singh, A., Ghazaryan, K., Minkina, T. M., & Al-Tawaha, A. R. M. (Eds.). (2024). *Sustainable agriculture: nanotechnology and biotechnology for crop production and protection*. Walter de Gruyter GmbH & Co KG.

Rajput, V. D., Singh, A., Minkina, T. M., Verma, K. K., & Singh, A. K. (2023). *Nanotechnology for sustainable agriculture: An innovative and eco-friendly approach* (1st ed.). Apple Academic Press., DOI: 10.1201/9781003333128

Ramezani, M., Ramezani, F., & Gerami, M. (2019). Nanoparticles in Pest Incidences and Plant Disease Control. In *Nanotechnology for Agriculture: Crop Production & Protection* (pp. 233–272). Springer Singapore. DOI: 10.1007/978-981-32-9374-8_12

Rani, M., Kaushik, P., Bhayana, S., & Kapoor, S. (2023). Impact of organic farming on soil health and nutritional quality of crops. *Journal of the Saudi Society of Agricultural Sciences*, *22*(8), 560–569. DOI: 10.1016/j.jssas.2023.07.002

Rani, N., Kumari, K., Sangwan, P., Barala, P., Yadav, J., Vijeta, , Rahul, , & Hooda, V. (2022). Nano-Iron and Nano-Zinc Induced Growth and Metabolic Changes in Vigna radiata. *Sustainability (Basel)*, *14*(14), 8251. DOI: 10.3390/su14148251

Rao, S., & Shekhawat, G. S. (2016). Phytotoxicity and oxidative stress of nanoparticles in *Brassica juncea. 3 Biotech, 6*(2), 244. https://doi.org/DOI: 10.1007/s13205-016-0565-0

Rashid, A., Schutte, B. J., Ulery, A., Deyholos, M. K., Sanogo, S., Lehnhoff, E. A., & Beck, L. (2023). Heavy metal contamination in agricultural soil: Environmental pollutants affecting crop health. *Agronomy (Basel), 13*(6), 1521. DOI: 10.3390/agronomy13061521

Raza, A., Charagh, S., Salehi, H., Abbas, S., Saeed, F., Poinern, G. E. J., Siddique, K. H. M., & Varshney, R. K. (2023). Nano-enabled stress-smart agriculture: Can nanotechnology deliver drought and salinity-smart crops? *Journal of Sustainable Agriculture and Environment, 2*(3), 189–214. DOI: 10.1002/sae2.12061

Raza, M. A. S., Amin, J., Valipour, M., Iqbal, R., Aslam, M. U., Zulfiqar, B., & Elsalahy, H. H. (2024). Copper nanoparticles enhance drought tolerance in wheat. *Scientific Reports, 14*, 14254. DOI: 10.1038/s41598-024-62680-1 PMID: 38902296

Reidy, B., Haase, A., Luch, A., Dawson, K. A., & Lynch, I. (2013). Mechanisms of silver nanoparticle release, transformation and toxicity: A critical review of current knowledge and recommendations for future studies and applications. *Materials (Basel), 6*(6), 2295–2350. DOI: 10.3390/ma6062295 PMID: 28809275

Reinhold-Hurek, B., Bünger, W., Burbano, C. S., Sabale, M., & Hurek, T. (2015). Roots shaping their microbiome. *Annual Review of Phytopathology, 53*(1), 403–424. DOI: 10.1146/annurev-phyto-082712-102342 PMID: 26243728

Rico, C. M., Lee, S. C., Rubenecia, R., Mukherjee, A., Hong, J., Peralta-Videa, J. R., & Gardea-Torresdey, J. L. (2014). Cerium oxide nanoparticles impact yield and modify nutritional parameters in wheat (Triticum aestivum L.). *Journal of Agricultural and Food Chemistry, 62*(40), 9669–9675. DOI: 10.1021/jf503526r PMID: 25220448

Riding, M. J., Martin, F. L., Jones, K. C., & Semple, K. T. (2015). Carbon nanomaterials in clean and contaminated soils: Environmental implications and applications. *Soil (Göttingen), 1*(1), 1–21. DOI: 10.5194/soil-1-1-2015

Rillig, M. C., Wagner, M., Salem, M., Antunes, P. M., George, C., Ramke, H. G., & Zhl, J. (2010). Material derived from hydrothermal carbonization: Effects on plant growth and arbuscular mycorrhiza. *Applied Soil Ecology, 45*(3), 199–204. DOI: 10.1016/j.apsoil.2010.04.011

Rios, J. J., Martínez-Ballesta, M. C., Ruiz, J. M., Blasco, B., & Carvajal, M. (2017). Silicon-mediated salinity tolerance. *Frontiers in Plant Science, 8*, 948. DOI: 10.3389/fpls.2017.00948 PMID: 28642767

Rizwan, A., Zia-ur-Rehman, M., Rizwan, M., Usman, M., Anayatullah, S., Areej, , Alharby, H. F., Bamagoos, A. A., Alharbi, B. M., & Ali, S. (2023). Effects of silicon nanoparticles and conventional Si amendments on growth and nutrient accumulation by maize (Zea mays L.) grown in saline-sodic soil. *Environmental Research, 227*, 115740. DOI: 10.1016/j.envres.2023.115740 PMID: 36997044

Roberts, K. G., Gloy, B. A., Joseph, S., Scott, N. R., & Lehmann, J. (2010). Life cycle assessment of biochar systems: Estimating the energetic, economic, and climate change potential. *Environmental Science & Technology*, *44*(2), 827–833. DOI: 10.1021/es902266r PMID: 20030368

Rondon, M. A., Lehmann, J., Ramírez, J., & Hurtado, M. (2007). Biological nitrogen fixation by common beans (Phaseolus vulgaris L.) increases with bio-char additions. *Biology and Fertility of Soils*, *43*(6), 699–708. DOI: 10.1007/s00374-006-0152-z

Ronsse, F., Van Hecke, S., Dickinson, D., & Prins, W. (2013). Production and characterization of slow pyrolysis biochar: Influence of feedstock type and pyrolysis conditions. *Global Change Biology. Bioenergy*, *5*(2), 104–115. DOI: 10.1111/gcbb.12018

Rubin, R., Oldfield, E., Lavallee, J., Griffin, T., Mayers, B., & Sanderman, J. (2023). Climate mitigation through soil amendments: Quantification, evidence, and uncertainty. *Carbon Management*, *14*(1), 1. DOI: 10.1080/17583004.2023.2217785

Rudrappa, T., Biedrzycki, M. L., & Bais, H. P. (2008). Causes and consequences of plant-associated biofilms. *FEMS Microbiology Ecology*, *64*(2), 153–166. DOI: 10.1111/j.1574-6941.2008.00465.x PMID: 18355294

Rui, M., Ma, C., Hao, Y., Guo, J., Rui, Y., Tang, X., & Zhu, S. (2016). Iron oxide nanoparticles as iron fertilizer for peanut. *Frontiers in Plant Science*, *7*, 815. DOI: 10.3389/fpls.2016.00815 PMID: 27375665

Ruttkay-Nedecky, B., Krystofova, O., Nejdl, L., & Adam, V. (2017). Nanoparticles based on essential metals and their phytotoxicity. *Journal of Nanobiotechnology*, *15*(1), 33. DOI: 10.1186/s12951-017-0268-3 PMID: 28446250

Saadi, M. A. S. R., Uluutku, B., Parvini, C. H., & Solares, S. D. (2020). Soft sample deformation, damage and induced electromechanical property changes in contact- A nd tapping-mode atomic force microscopy. *Surface Topography : Metrology and Properties*, *8*(4), 045004. Advance online publication. DOI: 10.1088/2051-672X/abb888

Saikanth, D. R. K., Singh, B. V., Rai, A. K., Kumar, U. S., Surender, , Yadav, B., & Singh, O. (2023). Biochar implementation in rice paddies for addressing greenhouse gas emissions and nutrient loss: A review. *International Journal of Plant and Soil Science*, *35*(18), 610–623. DOI: 10.9734/ijpss/2023/v35i183326

Sajid, M. (2022). Nanomaterials: Types, properties, recent advances, and toxicity concerns. *Current Opinion in Environmental Science & Health*, *25*, 100319. DOI: 10.1016/j.coesh.2021.100319

Salama, H. M. H. (2012). Effects of silver nanoparticles in some crop plants, Common bean (*Phaseolus vulgaris* L.) and corn *(Zea mays L.)*. In *International. Research Journal of Biotechnology*, *3*, 10.

Saleeb, N., Robinson, B., Cavanagh, J., Hossain, A. K. M., & Gooneratne, R. (2019). Biochemical changes in sunflower plant exposed to silver nanoparticles/silver ions.

Salonen, V. P., & Korkka-Niemi, K. (2007). Influence of parent sediments on the concentration of heavy metals in urban and suburban soils in Turku, Finland. *Applied Geochemistry*, *22*(5), 906–918. DOI: 10.1016/j.apgeochem.2007.02.003

Santos-Francés, F., Martínez-Graña, A., Rojo, P. A., and Sánchez, A. G. (2017). Geochemical Background and Baseline Values Determination and Spatial Distribution of Heavy Metal Pollution in Soils of the Andes Mountain Range (Cajamarca-Huancavelica, Peru). *International Journal of Environmental Research and Public Health 2017, Vol. 14, Page 859* 14, 859. DOI: 10.3390/ijerph14080859

Santos-Francés, F., García-Sánchez, A., Alonso-Rojo, P., Contreras, F., & Adams, M. (2011). Distribution and mobility of mercury in soils of a gold mining region, Cuyuni river basin, Venezuela. *Journal of Environmental Management*, *92*(4), 1268–1276. DOI: 10.1016/j.jenvman.2010.12.003 PMID: 21215510

Schimmelpfennig, S., & Glaser, B. (2012). One step forward toward characterization: Some important material properties to distinguish biochars. *Journal of Environmental Quality*, *41*(4), 1001–1013. DOI: 10.2134/jeq2011.0146 PMID: 22751042

Schlich, K., & Hund-Rinke, K. (2015). Influence of soil properties on the effect of silver nanomaterials on microbial activity in five soils. *Environmental Pollution*, *196*, 321–330. DOI: 10.1016/j.envpol.2014.10.021 PMID: 25463729

Schmidt, H. P., Pandit, B. H., Cornelissen, G., & Kammann, C. I. (2017). Biochar-based fertilization with liquid nutrient enrichment: 21 field trials covering 13 crop species in Nepal. *Land Degradation & Development*, *28*(8), 2324–2342. DOI: 10.1002/ldr.2761

Schwab, F., Zhai, G., Kern, M., Turner, A., Schnoor, J. L., & Wiesner, M. R. (2016). Barriers, pathways and processes for uptake, translocation and accumulation of nanomaterials in plants – Critical review. *Nanotoxicology*, *10*(3), 257–278. DOI: 10.3109/17435390.2015.1048326 PMID: 26067571

Sedlak, H. R., Hnilova, M., Grosh, C., Fong, H., Baneyx, F., Schwartz, D., Sarikaya, M., Tamerler, C., & Traxler, B. (2012). Engineered Escherichia coli silver-binding periplasmic protein that promotes silver tolerance. *Applied and Environmental Microbiology*, *78*(7), 2289–2296. DOI: 10.1128/AEM.06823-11 PMID: 22286990

Sekhon, B. (2014). Nanotechnology in agri-food production: An overview. *Nanotechnology, Science and Applications*, *31*, 31. Advance online publication. DOI: 10.2147/NSA.S39406 PMID: 24966671

Selvaganapathi, K., Kumar, K. A., Bhople, B. S., Singh, V. K., Jayanthi, J., Singh, O., Bárek, V., Alshehri, M. A., Gaber, A., & Hossain, A. (2025, September). **Singh, O.**, Bárek, V., Alshehri, M. A., Gaber, A., & Hossain, A. (2025). Boosting soil nutrients, carbon storage, and wheat yield with co-compost organic fertilizers and growth-promoting microorganisms in Typic Haplustept soils. *Journal of Soil Science and Plant Nutrition*, *25*(3), 6592–6607. Advance online publication. DOI: 10.1007/s42729-025-02551-4

Sembada, A. A., & Lenggoro, I. W. (2024). Transport of Nanoparticles into Plants and Their Detection Methods. *Nanomaterials (Basel, Switzerland)*, *14*(2), 131. DOI: 10.3390/nano14020131 PMID: 38251096

Seneviratne, M., Gunaratne, S., Bandara, T., Weerasundara, L., Rajakaruna, N., Seneviratne, G., & Vithanage, M. (2016). Plant growth promotion by Bradyrhizobium japonicum under heavy metal stress. *South African Journal of Botany*, *105*, 19–24. DOI: 10.1016/j.sajb.2016.02.206

Serri, C., Cruz-Maya, I., Bonadies, I., Rassu, G., Giunchedi, P., Gavini, E., & Guarino, V. (2023). Green Routes for Bio-Fabrication in Biomedical and Pharmaceutical Applications. *Pharmaceutics*, *15*(6), 1744. DOI: 10.3390/pharmaceutics15061744 PMID: 37376192

Sevilla, M., & Fuertes, A. B. (2009). The production of carbon materials by hydrothermal carbonization of cellulose. *Carbon*, *47*(9), 2281–2289. DOI: 10.1016/j.carbon.2009.04.026

Shabangu, S., Woolf, D., Fisher, E. M., Angenent, L. T., & Lehmann, J. (2014). Techno-economic assessment of biomass slow pyrolysis into different biochar and methanol concepts. *Fuel*, *117*, 742–748. DOI: 10.1016/j.fuel.2013.08.053

Shackley, S., Hammond, J., Gaunt, J., & Ibarrola, R. (2011). The feasibility and costs of biochar deployment in the UK. *Carbon Management*, *2*(3), 335–356. DOI: 10.4155/cmt.11.22

Shahi, U. P., Singh, O., Singh, V. K., Shivangi, P. K., Singh, R., & Singh, V. K. Anand Singh, Rajput, V. D., Rajput, V. D., Singh, A., Ghazaryan, K. A., & Al Tawaha, A. R. M. (2024). Nanotechnology in rice farming: Optimizing nutrient management with nanofertilizers. In: Sustainable Agriculture: Nanotechnology and Biotechnology for Crop Production and Protection. Walter de Gruyter GmbH. DOI: DOI: 10.1515/9783111234694-003

Shah, S. S., Niaz, F., Ehsan, M. A., Das, H. T., Younas, M., Khan, A. S., Rahman, H. U., Nayem, S. M. A., Oyama, M., & Aziz, M. A. (2024). Advanced strategies in electrode engineering and nanomaterial modifications for supercapacitor performance enhancement: A comprehensive review. *Journal of Energy Storage*, *79*, 110152. DOI: 10.1016/j.est.2023.110152

Shanmugavel, D., Rusyn, I., Solorza-Feria, O., & Kamaraj, S.-K. (2023). Sustainable SMART fertilizers in agriculture systems: A review on fundamentals to in-field applications. *The Science of the Total Environment*, *904*, 166729. DOI: 10.1016/j.scitotenv.2023.166729 PMID: 37678530

Sharmaa, K., Khurana, P., & Uddina, Z. (2025). MetaInfoSci: An Integrated Web Tool for Scholarly Data Analysis. Available at: https://arxiv.org/pdf/2506.09056 (Accessed January 18, 2026).

Sharma, A., Singh, R. N., Song, X. P., Singh, R. K., Guo, D. J., Singh, P., Verma, K. K., & Li, Y.-R. (2023). Genome analysis of a halophilic *Virgibacillus halodenitrificans* ASH15. *Frontiers in Microbiology*, *14*, 1229955. DOI: 10.3389/fmicb.2023.1229955 PMID: 37808307

Sharma, R., & Kumar, V. (2024). Nano enabled agriculture for sustainable soil. *Waste Management Bulletin*, *2*(1), 152–161. DOI: 10.1016/j.wmb.2024.01.002

Shen, M., Zhang, H., Ding, Y., & Tang, Z. (2022). Biochar nanoparticles as a potential tool to mitigate abiotic stress in plants: A review. *Environmental Pollution*, *293*, 118536.

Shen, Z., Chen, Z., Hou, Z., Li, T., & Lu, X. (2015). Ecotoxicological effect of zinc oxide nanoparticles on soil microorganisms. *Frontiers of Environmental Science & Engineering*, *9*(5), 912–918. DOI: 10.1007/s11783-015-0789-7

Shen, Z., Som, A. M., Wang, F., Jin, F., McMillan, O., & Al-Tabbaa, A. (2016). Long-term impact of biochar on the immobilisation of nickel (II) and zinc (II) and the revegetation of a contaminated site. *The Science of the Total Environment*, *542*, 771–776. DOI: 10.1016/j.scitotenv.2015.10.057 PMID: 26551277

Shi, H., Magaye, R., Castranova, V., & Zhao, J. (2013). Titanium dioxide nanoparticles: A review of current toxicological data. *Particle and Fibre Toxicology*, *10*(1), 15. DOI: 10.1186/1743-8977-10-15 PMID: 23587290

Siddiqui, H., Ahmed, K. B. M., Faizan, M., Baran, M. F., Alam, P., & Hayat, S. (2025). Pre-Sowing Seed Treatment with Glucose Boosts the Photosynthetic Efficiency, Metabolic Enzymes and Antioxidant Defense System to increase the Growth of <i>Abelmoschus esculentus</i> (L.) Moench. *Journal of Innovative Solutions for Eco-Environmental Sustainability*, *1*(1), 13854. DOI: 10.46991/JISEES.2025.01.1.13854

Singh, A., Bol, R., Lovynska, V., Singh, S., Singh, R. K., Sousa, J. R., Elshikh, M. S., Khachatryan, H., & Ghazaryan, K. (2025c). Nanoparticles: Dual role in alleviating abiotic stresses and boosting nutrient efficiency for sustainable agriculture. In *Advances in agronomy* (pp. 117–180). https://doi.org/DOI: 10.1016/bs.agron.2025.08.001

Singh, A., Sengar, R. S., Rajput, V. D., Minkina, T., & Singh, R. K. (2022). Zinc Oxide Nanoparticles Improve Salt Tolerance in Rice Seedlings by Improving Physiological and Biochemical Indices. *Agriculture 2022, Vol. 12, Page 1014, 12*(7), 1014. DOI: 10.3390/agriculture12071014

Singh, S., Singh, A., Chakhmakhchyan, A., Hayrapetyan, A., Darbinyan, N., & Ghazaryan, K. (2025). Global Research Trends on Drought Stress and Nanoparticles (2012–2025): A Bibliometric Analysis. *Journal of Innovative Solutions for Eco-Environmental Sustainability, 18*, 018–018. DOI: 10.46991/JISEES.2025.SI1.018

Singh, A., Bol, R., Lovynska, V., Singh, R. K., Sousa, J. R., & Ghazaryan, K. (2025). Application of nanoparticles for salinity stress management and biofortification in wheat: A review of dual approaches and insights. *Frontiers in Plant Science*, *16*, 1592866. Advance online publication. DOI: 10.3389/fpls.2025.1592866 PMID: 40688693

Singh, A., Rawat, S., Rajput, V. D., Minkina, T., Mandzhieva, S., Eloyan, A. S., Singh, R. K., Singh, O., El-Ramady, H., & Ghazaryan, K. (2024a). Nanotechnology products in agriculture and environmental protection: Advances and challenges. *Egyptian Journal of Soil Science*, *64*(4), 1355–1378. DOI: 10.21608/ejss.2024.300047.1802

Singh, A., Sharma, A., Singh, O., Rajput, V. D., Movsesyan, H., Minkina, T., Alexiou, A., Papadakis, M., Singh, R. K., Singh, S., Sousa, J. R., El-Ramady, H. R., Zulfiqar, F., Kumar, R., Al-Ghamdi, A. A., & Ghazaryan, K. (2024). In-depth exploration of nanoparticles for enhanced nutrient use efficiency and abiotic stresses management: Present insights and future horizons. *Plant Stress*, *14*, 100576. DOI: 10.1016/j.stress.2024.100576

Singh, B. P., Cowie, A. L., & Smernik, R. J. (2015). Biochar carbon stability in a clayey soil as a function of feedstock and pyrolysis temperature. *Environmental Science & Technology*, *46*(21), 11770–11778. DOI: 10.1021/es302545b PMID: 23013285

Singh, B., Singh, B. P., & Cowie, A. L. (2010). Characterisation and evaluation of biochars for their application as a soil amendment. *Soil Research (Collingwood, Vic.)*, *48*(7), 516–525. DOI: 10.1071/SR10058

Singh, M., Goswami, S. P., Sachan, P., Sahu, D. K., Beese, S., & Pandey, S. K. (2024). Nanotech for Fertilizers and Nutrients-Improving Nutrient use Efficiency with Nano-Enabled Fertilizers. *Journal of Experimental Agriculture International*, *46*(5), 220–247. DOI: 10.9734/jeai/2024/v46i52372

Singh, N., Joshi, E., Singh, D., & Sasode, N. C. (2020). Application of nanotechnology in agriculture. *Research Today*, *2*, 163–165.

Singh, O. (2023). Nanotechnology for sustainability and food security in agriculture: A nanopriming long story in short. In Singh, A., Rajput, V., Ghazaryan, K., Gupta, S., & Minkina, T. (Eds.), *Nanopriming Approach to Sustainable Agriculture* (pp. 315–339). IGI Global., DOI: 10.4018/978-1-6684-7232-3.ch014

Singh, O., Pandey, D., Parmar, S. S., Minkina, T., & Rajput, P. (2025d). Emerging role of nanotechnology in biofortification, plant growth, and crop protection. *Journal of Innovative Solutions for Eco-Environmental Sustainability*, *1*(1), 13490. DOI: 10.46991/JISEES.2025.01.1.13490

Singh, O., Shahi, U.P., Dhyani, B.P., Kumar, S., Vivek, S., & Sengar, R.S., Shivangi, and Singh, A. (2023). Effect of zinc oxide nanoparticles application on growth and yield of basmati rice (Oryza sativa) in alkaline soil. *Indian Journal of Agronomy*, *68*(3), 1–6.

Singh, O., Singh, S., Shahi, U. P., Singh, R., Singh, P. K., Grigoryan, K. K., Khachatryan, H., Rajput, P., Perelomov, L., Kozmenko, S., & Burachevskaya, M. (2024). Nanotechnological Innovations for Soil Pollution Remediation and Environmental Conservation. In Rajput, V., Singh, A., Ghazaryan, K., Alexiou, A., & Said Al-Tawaha, A. (Eds.), *Harnessing NanoOmics and Nanozymes for Sustainable Agriculture* (pp. 237–255). IGI Global., DOI: 10.4018/979-8-3693-1890-4.ch012

Singh, O., Singh, S., Singh, V. K., & Singh, A. (2023). Biochar: An Organic Amendment for Sustainable Soil Health. In Baskar, C., Ramakrishna, S., & Daniela La Rosa, A. (Eds.), *Encyclopedia of Green Materials*. Springer., DOI: 10.1007/978-981-16-4921-9_265-1

Singh, S., Singh, O., & Shahi, U. P. (2023). Nanotechnology: A New Innovation in Ghazaryan, K., Pandey, D., Singh, S., Varagyan, V., Alexiou, A., Petropoulos, D., Kriemadis, A., Rajput, V., Minkina, T., Singh, R., Sousa, J., Kumar, S., El-Ramady, H., Singh, O., & Singh, A. (2024). Enhancing Crop Production: Unveiling the Role of Nanofertilizers in Sustainable Agriculture and Precision Nutrient Management. *Egyptian Journal of Soil Science*, *64*(3), 981–1007.

Singh, S., Singh, O., Shahi, U. P., Singh, P. K., Singh, A., Rajput, V. D., Minkina, T., El-Ramady, H., & Ghazaryan, K. (2024a). Carbon sequestration through organic amendments, clay mineralogy and agronomic practices: A review. *Egyptian Journal of Soil Science*, *64*(2), 581–598. DOI: 10.21608/ejss.2024.260719.1707

Smolders, E., & Degryse, F. (2002). Fate and effect of zinc from tire debris in soil. *Environmental Science & Technology*, *36*(17), 3706–3710. DOI: 10.1021/es025567p PMID: 12322741

Sohi, S. P., Krull, E., Lopez-Capel, E., & Bol, R. (2010). A review of biochar and its use and function in soil. *Advances in Agronomy*, *105*, 47–82. DOI: 10.1016/S0065-2113(10)05002-9

Sondi, I., & Salopek-Sondi, B. (2004). Silver nanoparticles as antimicrobial agent: A case study on E. coli as a model for Gram-negative bacteria. *Journal of Colloid and Interface Science*, *275*(1), 177–182. DOI: 10.1016/j.jcis.2004.02.012 PMID: 15158396

Soni, D., Gandhi, D., Tarale, P., Bafana, A., Pandey, R. A., & Sivanesan, S. (2017). Oxidative stress and genotoxicity of zinc oxide nanoparticles to Pseudomonas species, human promyelocytic leukemic (HL-60), and blood cells. *Biological Trace Element Research*, *178*(2), 218–227. DOI: 10.1007/s12011-016-0921-y PMID: 28058665

Son, K. H., Hong, J. H., & Lee, J. W. (2016). Carbon nanotubes as cancer thera-peutic carriers and mediators. *International Journal of Nanomedicine*, *11*, 5163–5185. DOI: 10.2147/IJN.S112660 PMID: 27785021

Sousa, J. R., Ferreira, J., Saraiva, F., Singh, R. K., Singh, A., Ghazaryan, K., & Trindade, H. (2025). Soil Amendments for the Remediation of Cu-Contaminated Vineyard Soil: Effects of Biochar, Nanobiochar and Chitosan on Copper Bioavailability and Ecotoxicity. *Journal of Innovative Solutions for Eco-Environmental Sustainability*, 033–033. DOI: 10.46991/JISEES.2025.SI1.033

Spokas, K. A. (2010). Review of the stability of biochar in soils: predictability of O: C molar ratios. *Carbon Management*, *1*(2), 289–303. DOI: 10.4155/cmt.10.32

Spokas, K. A., Novak, J. M., Stewart, C. E., Cantrell, K. B., Uchimiya, M., DuSaire, M. G., & Ro, K. S. (2011). Qualitative analysis of volatile organic compounds on biochar. *Chemosphere*, *85*(5), 869–882. DOI: 10.1016/j.chemosphere.2011.06.108 PMID: 21788060

Steffan, J. J., Derby, J. A., & Brevik, E. C. (2020). Soil pathogens that may potentially cause pandemics, including severe acute respiratory syndrome (SARS) coronaviruses. *Current Opinion in Environmental Science & Health*, *17*, 35–40. DOI: 10.1016/j.coesh.2020.08.005 PMID: 33521411

Steiner, C., Glaser, B., Teixeira, W. G., Lehmann, J., Blum, W. E., & Zech, W. (2008). Nitrogen retention and plant uptake on a highly weathered central Amazonian Ferralsol amended with compost and charcoal. *Journal of Plant Nutrition and Soil Science*, *171*(6), 893–899. DOI: 10.1002/jpln.200625199

Stiegler, L. M., Luchs, T., & Hirsch, A. (2020). Shell-by-Shell Functionalization of Inorganic Nanoparticles. *Chemistry (Weinheim an der Bergstrasse, Germany)*, *26*(39), 8483–8498. DOI: 10.1002/chem.202000195 PMID: 32167598

Stopar, K. (2016). Presence of nanotechnology in agriculture: Bibliometric approach. *Acta Agriculturae Slovenica*, *107*(2), 497–507. DOI: 10.14720/aas.2016.107.2.20

Subash, N., Dutta, D., Ghasal, P., Ravisankar, N., Chaudhary, V. P., Kumar, S., Meena, L. R., & Singh, O. (2023). Brahmdutt, Singh, S., & Sunita, K. (2023). A Composite Index to Assess the Climate-Carbon-Yield-Sustainability of Cereal Based Cropping System. *International Journal of Plant Production*, *17*(4), 729–755. DOI: 10.1007/s42106-023-00268-x

Subhan, M. A., & Subhan, T. (2022). Safety and global regulations for application of nanomaterials. In *Nanomaterials Recycling* (pp. 83–107). Elsevier. DOI: 10.1016/B978-0-323-90982-2.00005-6

Subramanian, V., Peijnenburg, W. J., Vijver, M. G., Blanco, C. F., Cucurachi, S., & Guinée, J. B. (2023). Approaches to implement safe by design in early product design through combining risk assessment and life cycle assessment. *Chemosphere*, *311*, 137080. DOI: 10.1016/j.chemosphere.2022.137080 PMID: 36328317

Sukhanova, A., Bozrova, S., Sokolov, P., Berestovoy, M., Karaulov, A., & Nabiev, I. (2018). Dependence of nanoparticle toxicity on their physical and chemical properties. *Nanoscale Research Letters*, *13*(1), 1–21. DOI: 10.1186/s11671-018-2457-x PMID: 29417375

Sun, K., Kang, M., Zhang, Z., Jin, J., Wang, Z., Pan, Z., & Xing, B. (2017). A new insight into the adsorption of wastewater-derived organic contaminants on biochar: A review. *Journal of Cleaner Production*, *148*, 590–602.

Sun, M., Yu, Q., Hu, M., Hao, Z., Zhang, C., & Li, M. (2014). Lead sulfide nanoparticles increase cell wall chitin content and induce apoptosis in Saccharomyces cerevisiae. *Journal of Hazardous Materials*, *273*, 7–16. DOI: 10.1016/j.jhazmat.2014.03.008 PMID: 24704549

Sun, X. D., Ma, J. Y., Feng, L. J., Duan, J. L., Xie, X. M., Zhang, X. H., Kong, X., Ding, Z., & Yuan, X. Z. (2023). Magnetite nanoparticle coating chemistry regulates root uptake. *Proceedings of the National Academy of Sciences of the United States of America*, *120*(27), e2304306120. DOI: 10.1073/pnas.2304306120 PMID: 37364127

Sun, Y., Gao, B., Yao, Y., Fang, J., Zhang, M., Zhou, Y., Chen, H., & Yang, L. (2014). Effects of feedstock type, production method, and pyrolysis temperature on biochar and hydrochar properties. *Chemical Engineering Journal*, *240*, 574–578. DOI: 10.1016/j.cej.2013.10.081

Tang, L., Huang, H., Hao, H., & Zhao, K. (2013). Development of plasma pyrolysis/gasification systems for energy efficient and environmentally sound waste disposal. *Journal of Electrostatics*, *71*(5), 839–847. DOI: 10.1016/j.elstat.2013.06.007

Tan, X., Liu, Y., Zeng, G., Wang, X., Hu, X., Gu, Y., & Yang, Z. (2015). Application of biochar for the removal of pollutants from aqueous solutions. *Chemosphere*, *125*, 70–85. DOI: 10.1016/j.chemosphere.2014.12.058 PMID: 25618190

Tarafdar, J. C. (2015). *Tapan Adhikari. Nanotechnology in Soil Science*. Book., https://www.researchgate.net/publication/292539409

Tharini, C., Sachdev, A., & Manikandan, E. (2024). An efficient sensing system using ion-selective membrane on Ni_2O_3/rGO nanocomposite for electrochemical detection of nitrate ions. *Journal of Alloys and Compounds*, *980*, 173414. DOI: 10.1016/j.jallcom.2023.173414

Thio, B. J., Montes, M. O., Mahmoud, M. A., Lee, D. W., Zhou, D., & Keller, A. A. (2012). Mobility of capped silver nanoparticles under environmen-tally relevant conditions. *Environmental Science & Technology*, *46*(13), 6985–6991. DOI: 10.1021/es203596w PMID: 22133047

Tilman, D., Cassman, K. G., Matson, P. A., Naylor, R., & Polasky, S. (2002). Agricultural sustainability and intensive production practices. *Nature*, *418*(6898), 671–677. DOI: 10.1038/nature01014 PMID: 12167873

Titirici, M. M., Thomas, A., & Antonietti, M. (2007). Back in the black: Hydrothermal carbonization of plant material as an efficient chemical process to treat the C_O_2 problem? *New Journal of Chemistry*, *31*(6), 787–789. DOI: 10.1039/b616045j

Tiwari, S., Upadhyay, N., Singh, B. K., Singh, V. K., & Dubey, N. K. (2022). Facile Fabrication of Nanoformulated Cinnamomum glaucescens Essential Oil as a Novel Green Strategy to Boost Potency Against Food Borne Fungi, Aflatoxin Synthesis, and Lipid Oxidation. *Food and Bioprocess Technology*, *15*(2), 319–337. DOI: 10.1007/s11947-021-02739-3

Toksha, B., Sonawale, V. A. M., Vanarase, A., Bornare, D., Tonde, S., Hazra, C., Kundu, D., Satdive, A., Tayde, S., & Chatterjee, A. (2021). Nanofertilizers: A review on synthesis and impact of their use on crop yield and environment. *Environmental Technology & Innovation*, *24*, 101986. DOI: 10.1016/j.eti.2021.101986

Tomczyk, A., Sokołowska, Z., & Boguta, P. (2020). Biochar as a soil amendment for improving soil quality. *International Agrophysics*, *34*(2).

Tourinho, P. S., van Gestel, C. A. M., Lofts, S., Svendsen, C., Soares, A. M. V. M., & Loureiro, S. (2012). Metal-based nanoparticles in soil: Fate, behavior, and effects on soil invertebrates. *Environmental Toxicology and Chemistry*, *31*(8), 1679–1692. DOI: 10.1002/etc.1880 PMID: 22573562

Tran, T. K., Nguyen, M. K., Lin, C., Hoang, T. D., Nguyen, T. C., Lone, A. M., & Nguyen, D. D. (2023). Review on fate, transport, toxicity and health risk of nanoparticles in natural ecosystems: Emerging challenges in the modern age and solutions toward a sustainable environment. *The Science of the Total Environment*, •••, 169331. PMID: 38103619

Tripathi, D. K., Singh, S., Singh, S., Pandey, R., Singh, V. P., Sharma, N. C., & Chauhan, D. K. (2017). Manufactured nanoparticles in plants. *Plant Physiology and Biochemistry*, *110*, 2–12. DOI: 10.1016/j.plaphy.2016.07.030 PMID: 27601425

Tripathi, S., Bahuguna, R. N., Shrivastava, N., Singh, S., Chatterjee, A., Varma, A., & Jagadish, S. K. (2022). Microbial biofortification: A sustainable route to grow nutrient-rich crops under changing climate. *Field Crops Research*, *287*, 108662. DOI: 10.1016/j.fcr.2022.108662

Trivedi, P., Leach, J. E., Tringe, S. G., Sa, T., & Singh, B. K. (2020). Plant–microbiome interactions: From community assembly to plant health. *Nature Reviews. Microbiology*, *18*(11), 607–621. DOI: 10.1038/s41579-020-0412-1 PMID: 32788714

Tsai, W. T., Liu, S. C., Chen, H. R., Chang, Y. M., & Tsai, Y. L. (2012). Textural and chemical properties of swine-manure-derived biochar pertinent to its potential use as a soil amendment. *Chemosphere*, *89*(2), 198–203. DOI: 10.1016/j.chemosphere.2012.05.085 PMID: 22743180

Tyagi, S., Rajput, P., Pramanik, A., Dixit, V., Sinha, A., Pal, A. K., Chaubey, K. K., Manjunath, B. T., Verma, D. K., Bachheti, R. K., & Bachheti, A. (2023). *Plant Growth-Promoting Rhizobacteria Nanoemulsion and Their Applications*., DOI: 10.1007/978-981-99-5398-1_7

Uchimiya, M., Lima, I. M., Klasson, K. T., Chang, S., & Wartelle, L. H. (2011). Immobilization of heavy metal, lead, in contaminated soils by biochar and manuscripts from different biomass. *Chemosphere*, *80*(8), 923–930.

Uchimiya, M., Lima, I. M., Klasson, K. T., & Wartelle, L. H. (2010). Contaminant immobilization and nutrient release by biochar soil amendment: Roles of natural organic matter. *Chemosphere*, *80*(8), 935–940. DOI: 10.1016/j.chemosphere.2010.05.020 PMID: 20542314

Uddin, M., Desai, F., & Asmatulu, E. (2020). Engineered nanomaterials in the environment: Bioaccumulation, biomagnification and biotransformation. *Environmental Chemistry Letters*, *18*(4), 1073–1083. DOI: 10.1007/s10311-019-00947-0

Urra, J., Alkorta, I., & Garbisu, C. (2019). Potential benefits and risks for soil health derived from organic amendments. *Agronomy (Basel)*, *9*(9), 542. DOI: 10.3390/agronomy9090542

Usman, A. R. A., Abduljabbar, A., Vithanage, M., Ok, Y. S., Ahmad, M., Ahmad, M., & Al-Wabel, M. I. (2016). Biochar production from date palm waste: A review of technologies, characterization, and applications. *Waste and Biomass Valorization*, *7*(5), 1187–1200.

Uzoma, K. C., Inoue, M., Andry, H., Fujimaki, H., Zahoor, A., & Nishihara, E. (2011). Effect of cow manure biochar on maize productivity under sandy soil condition. *Soil Use and Management*, *27*(2), 205–212. DOI: 10.1111/j.1475-2743.2011.00340.x

Valdés, A., Guillermo, D., Katja, A., Cyrus, D., Dante, K., Nario, C. A., et al. (2021). The environmental geochemical baseline, background and sources of metal and metalloids present in urban, peri-urban and rural soils in the O´Higgins region, Chile. *Environmental Geochemistry and Health 2021 44:10* 44, 3173–3189. DOI: 10.1007/s10653-021-01098-4

van Nunen, K., Li, J., Reniers, G., & Ponnet, K. (2018). Bibliometric analysis of safety culture research. *Safety Science*, *108*, 248–258. DOI: 10.1016/j.ssci.2017.08.011

Van Zwieten, L., Kimber, S., Morris, S., Chan, K. Y., Downie, A., Rust, J., Joseph, S., & Cowie, A. (2010). Effects of biochar from slow pyrolysis of papermill waste on agronomic performance and soil fertility. *Plant and Soil, 327*(1), 235–246. DOI: 10.1007/s11104-009-0050-x

Vardharajula, S., Ali, S. Z., Tiwari, P. M., Eroğlu, E., Vig, K., & Dennis, V. A.. (2012). Functionalized carbon nanotubes: Biomedical applications. *International Journal ofNanomedicine, 7*, 5361–5374. DOI: 10.2147/IJN.S35832 PMID: 23091380

Vasuki, A., Paulpandi, V. K., Singh, R. D., Gurusamy, A., Mahendran, P. P., Sivakumar, T., & Prabakaran, K. (2023). Influence of Irrigation Methods and Nano-fertilizers Application on the Yield of Transplanted Lowland Rice (Oryza sativa L.) in Periyar Vaigai Command Area of Madurai. *Agricultural Science Digest - A Research Journal, Of.* DOI: 10.18805/ag.D-5840

Vega-Vásquez, P., Mosier, N. S., & Irudayaraj, J. (2020). Nanoscale Drug Delivery Systems: From Medicine to Agriculture. *Frontiers in Bioengineering and Biotechnology, 8*, 79. Advance online publication. DOI: 10.3389/fbioe.2020.00079 PMID: 32133353

Verma, K. K., Li, D. M., Singh, M., Rajput, V. D., Malviya, M. K., & Minkina, T.. (2020). Interactive role of silicon and plant–rhizobacteria mitigating abiotic stresses. *Plants, 9*(9), 1055. DOI: 10.3390/plants9091055 PMID: 32824916

Verma, K. K., Song, X. P., Joshi, A., Rajput, V. D., Singh, M., Sharma, A., Singh, R. K., Li, D.-M., Arora, J., Minkina, T., & Li, Y.-R. (2022a). Nanofertilizer possibilities for healthy soil, water, and food in the future. *Frontiers in Plant Science, 13*, 865048. DOI: 10.3389/fpls.2022.865048 PMID: 35677230

Verma, K. K., Song, X. P., Joshi, A., Tian, D. D., Rajput, V. D., & Singh, M. (2022b). Recent trends in nanofertilizers for advancing sustainable agriculture. *Nanomaterials (Basel, Switzerland), 12*, 173. DOI: 10.3390/nano12010173 PMID: 35010126

Verma, K. K., Song, X. P., Singh, M., Tian, D. D., Rajput, V. D., & Minkina, T.. (2023). Association of silicon and soil microorganisms induces stress mitigation. In de Mello Prado, R. (Ed.), *Benefits of silicon in the nutrition of plants* (pp. 299–328). Springer., DOI: 10.1007/978-3-031-26673-7_17

Vijayakumar, P. S., Abhilash, O. U., Khan, B. M., & Prasad, B. L. V. (2010). Nanogold-loadedsharp-edged carbon bullets as plant-gene carriers. *Advanced Functional Materials, 20*(15), 2416–2423. DOI: 10.1002/adfm.200901883

Vimbela, G. V., Ngo, S. M., Fraze, C., Yang, L., & Stout, D. A. (2017). Antibacterial properties and toxicity from metallic nanomaterials. *International Journal of Nanomedicine, 12*, 3941–3955. DOI: 10.2147/IJN.S134526 PMID: 28579779

Vimercati, L., Cavone, D., Caputi, A., De Maria, L., Tria, M., Prato, E., & Ferri, G. M. (2020). Nanoparticles: An experimental study of zinc nanoparticles toxicity on marine crustaceans and a general overview on the health implications in humans. *Frontiers in Public Health, 8*, 192. DOI: 10.3389/fpubh.2020.00192 PMID: 32509719

Vithanage, M., Herath, I., Joseph, S., Bundschuh, J., Bolan, N., Ok, Y. S., Kirkham, M. B., & Rinklebe, J. (2017). Interaction of arsenic with biochar in soil and water: A critical review. *Carbon, 113*, 219–230. DOI: 10.1016/j.carbon.2016.11.032

Vogel, H. J., Wollschläger, U., Helming, K., Heinrich, U., Willms, M., Wiesmeier, M., Russell, D., & Franko, U. (2019). Assessment of Soil Functions Affected by Soil Management. *Atlas of Ecosystem Services: Drivers, Risks, and Societal Responses*, 77–82. DOI: 10.1007/978-3-319-96229-0_13

Vurro, M., Miguel-Rojas, C., & Pérez-de-Luque, A. (2019). Safe nanotechnologies for increasing the effectiveness of environmentally friendly natural agrochemicals. *Pest Management Science, 75*(9), 2403–2412. DOI: 10.1002/ps.5348 PMID: 30672106

Waalewijn Kool, P. L., Diez Ortiz, M., & van Gestel, C. A. M. (2014). Effect of soil organic matter content and pH on ZnO nanoparticle toxicity to *Folsomia candida. Ecotoxicology and Environmental Safety, 108*, 9–15. DOI: 10.1016/j.ecoenv.2014.06.031 PMID: 25038266

Wahab, A., Muhammad, M., Ullah, S., Abdi, G., Shah, G. M., Zaman, W., & Ayaz, A. (2024). Agriculture and environmental management through nanotechnology: Eco-friendly nanomaterial synthesis for soil-plant systems, food safety, and sustainability. *The Science of the Total Environment, 926*, 171862. DOI: 10.1016/j.scitotenv.2024.171862 PMID: 38527538

Wang, C., Jiang, Y., Shao, Y., Chen, Z., Gao, Y., Liang, J., Gao, J., Fang, F., & Guo, J. (2024). The influence and risk assessment of multiple pollutants on the bacterial and archaeal communities in agricultural lands with different climates and soil properties. *Applied Soil Ecology, 193*, 105130. DOI: 10.1016/j.apsoil.2023.105130

Wang, L., Hu, C., & Shao, L. (2017). The antimicrobial activity of nanoparticles: Present situation and prospects for the future. *International Journal of Nanomedicine, 12*, 1227–1249. DOI: 10.2147/IJN.S121956 PMID: 28243086

Wang, L., Wang, Y., Ma, F., & Ruan, R. (2020). Synthesis and characterization of a novel magnetic biochar from rice husk for the removal of Cd2+ from aqueous solution. *Bioresource Technology*, *308*, 123287.

Wang, P., Menzies, N. W., Lombi, E., McKenna, B. A., Johannessen, B., Glover, C. J., Kappen, P., & Kopittke, P. M. (2013). Fate of ZnO nanoparticles in soils and cowpea (*Vigna unguiculata*). *Environmental Science & Technology*, *47*(23), 13822–13830. DOI: 10.1021/es403466p PMID: 24195448

Wang, T., Camps-Arbestain, M., Hedley, M., & Bishop, P. (2012). Predicting phosphorus bioavailability from high-ash biochars. *Plant and Soil*, *357*(1), 173–187.

Wang, X., Ou, G., Wang, N., & Wu, H. (2016). Graphene-based recyclable photo-absorbers for high-efficiency seawater desalination. *ACS Applied Materials & Interfaces*, *8*(14), 9194–9199. DOI: 10.1021/acsami.6b02071 PMID: 27019007

Wang, X., Xie, H., Wang, P., & Yin, H. (2023). Nanoparticles in Plants: Uptake, Transport and Physiological Activity in Leaf and Root. *Materials (Basel)*, *16*(8), 3097. DOI: 10.3390/ma16083097 PMID: 37109933

Wang, Z., Fu, S., Guo, Y., Han, Y., Ma, C., Li, R., & Yang, X. (2024). Classification and design strategies of polysaccharide-based nano-nutrient delivery systems for enhanced bioactivity and targeted delivery: A review. *International Journal of Biological Macromolecules*, *256*, 128440. DOI: 10.1016/j.ijbiomac.2023.128440 PMID: 38016614

Warnock, D. D., Lehmann, J., Kuyper, T. W., & Rillig, M. C. (2007). Mycorrhizal responses to biochar in soil–concepts and mechanisms. *Plant and Soil*, *300*(1), 9–20. DOI: 10.1007/s11104-007-9391-5

Waseem Ahmad Bhat, N. L. K. A. M. Z. A. D. R. A. Q. (2023). How to Conduct Bibliometric Analysis Using R- Studio: A Practical Guide. [EEL]. *European Economic Letters*, *13*(3), 681–700. DOI: 10.52783/eel.v13i3.350

Watts, N., Amann, M., Ayeb-Karlsson, S., Belesova, K., Bouley, T., Boykoff, M., Byass, P., Cai, W., Campbell-Lendrum, D., Chambers, J., Cox, P. M., Daly, M., Dasandi, N., Davies, M., Depledge, M., Depoux, A., Dominguez-Salas, P., Drummond, P., Ekins, P., & Costello, A. (2018). The Lancet Countdown on health and climate change. *Lancet*, *391*(10120), 581–630. DOI: 10.1016/S0140-6736(17)32464-9 PMID: 29096948

Weber, K., & Quicker, P. (2018). Properties of biochar. *Fuel*, *217*, 240–261. DOI: 10.1016/j.fuel.2017.12.054

Wei, C., & Wen, H. (2012a). Geochemical baselines of heavy metals in the sediments of two large freshwater lakes in China: Implications for contamination character and history. *Environmental Geochemistry and Health*, *34*(6), 737–748. DOI: 10.1007/s10653-012-9492-9 PMID: 23001472

Wen, W., Yang, F., Xie, S., Wang, C., Song, Y., Zhang, Y., et al. (2024). Determination of Geochemical Background and Baseline and Research on Geochemical Zoning in the Desert and Sandy Areas of China. *Applied Sciences 2024, Vol. 14, Page 10612* 14, 10612. DOI: 10.3390/app142210612

Wen, S., Su, Y., Dai, C., Jia, J , Fan, G.-C., Jiang, L.-P., Song, R.-B., & Zhu, J.-J. (2019). Plasmon coupling-enhanced Raman sensing platform integrated with exonuclease-assisted target recycling amplification for ultrasensitive and selective detection of microRNA-21. *Analytical Chemistry*, *91*(19), 12298–12306. DOI: 10.1021/acs.analchem.9b02476 PMID: 31486639

Wheeler, K. E., Chetwynd, A. J., Fahy, K. M., Hong, B. S., Tochihuitl, J. A., Foster, L. A., & Lynch, I. (2021). Environmental dimensions of the protein corona. *Nature Nanotechnology*, *16*(6), 617–629. DOI: 10.1038/s41565-021-00924-1 PMID: 34117462

Wiedner, K., Rumpel, C., Steiner, C., Pozzi, A., Maas, R., & Glaser, B. (2013). Chemical evaluation of chars produced by thermochemical conversion (gasification, pyrolysis and hydrothermal carbonization) of agro-industrial biomass on a commercial scale. *Biomass and Bioenergy*, *59*, 264–278. DOI: 10.1016/j.biombioe.2013.08.026

Wilson, K. (2014). How biochar works in soil. *Biochar Journal*, *32*, 25–33.

Wilson, M. A., Tran, N. H., Milev, A. S., Kannangara, G. S. K., Volk, H., & Lu, G. H. M. (2008). Nanomaterials in soils. *Geoderma*, *146*(1-2), 291–302. DOI: 10.1016/j.geoderma.2008.06.004

Wong, S. W., Leung, K. M., & Djurišić, A. B. (2013). A comprehensive review on the aquatic toxicity of engineered nanomaterials. *Reviews in Nanoscience and Nanotechnology*, *2*(2), 79–105. DOI: 10.1166/rnn.2013.1025

Woods, W. I., & Denevan, W. M. (2009). Amazonian dark earths: the first century of reports. In *Amazonian dark earths: Wim Sombroek's vision* (pp. 1–14). Springer. DOI: 10.1007/978-1-4020-9031-8_1

Woolf, D., Amonette, J. E., Street-Perrott, F. A., Lehmann, J., & Joseph, S. (2010). Sustainable biochar to mitigate global climate change. *Nature Communications*, *1*(1), 1–9. DOI: 10.1038/ncomms1053 PMID: 20975722

Woolf, D., Lehmann, J., & Lee, D. R. (2016). Optimal bioenergy power generation for climate change mitigation with or without carbon sequestration. *Nature Communications*, *7*(1), 1–11. DOI: 10.1038/ncomms13160 PMID: 27767177

Wu, H., Tito, N., & Giraldo, J. P. (2017). Cerium oxide nanoparticles protect photosynthesis. *ACS Nano*, *11*(11), 11283–11297. DOI: 10.1021/acsnano.7b05723 PMID: 29099581

Wu, W., Yang, M., Feng, Q., McGrouther, K., Wang, H., Lu, H., & Chen, Y. (2012). Chemical characterization of rice straw-derived biochar for soil amendment. *Biomass and Bioenergy*, *47*, 268–276. DOI: 10.1016/j.biombioe.2012.09.034

Xie, P., Yang, S. T., He, T., Yang, S., & Tang, X. H. (2017). Bioaccumulationand toxicity of carbon nanoparticles suspension injection in intra-venously exposed mice. *International Journal of Molecular Sciences*, *18*(12), 2562. DOI: 10.3390/ijms18122562 PMID: 29186019

Xing, B., Vecitis, C., & Senesi, N. (2016). *Engineered nanoparticles and theenvironment: physicochemical processes and toxicity. IUPACseries on biophysicochemical processes in environmental systems* (Vol. 4). Wiley. DOI: 10.1002/9781119275855

Xin, X., Nepal, J., Wright, A. L., Yang, X., & He, Z. (2022). Carbon nanoparticles improve corn (Zea mays L.) growth and soil quality: Comparison of foliar spray and soil drench application. *Journal of Cleaner Production*, *363*, 132630. DOI: 10.1016/j.jclepro.2022.132630

Xin, X., Zhao, F., Zhao, H., Goodrich, S. L., Hill, M. R., Sumerlin, B. S., Stoffella, P. J., Wright, A. L., & He, Z. (2020). Comparative assessment of polymeric and other nanoparticles impacts on soil microbial and biochemical properties. *Geoderma*, *367*, 114278. DOI: 10.1016/j.geoderma.2020.114278

Xu, C., Peng, C., Sun, L., Zhang, S., Huang, H., Chen, Y., & Shi, J. (2015). Distinctive effects of TiO2 and CuO nanoparticles on soil microbes and their community structures in flooded paddy soil. *Soil Biology & Biochemistry*, *86*, 24–33. DOI: 10.1016/j.soilbio.2015.03.011

Xue, Z., Wang, Z., Tan, L., and Fan, M. (2022). Geochemical baseline assessment and ecological risk evaluation of potentially toxic elements in soils: a case in Han River, China. *Journal of Soils and Sediments 2022 23:2* 23, 745–764. DOI: 10.1007/s11368-022-03339-0

Xu, G., Sun, J. N., Shao, H. B., & Chang, S. X. (2014). Biochar had effects on phosphorus sorption and desorption in three soils with differing acidity. *Ecological Engineering*, *62*, 54–60. DOI: 10.1016/j.ecoleng.2013.10.027

Yadav, A., Yadav, K., & Abd-Elsalam, K. A. (2023a). Nanofertilizers: Types, Delivery and Advantages in Agricultural Sustainability. *Agrochemicals*, *2*(2), 296–336. DOI: 10.3390/agrochemicals2020019

Yadav, A., Yadav, K., Ahmad, R., & Abd-Elsalam, K. A. (2023). Emerging frontiers in nanotechnology for precision agriculture: Advancements, hurdles and prospects. *Agrochemicals*, *2*(2), 220–256. DOI: 10.3390/agrochemicals2020016

Yadav, N., Garg, V. K., Chhillar, A. K., & Rana, J. S. (2023c). Recent advances in nanotechnology for the improvement of conventional agricultural systems: A review. *Plant Nano Biology*, *4*, 100032. DOI: 10.1016/j.plana.2023.100032

Yadav, S., Sharma, A. K., & Kumar, P. (2020). Nanoscale Self-Assembly for Therapeutic Delivery. *Frontiers in Bioengineering and Biotechnology*, *8*, 127. Advance online publication. DOI: 10.3389/fbioe.2020.00127 PMID: 32158749

Yamini, V., Shanmugam, V., Rameshpathy, M., Venkatraman, G., Ramanathan, G., & Garalleh, A. L. (2023). Environmental effects and interaction of nanoparticles on beneficial soil and aquatic microorganisms. *Environmental Research*, *236*, 116776. DOI: 10.1016/j.envres.2023.116776 PMID: 37517486

Yang, R., Gao, L., Liu, Z., Li, C., & Shen, Y. (2025). Sulfur nanoparticles enhance watermelon growth. *Plant Physiology and Biochemistry*, *229*, 110359. DOI: 10.1016/j.plaphy.2025.110359 PMID: 40811908

Yang, W., Li, T., Liao, S., Zhou, J., & Huang, L. (2024). Organ-on-a-chip platforms integrated with biosensors for precise monitoring of the cells and cellular microenvironment. *Trends in Analytical Chemistry*, *172*, 117569. DOI: 10.1016/j.trac.2024.117569

Yan, L., Li, P., Zhao, X., Ji, R., & Zhao, L. (2020). Physiological and metabolic responses of maize (Zea mays) plants to Fe3O4 nanoparticles. *The Science of the Total Environment*, *718*, 137400. DOI: 10.1016/j.scitotenv.2020.137400 PMID: 32105936

Yao, Y., Gao, B., Zhang, M., Inyang, M., & Zimmerman, A. R. (2012). Effect of biochar amendment on sorption and leaching of nitrate, ammonium, and phosphate in a sandy soil. *Chemosphere*, *89*(11), 1467–1471. DOI: 10.1016/j.chemosphere.2012.06.002 PMID: 22763330

Yashveer, S., Singh, V., Kaswan, V., Kaushik, A., & Tokas, J. (2014). Green biotechnology, nan-otechnology and bio-fortification: Perspec-tives on novel environment-friendly crop im-provement strategies. *Biotechnology & Genetic Engineering Reviews*, *30*(2), 113–126. DOI: 10.1080/02648725.2014.992622 PMID: 25598358

Yayehrad, A. T., Siraj, E. A., Wondie, G. B., Alemie, A. A., Derseh, M. T., & Ambaye, A. S. (2021). Could Nanotechnology Help to End the Fight Against COVID-19? Review of Current Findings, Challenges and Future Perspectives. *International Journal of Nanomedicine*, *16*, 5713–5743. DOI: 10.2147/IJN.S327334 PMID: 34465991

Yeo, M. K., & Pak, S. W. (2008). Exposing zebrafish to silver nanoparticlesduring caudal fin regeneration disrupts caudal fin growth and p53signaling. *Molecular & Cellular Toxicology*, *4*(4), 311–317.

Yin, H., Cao, Y., Marelli, B., Zeng, X., Mason, A. J., & Cao, C. (2021). Soil Sensors and Plant Wearables for Smart and Precision Agriculture. *Advanced Materials*, *33*(20), 2007764. Advance online publication. DOI: 10.1002/adma.202007764 PMID: 33829545

You, Y., Kerner, P., Shanmugam, S., & Khodakovskaya, M. V. (2023). Differential effects of carbon nanotubes and graphene on the tomato rhizosphere microbiome. *Environmental Science. Nano*, *10*(6), 1570–1584. DOI: 10.1039/D2EN01026G

Yuan, J. H., & Xu, R. K. (2011). The amelioration effects of low temperature biochar generated from nine crop residues on an acidic Ultisol. *Soil Use and Management*, *27*(1), 110–115. DOI: 10.1111/j.1475-2743.2010.00317.x

Yuan, J. H., Xu, R. K., & Zhang, H. (2011). The forms of alkalis in the biochar produced from crop residues at different temperatures. *Bioresource Technology*, *102*(3), 3488–3497. DOI: 10.1016/j.biortech.2010.11.018 PMID: 21112777

Zahedi, S. M., Abdelrahman, M., Hosseini, M. S., Hoveizeh, N. F., & Tran, L. S. P. (2019). Alleviation of the effect of salinity on growth and yield of strawberry by foliar spray of selenium-nanoparticles. *Environmental Pollution*, *253*, 246–258. DOI: 10.1016/j.envpol.2019.04.078 PMID: 31319241

Zain, M., Ma, H., Ur Rahman, S., Nuruzzaman, M., Chaudhary, S., Azeem, I., Mehmood, F., Duan, A., & Sun, C. (2024). Nanotechnology in precision agriculture: Advancing towards sustainable crop production. *Plant Physiology and Biochemistry*, *206*, 108244. DOI: 10.1016/j.plaphy.2023.108244 PMID: 38071802

Zakaria, Z., Zulkafflee, N. S., Mohd Redzuan, N. A., Selamat, J., Ismail, M. R., Praveena, S. M., & Abdull Razis, A. F. (2021). Understanding potential heavy metal contamination, absorption, translocation and accumulation in rice and human health risks. *Plants*, *10*(6), 1070. DOI: 10.3390/plants10061070 PMID: 34073642

Zaki, N. M., & Tirelli, N. (2010). Gateways for the intracellular access of nanocarriers: A review of receptor-mediated endocytosis mechanisms and of strategies in receptor targeting. *Expert Opinion on Drug Delivery*, *7*(8), 895–913. DOI: 10.1517/17425247.2010.501792 PMID: 20629604

Zaki, S. A., Ouf, S. A., Abd-Elsalam, K. A., Asran, A. A., Hassan, M. M., Kalia, A., & Albarakaty, F. M. (2022). Trichogenic silver-based nanoparticles for suppression of fungi involved in damping-off of cotton seedlings. *Microorganisms*, *10*(2), 344. DOI: 10.3390/microorganisms10020344 PMID: 35208801

Zamora-Justo, J. A., Abrica-González, P., Vázquez-Martínez, G. R., Muñoz-Diosdado, A., Balderas-López, J. A., & Ibáñez-Hernández, M. (2019). Polyethylene glycol-coated gold nanoparticles as DNA and atorvastatin delivery systems and cytotoxicity evaluation. *Journal of Nanomaterials*, *5982047*, 1–11. Advance online publication. DOI: 10.1155/2019/5982047

Zareabyaneh, H., & Bayatvarkeshi, M. (2015). Effects of slow-release fertilizers on nitrate leaching, its distribution in soil profile, N-use efficiency, and yield in potato crop. *Environmental Earth Sciences*, *74*(4), 3385–3393. DOI: 10.1007/s12665-015-4374-y

Zargar, M., Bayat, M., Saquee, F. S., Diakite, S., Ramzanovich, N. M., & Akhmadovich, K. A. S. (2023). New Advances in Nano-Enabled Weed Management Using Poly(Epsilon-Caprolactone)-Based Nanoherbicides. *Agriculture*, *13*(10), 2031. DOI: 10.3390/agriculture13102031

Zeng, K., Gauthier, D., Soria, J., Mazza, G., & Flamant, G. (2017). Solar pyrolysis of carbonaceous feedstocks: A review. *Solar Energy*, *156*, 73–92. DOI: 10.1016/j.solener.2017.05.033

Zhang, A., Cui, L., Pan, G., Li, L., Hussain, Q., Zhang, X., & Crowley, D. (2012). Effect of biochar amendment on yield and methane and nitrous oxide emissions from a rice paddy from Tai Lake plain, China. *Agriculture, Ecosystems & Environment*, *159*, 1–8.

Zhang, L., Wu, L., Si, Y., & Shu, K. (2018). Size-dependent cytotoxicity of silver nanoparticles to Azotobacter vinelandii: Growth inhibition, cell injury, oxidative stress and internalization. *PLoS One*, *13*(12), e0209020. DOI: 10.1371/journal.pone.0209020 PMID: 30566461

Zhang, M. X., Zhao, L. Y., He, Y. Y., Hu, J. P., Hu, G. W., Zhu, Y., & Zhang, J. L. (2024). Iron nanomaterials enhance growth and nitrogen fixation in alfalfa. *Bioresource Technology*, *391*, 129987. DOI: 10.1016/j.biortech.2023.129987 PMID: 37951551

Zhang, M., Tao, S., & Wang, X. (2020). Interactions between organic pollutants and carbon nanomaterials and the associated impact on microbial availability and degradation in soil: A review. *Environmental Science. Nano*, *7*(9), 2486–2508. DOI: 10.1039/D0EN00515K

Zhang, P., Guo, Z., Zhang, Z., Fu, H., White, J. C., & Lynch, I. (2020). Nanomaterial transformation in the soil–plant system: Implications for food safety and application in agriculture. *Small*, *16*(21), 2000705. DOI: 10.1002/smll.202000705 PMID: 32462786

Zhang, W., Pu, Z., Du, S., Chen, Y., & Jiang, L. (2016). Fate of engineeredcerium oxide nanoparticles in an aquatic environment andtheir toxicity toward 14 ciliated protist species. *Environmental Pollution*, *212*, 584–591. DOI: 10.1016/j.envpol.2016.03.011 PMID: 26986089

Zhang, X., Wang, H., He, L., Lu, K., Sarmah, A., Li, J., Bolan, N. S., Pei, J., & Huang, H. (2013). Using biochar for remediation of soils contaminated with heavy metals and organic pollutants. *Environmental Science and Pollution Research International*, *20*(12), 8472–8483. DOI: 10.1007/s11356-013-1659-0 PMID: 23589248

Zhao, F. J., Ma, J. F., Meharg, A. A., & McGrath, S. P. (2010). Cadmium uptake and accumulation in rice. [). Academic Press.]. *Advances in Agronomy*, *106*, 1–46.

Zhao, L., Cao, X., Mašek, O., & Zimmerman, A. (2013). Heterogeneity of biochar properties as a function of feedstock sources and production temperatures. *Journal of Hazardous Materials*, *256*, 1–9. DOI: 10.1016/j.jhazmat.2013.04.015 PMID: 23669784

Zhao, X., Yu, M., Xu, D., Liu, A., Hou, X., Hao, F., Long, Y., Zhou, Q., & Jiang, G. (2017). Distribution, bioaccumulation, trophic transfer, and influ-ences of CeO2 nanoparticles in a constructed aquatic food web. *Environmental Science & Technology*, *51*(9), 5205–5214. DOI: 10.1021/acs.est.6b05875 PMID: 28383254

Zhao, X., Zhang, J., Song, Z., Liu, H., Li, L., & Ma, C. (2017). Microwave pyrolysis of straw bale and energy balance analysis. *Journal of Analytical and Applied Pyrolysis*, *127*, 284–292.

Zimmerman, A. R., Gao, B., & Ahn, M. Y. (2011). Positive and negative carbon mineralization priming effects among a variety of biochar-amended soils. *Soil Biology & Biochemistry*, *43*(6), 1169–1179. DOI: 10.1016/j.soilbio.2011.02.005

About the Contributors

Abhishek Singh is a passionate Senior researcher (senior scientist) and researcher who is currently conducting cutting-edge research at the Institute of Biology, Applied Biology, and Ecology lab, Yerevan State University in the Republic of Armenia. Dr. Singh has a strong academic background in the fields of agriculture and biotechnology, having earned his doctoral degree from the esteemed Sardar Vallabh Bhai Patel University of Agriculture and Technology in Meerut, UP, India. Dr. Singh's academic journey started with an Integrated MSc in Biotechnology, with a specialization in Plant Studies, from the prestigious Sam Higginbottom University of Agriculture, Technology, and Sciences in Allahabad, India. Currently, Dr. Singh is focused on conducting research on the effects of salinity stress on various aspects of plant biology. His research covers a wide range of topics, including physiobiochemical characteristics, morphological transformations, anatomical variations, and the ultrastructure of cellular and subcellular organelles. Additionally, his research delves into the complexities of DNA, RNA, and protein damage, seeking to comprehensively understand the multifaceted impacts of salinity stress on plant life. Apart from his work on salinity stress, Dr. Singh is also delving into exploring the interactions between metallic nanoparticles and plants. His research covers the entire lifecycle of these nanoparticles, including their bioaccumulation, uptake, translocation, and the potential toxic effects they may cause. His contributions in this area are crucial for advancing sustainable agriculture by shedding light on the interactions between nanoparticles, plants, and soil, as well as their physiological and genomic responses. Dr. Singh's dedication to scholarship is evident through his extensive publication record, which includes 25 full-length research articles, 10 insightful books, and 30 chapters, all indexed in the prestigious Scopus database. Furthermore, he has presented his research findings through six articles at scientific conferences. Beyond academia, he has shared his expertise through over 30 articles related to agriculture in various magazines, further contributing to the dissemination of knowledge in his field. Dr. Singh actively participates in peer review activities for reputable publications on a global scale, and his contributions have earned him accolades, including a certificate as a peer reviewer from Springer and recognition for his valuable insights. Moreover, Dr. Singh generously lends his expertise to the editorial boards of numerous journals and societies, contributing to the advancement of knowledge in the fields of agriculture and biotechnology. The state government of Uttar Pradesh, India, recognized Dr. Abhishek Singh's outstanding contributions with the prestigious "Appreciation Award 2021," underscoring his significant achievements and his position as a distinguished researcher in the fields of agriculture and biotechnology.

Vishnu D. Rajput is working as Principal Scientist at Southern Federal University, Russia. Dr. Rajput has an exceptional academic background in Agriculture and Biological Sciences, and received a distinction in both his B.Sc. Ag and M.Sc. Ag degrees from CS Azad University of Agriculture & Technology, India, which laid the foundation for an illustrious career. Driven by an insatiable thirst for knowledge, he pursued a PhD at the esteemed Chinese Academy of Sciences, China, and further honed his expertise through a Post-doc at Southern Federal University, Russia. Dr. Rajput focuses on the study of soil contamination, specifically with potentially toxic elements and metallic nanoparticles. His research delves into the intricate web of bioaccumulation, bio/geo-transformations, uptake, translocation, and the toxic repercussions of metallic nanoparticles on various facets of plant physiology. Dr. Rajput's studies span from the macroscopic aspects of plant morphology to the microscopic intricacies of cellular and subcellular organelles, including cytomorphometric alterations and DNA damage. Recently, he is working on bioengineering soil rhizospheric microbiome. Dr Rajput has contributed significantly to the realm of environmental science, shining a spotlight on the intricate interactions between nanoparticles, heavy metals, plants, soil, and microbial communities and offering valuable insights into the broader ecosystem. He has an impressive record of 500+ scientific publications, including 375 peer-reviewed (indexed in Scopus), highly rated full-length articles, 30 enlightening books, and 101 chapters. He has also presented his findings in 55 conference articles. Recognized on an international scale for his profound expertise and contributions, Dr. Rajput has evaluated 250 manuscripts for esteemed journals and received outstanding reviewing certificates from prestigious publishers like Elsevier and Springer. He serves on the editorial boards of various high-impact journals, contributing to the advancement of academic discourse in his field. Moreover, Dr. Rajput has received the "Certificate for Appreciation 2019" and the "Certificate of Honor 2020" from Southern Federal University, Russia, in acknowledgment of his outstanding contributions to academia and creative research. He has also been recognized as a "Highly Qualified Specialist" by SFedU and the Russian Ministry of Internal Affairs, a testament to his exceptional expertise and achievements. In summary, Dr. Vishnu D. Rajput is a distinguished academic in the fields of Agriculture and Biological Sciences, making significant contributions to the study of soil contamination and environmental science. His dedication to scholarly excellence is reflected in his impressive portfolio of research publications, editorial board memberships, and awards.

Rupesh Kumar Singh is a passionate researcher with 10 years of post-PhD experience in diverse area of Agronomy, plant molecular biology, Vitiviniculture, plant physiology, Plant pathology, genetic engineering and modern genome editing technology, tissue culture, biochemistry, nanotechnology, biocontrol agent development and formulations, transcriptomics, proteomics, metabolomics. The research profile includes 3212 citations, H index-24 and 92 peer reviewed published scientific communications indexed in Scopus database including 1 granted patent and 2 provisional submissions recently. He gained academic and industrial research experience while working in national and international institutes and industries of good repute.

Karen Ghazaryan is a highly regarded academic figure who serves as the Dean of the Faculty of Biology at Yerevan State University. With 21 years of experience in scientific research across several domains, he has made significant contributions to critical areas such as Soil Ecology, Water Resources Monitoring and Management, Climate Change and its Environmental Consequences, Ecological Management of the Environment, Ecological Evaluation of the Environment, and Biodiversity Protection. Dr. Ghazaryan's expertise is reflected in his extensive portfolio of 96 scientific research articles, which demonstrates his unwavering commitment to advancing knowledge and understanding within his areas of specialization. He has also authored influential books, further establishing himself as a thought leader and authoritative voice in his field. In addition to his scholarly achievements, Dr. Ghazaryan is an active contributor to the progress of scientific publishing. He serves as a distinguished member of various editorial boards for peer-reviewed journals, and his commitment to advancing research and fostering academic excellence has had a lasting impact on the scientific community. His invaluable insights continue to shape the ever-evolving landscape of environmental sciences. Dr. Ghazaryan exemplary career, dedication, and remarkable contributions to academia position him as an outstanding and respected figure in the realm of Soil Ecology and Environmental Sciences. His profound commitment to advancing knowledge and preserving the delicate ecological balance of our planet serves as an inspiration to aspiring researchers and scientists worldwide, making him a true luminary in his field.

Tatiana Minkina is the esteemed Head of the Soil Science and Land Evaluation Department at Southern Federal University. She also holds the position of Head of the International Master's Degree Educational Program "Management and Estimation of Land Resources" from 2015 to 2022, which has received accreditation by ACQUIN. Her primary areas of scientific interest include Soil Science, Biogeochemistry of Trace Elements, and Environmental Soil Chemistry. Additionally, she specializes in soil monitoring, assessment, modeling, and remediation, utilizing physicochemical treatment methods.In recognition of her exceptional contributions to the educational process and the training of highly qualified specialists, she was awarded the Diploma of the Ministry of Education and Science of the Russian Federation in 2015. Currently, Professor Dr. Tatiana Minkina is actively involved in various projects funded by prestigious organizations like the Russian Scientific Foundation, Ministry of Education and Science of the Russian Federation, and the Russian Foundation of Basic Research.She holds memberships in several esteemed organizations, including the Expert Group of the Russian Academy of Science, the International Committee on Contamination Land, Eurasian Soil Science Societies, the International Committee on Protection of the Environment, and the International Scientific Committee of the International Conferences on Biogeochemistry of Trace Elements.Professor Dr. Tatiana Minkina's impressive research portfolio includes a total of 757 scientific publications, out of which 389 are in English. She serves as an invited editor for the Open Access Journal by MDPI, "Water," with ISSN 2073-4441 and an Impact Factor of 2.524. Additionally, she holds editorial board membership in esteemed journals such as "Geochemistry: Environment, Exploration, Analysis" with ISSN 1467-7873 and an Impact Factor of 1.109; 5yr IF: 1.769; SJR: 0.334; SNIP: 0.598, as well as "Eurasian Journal of Soil Science" with e-ISSN 2147 – 4249.Professor Dr. Tatiana Minkina's contributions to the field of Soil Science and Environmental Chemistry have earned her a well-deserved reputation as a respected and accomplished researcher in her domain.

***

E. Kannan, Registrar & Professor in School of Computing at Vel Tech University, Avadi, Chennai. He obtained his Ph.D from NIT, Trichy in the year 2006. His research interest spans across computer networking and parallel computing. Much of his work has been on improvising the understanding, design and the performance of parallel and networked computer systems, mainly through the application of data mining, statistics and performance evaluation. He has published more than 100 research publications in Scopus and SCI indexed international journals and conferences. He began his career in 1991 and has held various responsible positions in reputed institutions. He is also a member in many professional's society and a senior member in IEEE.

Vishal Eswaran is an accomplished Senior Big Data Engineer with an impressive career spanning over 6 years. His fervor for constructing robust data pipelines, unearthing insights from intricate datasets, identifying trends, and predicting future trajectories has fueled his journey. Throughout his tenure, Vishal has lent his expertise to empower numerous prominent US healthcare clients, including CVS Health, Aetna, and Blue Cross and Blue Shield of North Carolina, with informed business decisions drawn from expansive datasets. Vishal's ability to distill intricate data into comprehensive documents and reports stands as a testament to his proficiency in managing multifaceted internal and external data analysis responsibilities. His aptitude for synthesizing complex information ensures that insights are both accessible and impactful for strategic decision-making. Moreover, Vishal's distinction extends to his role as a co-author of the book "Internet of Things - Future Connected Devices." This book not only underscores his prowess in the field but also showcases his visionary leadership in the realm of Internet of Things (IoT). His insights resonate with a forward-looking perspective, emphasizing the convergence of technology and human life. As the author of "Secure Connections: Safeguarding the Internet of Things (IoT) with Cybersecurity," Vishal Eswaran's reputation as a thought leader is further solidified. His work is a manifestation of his commitment to ensuring the security of interconnected devices within the IoT landscape, a vital consideration in our digitally driven world. Vishal's dedication to enhancing the safety and integrity of IoT ecosystems shines through in his work.

Vivek Eswaran, with 8 years of experience as a Senior Software Engineer specializing in front-end development, he brings a vital perspective to securing the Internet of Things (IoT). At Medallia, Vivek played an instrumental role in crafting engaging user interfaces and optimized digital experiences. This profound expertise in front-end engineering equips them to illuminate the crucial synergy between usability and cybersecurity as IoT adoption accelerates. In the new book "Secure Connections: Safeguarding the Internet of Things with Cybersecurity," Vivek combines their real-world experiences building intuitive and secure software systems with cutting-edge insights into strengthening IoT ecosystems. Drawing parallels between front-end best practices and security imperatives, they offer readers an invaluable guide for fortifying IoT without compromising usability. As businesses and consumers continue rapidly connecting people, processes, and devices, Vivek's contribution provides timely insights. Blending user empathy with security proficiency, Vivek empowers audiences to realize the potential of IoT through resilient and human-centered systems designed for safety without friction.

Ushaa Eswaran is an esteemed author, distinguished researcher, and seasoned educator with a remarkable journey spanning over 34 years, dedicated to advancing academia and nurturing the potential of young minds. Currently serving as a Principal and Professor in Andhra Pradesh, India, her vision extends beyond imparting cutting-edge technical expertise to encompass the nurturing of universal human values. With a foundation in Electronics Engineering, Dr. Eswaran delved into the realm of biosensors, carving a pioneering path in nanosensor models, a remarkable achievement that earned her a well-deserved Doctorate. Her insights have been encapsulated in her acclaimed book, "Internet of Things: Future Connected Devices," offering profound insights into the evolving IoT landscape. Her expertise also finds its place in upcoming publications centered around computer vision and IoT technologies. Dr. Eswaran's commitment to literature is rooted in her unwavering passion to equip the younger generation with the latest knowledge fortified by ethical principles. Her book stands as a beacon of practical wisdom, providing a roadmap through the intricate IoT terrain while shedding light on its future societal impacts. Her forthcoming contributions unveil her interdisciplinary perspective, seamlessly integrating electronics, nanotechnology, and computing. Bolstering her scholarly contributions is her ORCID identifier, 0000-0002-5116-3403, a testament to her prolific research journey that encompasses over a hundred published papers. Dr. Eswaran thrives in merging her profound academic insights with her dedication to nurturing holistic student growth. Her tireless exploration of the dynamic interface between technology and human values continues to shape her works. As the author of "Secure Connections: Safeguarding the Internet of Things (IoT) with Cybersecurity," Dr. Ushaa Eswaran's voice emerges as a beacon of wisdom in the realm of IoT. Her work encapsulates her dedication to enhancing the interconnected world while ensuring its resilience against cyber threats.

Keerthana Murali, with over 5 years of experience as a Site Reliability Engineer at Dell, Keerthna Murali has honed an intricate expertise in maintaining and optimizing robust digital infrastructures. On the frontlines of ensuring seamless online experiences, Keerthna specialized in troubleshooting complex issues and proactively enhancing system performance and availability. These capabilities uniquely position them to tackle the critical challenge of security for rapidly emerging IoT ecosystems. In their new book "Secure Connections," Keerthna channels their real-world experiences maintaining enterprise-scale platforms into a compelling vision for building security into the foundation of IoT systems. Blending software engineering best practices with cybersecurity insights, they offer a prescient guide for developers, IT leaders, and security experts seeking to realize IoT's potential while mitigating its risks.

Omkar Singh, born on September 1, 1995, into a farming family in Uttar Pradesh, India, has emerged as a dedicated researcher and scholar in the field of Soil Science and Agricultural Chemistry. He earned his Ph.D. from S.V.P. University of Agriculture and Technology, Meerut, following an M.Sc. in Agriculture from N.D. University of Agriculture and Technology. Growing up in a farming environment, Omkar possesses a profound understanding of the challenges faced by the farming communities in the eastern Indo-Gangetic plains. His research focus includes the impact of climate change on agriculture, particularly in the context of sustainable and resilient cropping systems. As a Senior Research Fellow at SVPUA&T, Meerut, he actively engages in field experiments, soil, and plant sample analyses, and data interpretation. Omkar has published extensively, contributing to reputable journals with articles on topics ranging from the effects of nanoparticles on rice growth to sustainable soil management practices for climate resilience. He has also been a part of various national and international conferences, workshops, and training programs.

Mohit Yadav is a Research Scholar specializing in nanotechnology applications in soil science and sustainable agriculture. His research focuses on soil–plant interactions, nanomaterial dynamics, soil health improvement, and abiotic stress mitigation. He has authored research and review articles and book chapters related to nanomaterials, soil remediation, and crop performance.

Index

P

R

S

W

www.ingramcontent.com/pod-product-compliance
Lightning Source LLC
LaVergne TN
LVHW060004031026
849161LV00025B/1413
9798369370063